Ethics and Animals

CONTEMPORARY ISSUES IN BIOMEDICINE, ETHICS, AND SOCIETY

Ethics and Animals, edited by **Harlan B. Miller** and **William H. Williams,** *1983*

Profits and Professions, edited by **Wade L. Robison, Michael S. Pritchard,** and **Joseph Ellin,** *1983*

Visions of Women, edited by **Linda A. Bell,** *1983*

Medical Genetics Casebook, by **Colleen Clements,** *1982*

Who Decides? edited by **Nora K. Bell,** *1982*

The Custom-Made Child?, edited by **Helen B. Holmes, Betty B. Hoskins,** and **Michael Gross,** *1981*

Birth Control and Controlling Birth, edited by **Helen B. Holmes, Betty B. Hoskins,** and **Michael Gross,** *1980*

Medical Responsibility, edited by **Wade L. Robison** and **Michael S. Pritchard,** 1979

Contemporary Issues in Biomedical Ethics, edited by **John W. Davis, Barry Hoffmaster,** and **Sarah Shorten,** *1979*

Ethics
and
Animals

Edited by

Harlan B. Miller

and

William H. Williams

Humana Press · **Clifton, New Jersey**

Library of Congress Cataloging in Publication Data

Main entry under title:

Ethics and animals.

  (Contemporary issues in biomedicine, ethics, and
society)
  Bibliography: p.
  Includes index.
  1. Animals, Treatment of—Moral and ethical aspects—
Addresses, essays, lectures. 2. Bioethics—Addresses,
essays, lectures. 3. Medical ethics—Addresses, essays,
lectures. I. Miller, Harlan B. II. Williams,
William H. (William Hatton), 1934–   . III. Series.
HV4711.E87 1983   179'.3   82-21387
ISBN 0-89603-036-9

©1983 The HUMANA Press Inc.
Crescent Manor
PO Box 2148
Clifton, NJ 07015

Printed in the United States of America.

   A slightly different version of Professor Regan's essay has been published in *Environmental Ethics*
and, in Polish translation, in *Etyka*. Professor Narveson's paper is published both here and in *Animal
Regulation Studies* by agreement with Elsevier Scientific Publishing Company, Amsterdam.
   The poem by Peter Rabbit on pp. 321–322 is reprinted with the permission of Random House and
of the successors to the Portola Institute, The Next Whole Earth Catalog, Box 428, Sausalito,
California, 94966.
   The selections from Aldo Leopold's *Sand County Almanac* on pp. 185 through 194 are reprinted
with the permisson of Oxford University Press.
   The selections from Christopher Stone's *Should Trees Have Standing?* on pp. 183 through 184 are
reprinted with the permission of William Kaufman, Inc.
   The selections from Paul Shepard's *The Tender Carnivore and the Sacred Game* on pp. 191
through 192 are reprinted with the permission of Charles Scribner's Sons.

CONTENTS

Preface xi

Introduction: 'Platonists' and 'Aristotelians',
 Harlan B. Miller 1

SECTION I: Nonhumans in the Eyes and Minds of Humans

Introduction 17

Animal Rights, Human Wrongs, **Tom Regan** 19

Animal Rights Revisited, **Jan Narveson** 45

Knowing Our Place in the Animal World, **Annette C. Baier** 61

The Clouded Mirror: Animal Stereotypes and Human Cruelty,
 Thomas L. Benson 79

SECTION II: Morality, Legality, and Animals

Introduction 93

Moral Community and Moral Order: The Intensive and
 Extensive Limits of Interaction, **James M. Buchanan** 95

The Legal and Moral Bases of Animal Rights,
 Bernard E. Rollin 103

SECTION III: Humans and Other Animals—Killing

Introduction 121

Life, Death, and Animals, **Edward Johnson** 123

Killing Persons and Other Beings, **Dale Jamieson**   135
Interspecific Justice and Animal Slaughter,
   Donald VanDeVeer   147

SECTION IV: Humans and Other Animals— Linkages and Likenesses

Introduction   165
Humans, Animals, and 'Animal Behavior',
   Stephen R. L. Clark   169
Ecology, Morality, and Hunting, **Peter S. Wenz**   183
Humans as Hunting Animals, **Patrick F. Scanlon**   199
Apes and Language Research, **Duane M. Rumbaugh and**
   Sue Savage-Rumbaugh   207

SECTION V: Human Interests, Porcine Interests, and Chipmunk Interests

Introduction   221
The Priority of Human Interests, **Lawrence C. Becker**   225
Comments on "The Priority of Human Interests,"
   James Cargile   243
The Case Against Raising and Killing Animals for Food,
   Bart Gruzalski   251
Postscript, **James Cargile**   267

SECTION VI: Animal Rights?

Introduction   273
Do Animals Have a Right to Life?, **James Rachels**   275
On Why We Would Do Better To Jettison Moral Rights,
   R. G. Frey   285

SECTION VII: Breadth of Vision

Introduction   305
Philosophy, Ecology, Animal Welfare, and
   the 'Rights' Question, **Michael W. Fox**   307
Deciding What to Kill, **T. Nicolaus Tideman**   317

SECTION VIII: Facts and Acts

Introduction 325

Chicken–Environment Interactions, **W. B. Gross** 329

Against A Scientific Justification of Animal Experiments,
  Deborah G. Mayo 339

"Animal Liberation" as Crime: The Hawaii Dolphin Case,
  Gavan Daws 361

Fighting for Animal Rights: Issues and Strategies, **Henry
  Spira** 373

Epilog 379

Works Cited 381

Index 391

CONTRIBUTORS

ANNETTE C. BAIER · Department of Philosophy, University of Pittsburgh, Pittsburgh, Pennsylvania

LAWRENCE C. BECKER · Department of Philosophy and Religion, Hollins College, Hollins, Virginia

THOMAS L. BENSON · Department of Philosophy, University of Maryland, Baltimore County, Baltimore, Maryland

JAMES M. BUCHANAN · Center for the Study of Public Choice, Virginia Polytechnic Institute and State University, Blacksburg, Virginia

JAMES CARGILE · Corcoran Department of Philosophy, University of Virginia, Charlottesville, Virginia

STEPHEN R. L. CLARK · Department of Moral Philosophy, University of Glasgow, Glasgow, Scotland

GAVAN DAWS · Research School of Pacific Studies, The Australian National University, Canberra, Australia

MICHAEL W. FOX · Institute for the Study of Animal Problems, Washington, District of Columbia

R. G. FREY · Department of Philosophy, University of Liverpool, Liverpool, England

WALTER B. GROSS · College of Veterinary Medicine, Virginia Polytechnic Institute and State University, Blacksburg, Virginia

BART GRUZALSKI · Department of Philosophy and Religion, Northeastern University, Boston, Massachusetts

DALE JAMIESON · Department of Philosophy, University of Colorado, Boulder, Colorado

EDWARD JOHNSON · Department of Philosophy, University of New Orleans, New Orleans, Louisiana

DEBORAH G. MAYO · Department of Philosophy and Religion, Virginia Polytechnic Institute and State University, Blacksburg, Virginia

HARLAN B. MILLER · Department of Philosophy and Religion, Virginia Polytechnic Institute and State University, Blacksburg, Virginia

JAN NARVESON · Department of Philosophy, University of Waterloo, Waterloo, Canada

JAMES RACHELS · Department of Philosophy, University of Alabama in Birmingham, Alabama

TOM REGAN · Department of Philosophy and Religion, North Carolina State University, Raleigh, North Carolina

BERNARD E. ROLLIN · Department of Philosophy, Colorado State University, Fort Collins, Colorado

DUANE M. RUMBAUGH · Department of Psychology, Georgia State University, Atlanta, Georgia

SUE SAVAGE-RUMBAUGH · Yerkes Regional Primate Research Center, Emory University, Atlanta, Georgia

PATRICK F. SCANLON · Department of Fisheries and Wildlife Sciences, Virginia Polytechnic Institute and State University, Blacksburg, Virginia

HENRY SPIRA · New York, New York

T. NICOLAUS TIDEMAN · Center for the Study of Public Choice, Virginia Polytechnic Institute and State University, Blacksburg, Virginia

DONALD VANDEVEER · Department of Philosophy and Religion, North Carolina State University, Raleigh, North Carolina

PETER S. WENZ · Department of Philosophy, Sangamon State University, Springfield, Illinois

PREFACE

This volume is a collection of essays concerned with the morality of human treatment of nonhuman animals. The contributors take very different approaches to their topics and come to widely divergent conclusions. The goal of the volume as a whole is to shed a brighter light upon an aspect of human life—our relations with the other animals—that has recently seen a great increase in interest and in the generation of heat.

The discussions and debates contained herein are addressed by the contributors to each other, to the general public, and to the academic world, especially the biological, philosophical, and political parts of that world. The essays are organized into eight sections by topics, each section beginning with a brief introduction linking the papers and the sections to one another. There is also a general introduction and an Epilog that suggests alternate possible ways of organizing the material. The first two sections are concerned with the place of animals in the human world: Section I with the ways humans view animals in literature, philosophy, and other parts of human culture, and Section II with the place of animals in human legal and moral community. The next three sections concern comparisons between human and nonhuman animals: Section III on the rights and wrongs of killing, Section IV on the humanity of animals and the animality of humans, and Section V on questions of the conflict of human and animal interests. Section VI focuses on the notion of moral rights and the propriety and utility of ascribing such rights to nonhumans. Section VII consists of two papers, both of which argue that the perspective of the rest of the volume is too constrained; and Section VIII is concerned with two sorts of factual questions—those concerning scientific research with animals and those concerning the actions of 'animal liberationists.'

Although this volume was not designed as a textbook, it could serve very effectively as the main required work in a course on the morality of human treatment of animals. Instructors of such courses might find the alternate organizational suggestions of the Epilog useful.

All the essays are recent, and all but one are appearing here for the first time. In contrast to several other recent collections on this topic, this volume contains some spirited defenses of some human uses of other animals, as well as attacks on such uses. In addition it contains factual treatments of morally relevant research on ape language learning and the welfare of chickens, and a criticism of animal research primarily on epistemological rather than ethical grounds.

The essays presented here arose out of papers and comments delivered at, or were prepared by other participants in, our conference, The Moral Foundations of Public Policy: Ethics and Animals, which was held on the campus of Virginia Polytechnic Institute and State University, May 24–27, 1979. That conference was made possible by a grant from the Virginia Foundation for the Humanities and Public Policy and matching funds from VPI & SU. We are grateful to both for their support. We are indebted as well to the participants in that conference. We thank those whose essays appear here both for their contributions and for their patience during the lengthy period between the conference and the publication of this collection. In addition we wish to express our gratitude to those panelists whose contributions to the conference could not (for various reasons) be represented by essays in this volume: Thomas L. Carson, Joseph J. Franchina, and Peter Singer.

Our special thanks are due to our departmental secretaries, Betty Davis and Jeanne Keister, without whose efforts neither the conference nor this volume could have seen the light of day.

Harlan B. Miller
William H. Williams

Introduction

'Platonists' and 'Aristotelians'

Harlan B. Miller

> [P]lants exist for the sake of animals and the other animals for the
> good of man, the domestic species both for his service and for his
> good, and if not all at all events most of the wild ones for the sake of
> his good and of his supplies of other kinds, in order that they may fur-
> nish him both with clothing and with other appliances. If therefore na-
> ture makes nothing without purpose or in vain, it follows that nature
> has made all the animals for the sake of men. (*Politics* 1256b16–23,
> Loeb translation [H. Rackham])

Thus Aristotle. Yet, even though he can say that all (other) animals are
for the sake of man, Aristotle represents, or stands near the head of, that
tradition in Western thought most sympathetic to the claims and to the
standing of nonhuman animals. For Aristotle, as for Darwin, man is one
animal among the others, different surely, primary perhaps, but animal
certainly.

In contrast, for Plato man is not properly an animal at all, but a sort
of god trapped in an animal. For the body, and the lower two of the three
parts of the soul, is a *lower* animal for Plato, and the true soul is degraded
by its association with this 'beast'. When, late in the *Republic,* Plato produces yet another metaphor for his tripartite soul, it is 'reason'
(logistikon) that is a human, while the 'spirit' (thumos) is a lion, and the
'appetites' (epithumia) are a "many-headed beast" (588c–e).[1]

The identification of the body and that 'part' of the soul that operates
the body as 'animal' is a persistent theme in Western thought. It recurs in
the use of "you animal!" as a rebuke to sexual or (secondarily) nutritive
advances. (See Stephen Clark's essay in Section IV of this volume.)

---

[1]Citations in parentheses throughout the volume refer to the consolidated list of works
cited found at book's close.

1

Of course the Plato/Aristotle contrast is largely fictitious. With some care, one can find Aristotle saying Platonic-sounding things about nonhuman animals, and Plato occasionally recognizes the continuity between humans and nonhumans. At the end of the *Timaeus* (90e–92c) Plato presents a theory of evolution (or perhaps it should be called devolution) in which all other animal species are descended from man. In this story descent is definitely downward, and it should be noted that the first step downward from man is woman.

But we will follow many others in using caricatures of Plato and Aristotle as totems for opposing aspects of our cultural heritage. That no specific claims about the historical Athenian and Stagirite are intended will be indicated by the use of scare quotes and lower-case letters.

The 'platonic' tradition, then, though usually admitting that human beings have something of the animal about them, at least in the fallen state, holds that in essence man is not an animal at all, but a being of a higher sort. The 'aristotelian' tradition classifies man as an animal, perhaps even (with Aristotle) as the paradigm case of animality, though adding that man is superior to all 'lower' animals.

Has this 'platonic'/'aristotelian' opposition any practical significance? What difference does it make for our treatment of nonhumans whether we view them as our distant kin or as not our kin at all? Does not Aristotle conclude, in the quotation above "that nature has made all the animals for the sake of men"? Despite their theoretical differences may not 'platonic' and 'aristotelian' butchers, or cosmetics testers, or philosophers treat nonhumans in exactly the same way? Perhaps.

Perhaps not. In the very same part of Book I of the *Politics* from which our quotation comes, Aristotle defends his doctrine of natural slavery, and indicates his rank ordering of (in decreasing order of merit): (Greek) man, (Greek) woman, natural slave/barbarian, nonhuman animal. One of the ways that barbarians demonstrate their barbarity is in their inability to distinguish between inferiors, not grasping the difference between women, who should be persuaded if possible, and slaves, who should always be commanded (1252b5-8). "The ox is the poor man's slave" (1252b12). The point of recalling these passages is not to stimulate the adrenalin production of liberated woman readers, but to suggest that there is a difference, if only in potential status, between being entirely excluded from the realm of moral concern (as are animals in 'platonic' thought) and being included in a moral continuum, even near the bottom. Few now would dare publicly voice a view of foreigners (barbarians) as intrinsically morally inferior. Surely a majority of adult males in the English-speaking world would pay at least lip service to the moral equality of women. Once a subordinate group is granted any moral standing whatever, history seems to suggest that those who would maintain the group's subordination will be placed on the defensive. If all the human race were 'platonic,' the nonhuman animals could expect little, but if all humans were 'aristotelians' the 'beasts' could hope.

'Platonists' have not all regarded the treatment of nonhumans as a matter of moral indifference. Both St. Thomas Aquinas and Immanuel Kant held it wrong to abuse animals, because a human who so treats animals becomes more likely to mistreat humans, or shows a defective character. Here is it not that the animals are themselves given any moral standing, but that behavior in regard to them is important because of the analogy (not kinship) they bear to humans.

In Christianity the 'platonists' have generally had things their way. Aquinas is only one of the most distinguished (and more tender-hearted toward animals) representatives of the majority view. St. Francis of Assisi and Albert Schweitzer are probably the best known members of the minority. At present, if a member of the Christian clergy begins seriously to wonder if there is salvation for dogs and cats, he or she is quite likely to be either laughed at, sent for psychiatric examination, or excommunicated. (Just which alternative is selected will depend on the denomination and the age of the cleric.)

Both in his view of animals and in other matters, René Descartes is more 'platonic' than Plato. Animals, and human bodies, are machines reacting to environmental impacts in purely mechanical ways. The human body, *qua* body, is not different in kind from the horse body, but the human body has associated with it, in a strange and intimate way, something quite different—a soul. It is only souls that think, and 'thinking' includes all sentience (apparently), thus no nonhumans are sentient. By vivisecting nonhuman animals one may, therefore, learn much about the 'machine' common to humans and nonhumans, and may disregard the screams, since only humans can feel pain.

Despite Descartes and a long line of influential Cartesians, the 'aristotelian' tradition is in many ways dominant today. It was only a century from Descartes' 'demonstration' that animals are machines to La Mettrie's corollary that humans are machines in exactly the same way (1748). The Darwinian revolution consists in large part of stressing an 'aristotelian' view of nonhumans. Darwin's challenge to 'platonism' is not simply in his reconciliation of historical continuity with existing differences between species [in *The Origin of Species* (1859) and *The Descent of Man* (1871)], but in the systematic treatment of man as one animal among others even in 'mental' ways [as in *The Expression of The Emotions in Men and Animals* (1872)]. In the latter matter Darwin was, as he well knew, in the tradition of Aristotle, and both then and now the most vehement anti-Darwinists are usually the most dogmatically 'platonic'.

Today most reflective Westerners are probably both 'aristotelian' and 'platonic' in more or less unstable equilibrium. Convinced materialists (Marxist and non-Marxist) are officially 'aristotelian', as are many nonmaterialists, but they usually show by their actions and reactions that they are extreme human chauvinists. On the other hand, the avowed religious beliefs of a large part of the population of the Americas and Western Europe commit them to 'platonism', but only fundamentalists (and not all of them) would deny that humans are animals. (Several years ago, at a party attended primarily by academic couples, a woman declared passionately that "Man isn't an animal." The silence of the stunned audience was broken after a few seconds by a chemist's "What is he then? Vegetable or mineral?" Thus at the level of simple declaration, it seems clear that convinced 'platonists' and 'aristotelians' have little to say to one another. For genuine argument to be possible both parties must move to even more basic questions.)

WHY ANIMALS NOW?

There have always been some humans concerned about human treatment of nonhuman animals. Most of us, most of the time, 'platonist' or 'aristotelian', have taken exploitation of other animals for granted. There are scattered voices through the centuries protesting particular practices, but in general, in the Western world, widespread interest in the welfare of nonhumans is a recent development. Statutes forbidding cruelty to animals date from the late 19th century at the earliest.

Quite recently (since the Second World War and especially since about 1970) there has been a great increase in the extent of concern, at least expressed concern, for nonhumans. One way in which this has been manifested is in the considerable increase in the number, membership, activity, and visibility of 'animals' organizations such as the Fund for Animals, the Humane Society, Greenpeace, the Whale Protection Fund, Friends of Animals, Society for Animal Rights, National Wildlife Federation, International Primate Protection League, and so on. Not only has there been an increase in the quantity of expressed concern, but in addition there has been a shift in quality, the growth of considerable diversity. Alongside the continuing (and expanding) 'establishment' activities of running shelters, lobbying for protective legislation, sponsoring educational programs, and so on, there has developed a new 'activist' movement, sometimes under the banner of 'animal liberation.' The activists have confronted whalers on the seas and sealers on the ice, they have picketed research centers and 'liberated' dolphins in Hawaii and experimental subjects in New York (see the papers by Daws and Spira in Section VIII).

This same period (since about 1970) has seen an extraordinary proliferation of literary attention to nonhumans. The books produced have been of many sorts: sympathetic fiction such as Richard Adams's *Watership Down* and *Plague Dogs* (1972 and 1978), catalogs of abuses such as Hans Ruesch's *Slaughter of the Innocent* (1978) and Richard Ryder's *Victims of Science* (1975), and philosophical works such as Mary Midgley's *Beast and Man* (1978) and Tom Regan and Peter Singer's anthology *Animal Rights and Human Obligations* (1976). The best known of all these recent books is probably Peter Singer's *Animal Liberation* (1975) a moral argument with a catalog of abuses. One further result or symptom of recent interest in animals has been the new, or at least newly increased, respectability of such interest in the circles of professional philosophy.

To what may this burgeoning attention to nonhumans be ascribed? There are at least six distinguishable factors. In arbitrary order the first is the momentum of 'liberation'. Once national chauvinism, colonialism, racism, and sexism have been vanquished, at least intellectually, it seems only reasonable to attack 'speciesism.' Put less tendentiously, the social history of the last two centuries can be seen in part as a continuing struggle to enlarge the boundaries of moral community. This continuity of 'animal liberation' with the other 'liberation' movements has been used as a 'slippery slope' argument against animal liberation as leading inexorably to 'plant liberation' and then on toward 'rights for rocks.' Historically, this continuity has frequently been exploited in the opposite direction. In the late 18th century Thomas Taylor (apparently) wrote *A Vindication of the Rights of Brutes* (1792) in response to and as refutation-by-spoof of, Mary Wollstonecroft's *A Vindication of the Rights of Women* (1792). After all, as the argument goes, if one lets women vote, why not cats and dogs? The linkage between women's liberation and animal liberation by enemies of both still lives nearly two centuries later in 1977, as indicated by the following quote from Michael E. Levin (1977):

> And, in truth, the last decade has seen an eruption of irrationality combined with a deep respect accorded to protest and 'liberation' per se. Anyone who perceives women's liberation as the boundless self-assertion of its female protagonists and the grotesque self-abasement of its male protagonists will appreciate Singer's analogy [of animal liberation to women's liberation]. For what is women's liberation but an appeal to guilt about being a man and acting in such ways as manhood demands? And what is this new discovery of rights of animals other than guilt about being *human* and doing those things necessary for a satisfactory human life."

A second factor in the new attention to nonhumans has been the recently acquired knowledge of certain sorts of facts about some kinds of animals. The linguistic attainments of chimpanzees and more recently of other of the great apes are quite impressive. Descartes took the use of language to be a central criterion for the presence of genuine thought.

[I]t has never yet been observed that any brute animal reached the
stage of using real speech, that is to say, of indicating by word or sign
something pertaining to pure thought and not to natural impulse. Such
speech is the only certain sign of thought hidden in a body. All men use
it, however stupid or insane they may be, and though they may lack
tongue and organs of voice; but no animals do. Consequently it can be
taken as a real specific difference between men and dumb animals (letter from Descartes to Henry More, February 5, 1649).

Modern Cartesians are compelled either to admit that the apes 'think,'
and thus cease to be Cartesians, or to deny that the behavior of these apes
is truly linguistic. That the second maneuver becomes more and more
desperate will be apparent to the reader of Rumbaugh and Savage-
Rumbaugh's paper in Section IV. The recent demonstrations of the intelligence of whales and porpoises poses a challenge to human chauvinism
that is even more basic. It is easy enough to see the apes as quasi-humans,
or failed, defective, humans. But if we are compelled by the facts to accord full-fledged moral status to cetaceans, then we must admit that persons need not even resemble humans. Here 'aristotelians' may be driven
to move beyond Aristotle. Perhaps humans are not the ultimate animals,
perhaps animality is not to be seen as derivative from a single paradigm,
or (worse yet) perhaps humanity itself is to be seen as an imperfect form.

The third (in this list) factor drawing attention to the moral status of
animals is the abortion controversy. The connection between abortion and
animals is not immediately apparent. For quite some time there was a
practical consensus on the prohibition of abortion that masked theoretical
disagreement. As long as abortion was a risky procedure with a significant chance of maternal death, those who considered the fetus a person
could make common cause with those who were concerned only (in such
cases) with diminishing the threats to women's health. As the medical
risk to the mother decreased to the point that abortion became safer than
childbirth, this unity came apart. Now issues that were previously ignored
or swept under the rug became central. Is the fetus a person? The concept
of a person moves to center stage. What sorts of things need we be morally concerned about? What sorts of things do not matter? If 'human' is
not the decisive concept (human fetuses are unquestionably human), then
what is? If humanity cannot be taken as morally significant without
begging important questions, how can we justify our uses of nonhumans?

A fourth factor, less esoteric than it might initially appear, is the current state of the philosophy of mind and related fields such as cognitive
psychology and neurophysiology. Dualistic ('platonic') views of mind
and body become less and less plausible. If, as it seems we must, we are
to conclude that our mental life is somehow identical with or an aspect of
the activity of our central nervous system, then since there is no sharp
boundary between humans and other animals in the complexity of nerv-

ous systems, there seems no justification for assuming a sharp differentiation in the type or level of sentience or ability to reflect.

Connected with this is the fifth factor, a group of discoveries and trends in the biological sciences, including sociobiology and ethology, that lend themselves to an attempt to draw conclusions about human social patterns from observations of the behavior of other species. This line of thought can show humans much about humanity, but only if one is willing to look at our species as a species, and as one species among others. Here too 'aristotelian' views of humanity are dominant, and are being pressed toward a denial of human uniqueness. *Homo sapiens* is a species, and not even a very special species.

The sixth factor, perhaps in some ways the most important of all, is the widespread, amorphous social movement marked by terms such as 'environment,' 'ecology,' and 'organic.' Sometimes, of course, these are mere buzz-words, but often they mark something genuine even if confused. There is a generally increased awareness of nature, of humanity's interdependence with other species and with all of the natural and artificial world. Many have come to feel that our culture has ravished the environment and slaughtered the animals, and that this has been not merely a waste or an inefficiency, but rather a crime akin to matricide. That recognition of ecological order will bring with it respect for other species is one theme of Michael Fox's paper in Section VII.

(There may be a seventh factor that merits at least parenthetical mention. The new respectability of science fiction and the recent discoveries about the other planets of our system have brought popular consciousness to an acknowledgment that somewhere 'out there' there are probably other minds, other intelligences, other persons. Of course these other persons will not be humans, and thus it follows that humanity is not necessary for personhood. Perhaps if we could learn to communicate with the cetaceans of Earth, we would be better prepared to talk to the creatures of the Proximi Centauri system.)

Having rehearsed some of the factors that make us more concerned about nonhuman animals than were previous generations, one must say something about the curious inconsistencies that sometimes mark modern interest in animals. We are, almost all of us, much more distant from the animals we eat than were our grandparents or great-grandparents, and we think little, if at all, about the origin of the meat that somehow arrives, neatly wrapped and labeled, in our supermarkets. If a use of animals is 'efficient' or (especially) 'scientific,' we give it no further thought, as long as it is out of sight. But some species have special places in human hearts. When, in 1973, it was revealed that the US Army was testing chemical warfare agents on beagles, a storm of protest resulted. Yet very few people protest similar tests on rats. In 1979 there was extensive protest in Great Britain against the export of horses to the Continent for food, but not against the export of cattle. Why beagles and not rats, horses and not cattle? Well, beagles are, in general, our friends and allies, and wild rats our enemies and competitors, but not these beagles or these rats. I know of nothing to say in defense of a significant difference in status between horses and cattle.

It is probably best simply to admit that most of us are less than wholly consistent about animals. We have little tolerance for clear cruelty, but also little stomach for inquiry into animal uses beyond our sight. (In his paper in Section I, Tom Regan argues that cruelty is not really the important issue.)

USING THE OTHERS

We use other animals in many ways: we eat them, experiment upon them, test an enormous number of substances on them, hunt and kill them for entertainment, race and fight them for entertainment, produce a wide variety of products from the uneaten parts of their bodies, and keep them as pets and as servants.

In the United States, the land animals we eat primarily are cattle, swine, chickens, and turkeys. We rarely think of these animals as animals, least of all when parts of their bodies are delivered to us ready-to-eat with "special-sauce-on-a-sesame-seed-bun." Yet they are animals, and the way that they live (as well—as James Cargile insists in Section V—as whether or not these particular animals live at all) depends on our consumption of them. Are the restrictions of the veal calf, the feedlot steer, the battery egglaying hen, worth the gain to humans? Although some aspects of some of our current farming practices (veal-calf raising, for instance) seem pretty surely morally suspect, arguments such as those of Walter Gross in Section VIII should make one reluctant to rush to judgment.

We also use animals, large numbers of them, in research and testing. Some of this research and testing is directed solely to determining the effect of a compound or a procedure on the very species on which it is tested. About such research we can ask whether the hoped-for gain in knowledge is worth the cost in experimental suffering. (One might conceivably learn something of some interest from a comparative study of reactions to being burned on the eyeball with lighted cigarettes, with a variety of humans as experimental subject (Anglo-Saxons, blacks, Japanese, Jews, Iranians, say). Would the possible gain in knowledge justify conducting such research?) When, on the other hand, the research or testing on animals is directed to discovering something about the human species, a second sort of question arises. Questions of this epistemological sort are addressed by Deborah Mayo in Section VIII.

Some sorts of experiments on animals are not intended to increase the stock of human knowledge, but rather to contribute to the education of individual humans. Introductory biology instruction from the junior high school through the college level often involves dissection of animals. Dissection by beginners, many believe, is likely to teach the beginners less than the observation of a dissection by an expert, or than the disassembly of a model. The animals dissected in introductory classes are usually killed in advance, but in more advanced student 'research' (from which no new knowledge is expected) the infliction of suffering on live animals is not uncommon. Some particularly disturbing instances of live animal 'research' have taken place in science fairs. Many have argued that a revision of traditional methods in biology instruction would not only markedly decrease animal suffering, but also significantly improve student learning. (See, for example, Orlans, 1980.)

The various sorts of ways in which humans prey on wild (i.e., nondomesticated) nonhumans can be grouped under the head of 'hunting.' This includes sport hunting, subsistence hunting, almost all fishing, trapping, and whaling. Very few voices are raised in opposition to subsistence hunting (though some lament its necessity) or to fishing (including the collection of shellfish, crabs, etc.), probably in part because of the widespread conviction that the level of sentience of these harvested water creatures is significantly lower than that of mammals.

Sport hunting, trapping, and whaling, in contrast, elicit widespread protest. In the case of whaling, the primary cause of opposition is the perceived importance of the cetaceans—morally, aesthetically, and/or ecologically—and the lack of any necessity for their use. Trapping is seen to be a cruel and extremely painful method of satisfying a demand now provided only (or at least primarily) by vanity. Legislation to ban leg-hold trapping is regularly introduced in the state legislatures of the United States (usually dying without a hearing).

Sport hunting is not, as a rule, as cruel as trapping, nor are the 'harvested' deer, squirrel, bear, and so on, the obvious intellectual or moral peers of the great whales. But sport hunting is a much bigger business (in the United States) than either whaling or trapping, in numbers of participating hunters, in the direct effect on nonparticipants, in advertising, and in visibility. That hunting, in the United States and probably everywhere, has a ritual character, that it is a rite of autumn in which certain values are affirmed and sacral acts performed, cannot be denied. Opponents and (civilized) proponents of hunting do not disagree about 'slob hunting' or about the preferability of quick kills to slow ones. The disagreement, often quite inarticulate, is about the values affirmed and the worth of affirming these values at some cost in animal suffering. (There are further, empirical, disagreements about the net effect of hunting on animal suffering, but these are secondary.) Anti-hunters generally perceive the values affirmed to be those of exploitation and of 'macho' domination by violation, while reflective hunters such as Patrick Scanlon in his paper in Section IV see the affirmed values to be those of humanity's honest acceptance of itself as a predatory species. Is it better, Scanlon asks, to kill without admitting to oneself that one is doing so (as when one sprays a garden or drives a car) or to face killing, and ecological interdependence, directly?

There are other sports in which humans use animals. In the English-speaking world we have no bullfights, but we do have dog fights (even though they are almost everywhere prohibited), chicken fights (also prohibited), horse races, rodeos, dog races (with rabbits as prizes and training aids), and the quasi-sports of dog shows, horse shows, and so on. Rodeos have attracted considerable fire, both because of the way in which animals are handled in the actual events (steer wrestling is hard on steers) and because of the methods and devices used to induce violent reactions such as bucking. The bucking strap tight across the genitals is a main target of protest. The use of rabbits in greyhound racing and training is also an instance of clearly brutal treatment of animals with no justification other than human entertainment.

Two (at least) human uses of animals remain to be mentioned. Dogs, cats, parrots, gerbils, and many other sorts of animals are our pets, and dogs and horses are sometimes our servants. Is this wrong in itself? It is surely not obviously so. Yet many human wrongs to nonhumans are wrongs to pets or ex-pets. To discard an animal because it is no longer as cute as it was when a kitten or puppy, or because it no longer fits one's color scheme, is to treat a living, sentient being as if it were a lampshade or slipcover (for some examples of this, see the film *The Animals are Crying*). Pets are dependent upon us in ways that wild animals are not, and though one might conceivably be charged with paternalism or 'interference with nature' if one has one's pet spayed or neutered, the alternative is often litter after litter of puppies or kittens that find no homes and will die sooner or later, quickly or slowly. Finally, what of animals in zoos? Zoos have helped to preserve species that would otherwise have become extinct, and have provided the resources necessary to teach us much we would not otherwise have known. But does this give us a blanket justification for sentencing wild creatures to a life in captivity?

QUESTIONS OF THEORY

The status accorded nonhuman animals in ethical and political theories of course varies considerably from theory to theory. There are at least possible 'platonic' or Cartesian theories on which the treatment of nonhumans is of no moral significance whatever. More common, or at least more widely espoused, are theories such as those of Aquinas and Kant on which, although cruelty to animals is genuinely and sincerely deplored, animals themselves are ascribed no moral standing. To treat animals cruelly is wrong because it tends to harden one's heart toward one's fellow humans, or shows that one's heart is already hard.

The most prominent of the moral theories clearly committed to ascribing moral status to animals per se is utilitarianism. The founder of modern hedonistic utilitarianism, Jeremy Bentham, was characteristically clear about this. If, as Bentham (1789) holds, the right act is that act that gives the greatest net amount of pleasure (or least net amount of displeasure), then the only morally important question is whether or not nonhumans are capable of pleasure and displeasure:

> [A] full-grown horse or dog is beyond comparison a more rational, as well as a more conversable animal, than an infant of a day, or a week, or even a month old. But suppose the case were otherwise, what would it avail? the question is not, Can they *reason,* nor Can they *talk*? but, Can they *suffer*?"

Though John Stuart Mill rejected Bentham's exclusively quantitative comparison of pleasures, insisting that pleasures differ in quality, he still takes animal pleasure and suffering to be of moral importance on their own. According to Mill (1863), it is "better to be a human being dissatisfied than a pig satisfied," but this is quite compatible with holding that of two situations differing only in that one contains a satisfied pig and the other a dissatisfied pig (or a neutral pig, or no pig at all) the first is preferable, morally preferable.

The best known philosopher of modern animal liberation, Peter Singer, is an avowed utilitarian. Singer insists that equal quantities of pleasure or displeasure be weighed equally, regardless of the species of the being experiencing the pleasure or displeasure. But it would be easy to suppose that there is more difference from Mill than in fact there is. Singer is at pains not to deny that humans (or at least most of them) can enjoy sorts of pleasures (of anticipation, reflective self-satisfaction, and so on) and suffer sorts of displeasures (remorse, dread for the fate of one's grandchildren, and so on) that just are not possible for nonhuman animals (or at least most of them).

Other thinkers find the utilitarian emphasis on pleasure and displeasure inadequate. For example, if I am on balance unhappy, and would continue to be so, and if no one else cares whether I live or die, then must not a utilitarian hold that killing me painlessly (and unexpectedly) is not only permissible, but morally obligatory? (For a variety of objections to utilitarianism, and responses to these objections, see the papers in Miller and Williams, 1982.) In the present volume, Tom Regan, in Section I, and James Rachels, in Section VI, argue that rights must be ascribed to animals on the basis of some of the shared characteristics of humans and nonhumans. Another case of arguing for the ascription of rights to nonhumans is contained in the position statement issued in 1979 by the (US) National Coalition for Alternatives to Animal Experimentation:

> Animals, like people, are not mere things. People have lives that are valuable whether or not they are of use to others. Animals, too, have lives that are valuable whether or not they are of use to others. People have rights. Animals, too, therefore, have rights. To treat them as mere things, as mere tools to be used for human pleasure, profit or curiosity, is to violate their rights.
>
> What is it about our lives that we value? Pleasure, and a minimum of pain; companionship; the satisfaction of wants and needs. These, certainly, are among the things that give value to our lives.
>
> Animals, too, have lives that are valuable to them in these ways. Animals, too, experience pleasure and pain, have needs, wants, seek companionship. In these ways at least we share a common nature with them. In these ways at least the value of animal and human life has a common basis.
>
> Does this mean that animals have the same rights people do? Must we say that they have a right to vote because we have this right? No, we need not say this. What we must say is that there are certain rights we share based upon our common nature, especially the rights not to be made to suffer, or to be killed merely for the pleasure, profit, or curiosity of another species.
>
> Yet animals are exploited in laboratories, in zoos, in modern "factory" farms, in schools. They are trapped and hunted, they are abused in sports and entertainment. In all these ways and more, animals are treated as mere things, as if they had no value in themselves.
>
> Such treatments must be stopped, not only for the sake of our humanity, but because the animals in their own right are entitled to just treatment. When rights are violated, justice, not kindness, is at issue. Their defenseless state, their inability to speak out for themselves, makes it our duty to speak out for them.
>
> They will continue to suffer and be killed unless we act for them. And act we must. Respect for justice requires nothing less.

Such appeals to rights, R. G. Frey argues in Section VI, gain nothing. Frey's position is that the very notion of a moral right only obscures the landscape of moral argumentation, that to say that a person or animal has a right not to be treated in a certain way is only to say, with the postulation of a gratuitous abstract entity, that it is wrong to treat the creature in that way.

Those who talk of animal rights and those who (like Peter Singer) eschew talk of rights agree that what is in question is the moral standing (or lack thereof) of nonhuman animals. To have moral standing is to be either a moral agent or a moral patient (the 'or' is, of course, inclusive). To be a moral agent is to be an entity capable of actions that may appropriately be evaluated as right or wrong. To be a moral patient is to be an entity of such a sort that what is done to that entity by a moral agent is per se, subject to moral evaluation. Normal human adults are both moral agents and moral patients, as they can be praised or blamed for their own actions and other agents can properly be praised or blamed for what they do to adult humans. Human infants and very young children are clearly and indisputably moral patients (it is wrong to cause gratuitous pain to an infant), but surely not moral agents. Animal liberationists would put most (or all) nonhuman animals into this category. Some 'abstract' agents such as corporations or nations may be clearly moral agents, but either clearly not moral patients (most corporations could not suffer in any ethically important way) or debatably moral patients (nations, families). (Most of us would consider a corporation's loss of no moral importance per se, even though the losses to stockholders, officers, and employees may be quite important.)

Just as humans passing through childhood and adolescence are generally seen as increasingly morally responsible for their acts—that is, as becoming moral agents in a diminished sense over a limited range and then gradually having the range broaden and the diminishment diminish—so it is quite possible to ascribe at least some sort of moral agency to some nonhumans in some circumstances. The mature and trained sheepdog who abandons the flock to the wolves may be subject to moral condemnation (not just 'negative reinforcement') and the family pet who suffocates trying to drag the (human) baby from the flames (see Reeve, 1978) may deserve the same kind of moral praise as the babysitter or fireman who does the same. Surely some nonhumans may sometimes be seen as moral agents without absurdity. Perhaps some sorts of nonhumans may come to be seen as having a moral status quite comparable to (some or most) humans. (One thinks first of primates and cetaceans, but that may be to cast the net too narrowly.)

The fundamental theoretical question about our relations with other animals is one that can be conceputalized in two different but overlapping ways. What are, and what ought to be, the boundaries of our moral community? In his paper in Section II, James Buchanan argues that the innate capacities of the human psyche place limits on the number, but less on the species, of creatures about whom anyone can possibly be morally concerned. A very nearby dog's welfare may be much more emotionally important to me than that of a distant human. It is folly, Buchanan suggests, to prescribe moral or political forms that ignore the intrinsic limitations of the human heart and mind. 'Ought' must not exceed 'can.' One may agree with this without agreeing with Buchanan about the limits of 'can.'

The main alternative to 'moral community' talk involves the analysis of personhood. Normal adult humans are persons, but 'human' and 'person' are not synonymous. The three persons of the Christian Trinity are not (most of them) humans. Mr. Spock of *Star Trek* is clearly a person, but not a human. Perhaps some cetaceans are persons, and perhaps some nonhuman primates are. But however central the concept of a person may

turn out to be in ethics (and in the philosophy of mind), it is clear that
nonpersons ('quasi-persons'? 'sub-persons'?) may well be moral patients
even if they are not full-fledged moral agents. It might be wrong (or
wrong-headed) to condemn a hamster mother for devouring her newborn,
but it is neither wrong nor wrong-headed to condemn a human for sadist-
ically torturing that same hamster. Perhaps persons are the paradigm, the
central, members of the moral community, but they are not the only
members. It *does* matter how we treat animals.

WHY DOES MORAL PHILOSOPHY MATTER?

There are those humans who believe that questions about the propriety of
human treatment of nonhumans are to be resolved by reflective intuition,
that one only need engage both left brain and right brain and contemplate
the facts to arrive at the right answers. That the significance of the left
brain/right brain distinction is very dubious is probably the least important of the defects of this position. It seems quite clear that the majority of
English-speaking humans, on reflection, would agree that most of the
many sorts of present human treatments of nonhumans are morally acceptable. It is equally clear that a minority of humans believes that many
of those sorts of treatments are unacceptable. What, then, are we to do?
We could simply choose up sides and vote, or choose up sides and fight.

Whether either, neither, or both of these choices is acceptable depends not so much on one's primary, substantive beliefs about the rights
and wrongs of human treatment of nonhumans, as on one's commitments
to secondary, procedural rules about the proper organization of human
society. It is possible, of course, to choose that sort of society in which
whatever the majority (or the aristocracy or the workers and peasants)
chooses is *ipso facto* considered to be correct. But if one wishes to live in
a society in which one may reasonably be asked for one's reasons, in
which (imperfectly) rational animals attempt to reason one with another,
then one must come, soon or late, to recognize the fundamental importance of moral theory. If your moral intuitions and mine conflict in a particular case, simple assertions that "It's right" and "It's wrong" are
nearly certain to get us nowhere. But once you claim that I am mistaken
about the facts, or I argue that your position in this case is inconsistent
with your position in another case, or you point out that I am failing to
comply with principles I have previously espoused, we are engaged in
moral argumentation, the stuff of moral philosophy.

Moral argument is often quite complex and frequently includes a
large component of factual argument. The essays in this volume join with
one another and with an ill-defined body of earlier work to increase and
(one hopes) to forward the continuing human argument about the moral
status of nonhuman animals.

SECTION I

Nonhumans in the Eyes and Minds of Humans

Introduction

The four papers comprising this Section are concerned with choices between ways of thinking about nonhuman animals.

Tom Regan describes three general accounts of the morality of human treatment of animals, which he labels the 'Kantian,' 'cruelty,' and 'utilitarian' accounts. Kant's view and that of the utilitarians were sketched in the Introduction. On the cruelty account, the reason we must not mistreat animals is that cruelty is wrong. Regan argues that all three are inadequate, and proposes a rights account. Rights are grounded in 'inherent value,' and at least many sorts of animals possess such value because they are not just alive, but have a life. Regan suggests a set of necessary conditions for the justifiable overriding of a right and concludes that, given that animals possess a right not to be harmed, a great many current practices constitute the unjustified denial of rights. We are thus, he holds, obligated to try to stop such practices.

Jan Narveson also maps the terrain of possible views of our moral relation to animals, but his coordinate system is not the same as Regan's and he produces a different set of alternatives. There are, Narveson holds, three plausible candidate views about animals, corresponding to the three sorts of live option in modern moral theory. The three moral theories are utilitarianism, libertarianism, and contractarianism. Very roughly, contractarian moral theories hold that moral obligations arise only by agreement and obtain only among agreeing parties. Libertarian moral theories (roughly) are theories in which the rights of individuals are paramount, but individuals have only negative rights (to non-interference) and no positive rights (to assistance).

None of these theories, Narveson argues, holds much promise for nonhuman animals. In connection with utilitarianism and libertarianism, an important part of Narveson's argument is his challenge to the claim that our exploitation of animals renders them worse off than they would otherwise be. (This difficult matter moves to center stage in the papers by Cargile and Gruzalski in Section V.) On a contractarian view (to which Narveson inclines) there is no room for moral duties toward those who cannot be considered to have made agreements, i.e., to the vast majority of animals.

Annette Baier's paper first addresses the relation between moral theory and moral intuition. No coherent theory can incorporate genuinely irreconcilable intuitions, and concerning the treatment of animals the intuitions of some of us seem genuinely irreconcilable with those of others. Which intuitions are we to discount? Intuitions likely to be influenced by special interests, special tastes, and dogmatic commitments are ipso facto suspicious, she argues; and these factors are much less apparent in the friends of animals than in the defenders of the status quo. A moral theory, thus, should be able to account for many of the pro-animal intuitions.

Perhaps Narveson has shown that the three theories he discusses fail to provide much hope for nonhumans. In any case, writes Baier, there is an important moral theory not yet considered (in this volume) on which the treatment of animals is of moral importance in itself. That theory is David Hume's. Hume, on Baier's account, is both Aristotelian in his ethical theory and 'aristotelian' (in the sense of the Introduction) in his view of man's relation to the other animals. We have much in common with other animals and naturally sympathize with them. Further we make tacit 'conventions' with animals as with our fellow men and women, and these conventions are morally binding. (Libertarian thought here differs sharply from Hume.)

On Humean moral theory, as Baier interprets it, the boundaries of moral community extend beyond those of the human species (and, Baier insists, the members of this community vary greatly in status).

Thomas L. Benson's paper is concerned with ways of thinking about animals that are more pervasive, if less articulate, than the moral theories compared in the three previous papers. He is concerned not so much with how philosophers and scientists think about animals when engaged in their theoretic tasks as with how (almost) all of us, including philosophers and scientists when off duty, perceive the animal world. Our perceptions are, he claims, largely dominated by a variety of stereotypes. These stereotypes often conflict with one another, and not all of them are unfavorable to the stereotyped creatures.

The animal stereotypes Benson examines are those of alien, child, moral paragon, demon, and machine. (The Cartesian view of animals and human bodies as machines comes essentially to the claim that mechanical causation alone governs the operations of the animal or body. The machine stereotype that Benson discusses is quite different, for on this view one is not so much interested in *understanding* the operation of the 'machine' as in *operating* it efficiently in pursuit of some goal. It is in this sense that armies and factories are sometimes accused of treating human beings as machines.)

Animals, Benson believes, are fundamentally mysterious, but a mystery is not solved by hiding it behind a stereotype. If we are even to consider making a fundamental change in our dealings with other animals we must first try to rid ourselves of the stereotypes that constrain us.

Animal Rights,
Human Wrongs[1]

Tom Regan

At this moment workers on board the mother ship of a pirate whaling vessel are disassembling the carcass of a whale. Though officially protected by agreement of the International Whaling Commission, pirate whalers operate outside IWC regulations, and it is not too incredible to imagine them butchering a great blue whale, the largest creature ever known to have lived on the earth—larger than thirty elephants, larger even than three of the largest dinosaurs laid end to end. A good catch, this leviathan of the deep. And, increasingly, a rare one. For the great blue, like hundreds of other animal species, is endangered—may, in fact, already be beyond the point of recovery.

But the crew members have other things on their minds. It will take some hours of hard work to butcher the huge carcass, a process now carried out at sea rather than in port. And this is not the only thing in whaling that has changed. The fabled days of the hunt, the individual Ahab pitted against the treacherous whale, must remain the work of fiction now. Whaling is applied technology, from the use of the most sophisticated sonar, to on-board refrigeration; from tracking helicopters, to explosive harpoons, the later a technological advance that expedites a whale's death. Time to die? About five minutes; sometimes twenty. Here is one man's account of one whale's demise.[2]

---

[1]The main body of this essay was originally presented as one of a series of lectures at Muhlenberg College on the general topic of "Problems and Directions of the Post-Abundant Society," March 1979. It has appeared in a slightly revised form in *Environmental Ethics* (Winter 1980) in a Polish translation in *Etyka* (18, 1981), and, as one of ten essays, in Tom Regan, *All That Dwell Therein* (Berkeley: University of California Press, 1982).

[2]Captain W.R.D. McLaughlin, *Call to the South* (London: Harrap) quoted in Dowding, 1971, p. 35.

> The gun roars. The harpoon hurls through the air and the whale-line
> follows. There is a momentary silence, and then the muffled explosion
> as the time fuse functions and fragments the grenade . . . There is now
> a fight between the mammal and the crew of the catching vessel—a
> fight to the death. It is a struggle that can have only one
> result . . . Deep in the whale's vast body is the mortal wound, and
> even if it could shake off the harpoon it would be doomed . . . A sec-
> ond harpoon buries itself just behind the dorsal fin . . . There is an-
> other dull explosion in the whale's vitals. Then comes a series of
> convulsions—a last despairing struggle. The whale spouts blood, keels
> slowly over, and floats belly upward. It is dead.

And if we ask, For what? To what end? In the name of what purpose is
this being done to the last remaining members of an irreplaceable species,
not in remote, barbarian yesterdays, but possibly even at this very moment, by supposedly civilized men, the answer is: For candle wax. And
soap. And oil. For pet food, margarine, fertilizer. For perfume.

In Thailand, at this moment, another sort of hunt, less technologic-
ally advanced, is in progress. The Thai hunter has hiked two miles
through thick vegetation and now, with his keen vision, he spots a female
gibbon and her infant sleeping high up in a tree. Jean-Yves Domalain
(1977) describes what follows:

> Down below, the hunter rams the double charge of gun-powder
> down the barrel with a thin iron rod, then the lead shot. The spark
> flashes from two flints, and the gun goes off in a cloud of white
> smoke . . . Overhead there is an uproar. The female gibbon, mortally
> wounded, clings to life. She still has enough strength to make two
> gigantic leaps, her baby still clinging to the long hair of her left thigh.
> At the third leap she misses the branch she was aiming for, and in a
> final desperate effort manages to grasp a lower one; but her strength is
> ebbing away and she is unable to pull herself up. Slowly her fingers
> begin to loosen her grip. Death is there, staining her pale fur. The
> youngster flattens himself in terror against her bloodstained flank.
> Then comes the giddy plunge of a hundred feet or more, broken by a
> terrible rebound off a tree trunk.

The object of this hunt is not to kill the female gibbon, but to capture the
baby. Unfortunately, in this case the infant's neck is broken by the fall, so
the shots were wasted. The hunter will have to move on, seeking other
prospects.[3]
   We are not dealing in fantasies when, at this moment, we consider
the day's work of the Thai hunter. Domalain makes it clear that both the
method of capture (killing the mother to get the infant) and the results (the

---

[3] For additional information on endangered species, see Amory, 1974 and Regenstein, 1975.

death of both) are the rule rather than the exception in the case of gibbons. And chimpanzees. And tigers. And organgutans. And lions. Some estimate that for every one animal captured alive, ten have been killed; and Domalain states that, for every ten captured, only two will live a few months beyond. The mortality rate stemming from hunts that aim to bring animals back alive thus is considerable.

So, no, we do not fantasize when we regard the female gibbon's weakening grip, the infant's alarmed clutching, the bonds of surprise and terror that unite them one last time as they begin, at this moment, their final descent. And if we ask, For what? To what end? In the name of what purpose is this scene played out, not once, but again—and again—and again—the answer is, So that pet stores might sell "exotic animals." So that roadside zoos might offer "new attractions." So that the world's scientists might have "subjects" for their experiments.

Not far from here, at this moment, a rabbit makes a futile effort to escape from a restraining device, called a stock, which holds the creature in place by clamping down around its neck. Immediately we think of trapping. The stock must be another kind of trap, like the infamous leg-hold trap. We are viewing one of the barbarities of trapping animals in the wild.

This is not so. The stock is not part of the trapper's arsenal. The stock is a handmaiden of science, and the rabbit at this moment confined by it is not in the wild, but in some research laboratory, not far from here. If we look closely we will see that one of the rabbit's eyes is ulcerated. It is badly inflamed, an open, running sore. With the passage of hours the sore increases in size until, at this moment, barely half the eye is visible. In a few days the eye will become permanently blind. In some cases— perhaps in this one—the eye is literally burned out of its socket.

No, this is no animal trapped in the wild. The rabbit we consider is a research subject in what is known as the Draize test, named after its inventor. This rabbit, like the hundreds of others in the neighboring stocks, is being used because rabbits, among other reasons, happen not to have tear ducts and so cannot themselves flush substances from their eyes or dilute them. The Draize test procedes routinely as follows: Concentrated solutions of a substance are administered to one of the rabbit's eyes; the other eye, a sort of control group, is left untroubled. Swelling, redness, destruction of iris or cornea, loss of vision are measured and the substance's eye-irritancy is thereby scientifically established.

What is this substance which, at this moment, in concentrated form, invades the rabbit's eye? Chances are it is a cosmetic, a new variety of toothpaste, shampoo, mouthwash, talcum, hand lotion, eye cosmetic, face cream, hair conditioner, perfume, cologne. So, when we ask, as we must, Why? To what end? In the name of what purpose does this unanesthetized rabbit endure the slow burning destruction of its eye?, the answer is: So that researchers might establish the eye-irritancy of mouthwash and talc, toothpaste and cologne.[4]

One final individual bids for our attention at this moment. A bobby calf is a male calf born to a dairy herd. Since the calf cannot give milk, something must be done with it—some use must be found for it. A common practice is to raise and sell it as a source of veal, as in veal Parmigiana. To make this as commercially profitable as possible the calf must be raised in highly unnatural conditions. Otherwise the youngster would romp and play, as is its wont; equally bad, it would forage and consume roughage. From a businessman's point of view, this is detrimental to the product: The romping produces muscle, which makes for tough meat, and the roughage will contain natural sources of iron, which will turn the calf's flesh red, definitely to be avoided since studies show that consumers have a decided preference for pale veal. So, the calf is kept permanently indoors, in a stall too narrow for it to turn around (recommended size: 1 foot 10 inches wide by 4 feet 6 inches long), frequently tethered to confine it even further, its short life lived out mostly in the dark on a floor of wood slats, its only contact with other living beings coming when it is fed and when, ultimately, it is transported to the slaughter house.

At this moment, then, we can see the tethered calf, unable to turn around, unable even to sit down without hunching up, devoid of companionship, the satisfaction of its natural urges to romp and forage denied, fed a wholly liquid diet kept intentionally deficient in iron so that its pale flesh will not be compromised—intentionally kept, that is to say, in an anemic state. At this moment, if we look, as we must, we can see this solitary creature hunched up in its darkened stall. And if we ask, For what? To what end? In the name of what purpose does the calf live so?, the answer is: So that humans might satisfy their preference for pale veal.[5]

A few cases only—the great blue, now floating belly up, its white underside bobbing through the water like an uninhabited ice floe; the infant gibbon, still clutching its mother, their fused bodies barely perceptible on the jungle floor; the furious movement of the rabbit's feet as it seeks relief from the corrosive liquid which, as certain and painful as a knife, moment by moment cuts away at its optic nerve; the dank immobility of the bobby calf. Multiply these cases by the hundreds; rather, multiply them by the thousands and we approach, perhaps, the magnitude of the death and suffering animals are enduring, at this moment, at the hands

---

[4]Especially important sources on the use of animals in research are Ryder (1975), Ruesch (1978), Vyvyan (1971), and Westacott (1949).

[5]For information on modern factory farming methods, see in particular Harrison (1964) and Singer (1975).

of human beings. As for the numbers, they are like distances in astronomy. We can write them down, compare them, add and subtract them, but as in the case of light years stacked upon light years, we lack the intellectual or imaginative wherewithal to hold them in steady focus. Item: Two million whales killed in the last fifty years. Item: Once, in the mid-nineteenth century, there were upwards of two billion passenger pigeons in the United States; now there are none. Item: Once, when white settlers first set foot on North America, there were sixty million buffalo; now the largest wild herd numbers 600. Item: The bald eagle, the national symbol of the United States, now virtually extinct in all areas save Alaska.

But extinct or endangered species aside, consider these figures for legally imported and declared wildlife brought into the United States, just in 1970, to supply the needs of the pet industry, zoos, and scientific researchers (Regenstein, 1975, p. 121):

101,302 mammals, 85,151 of which were primates (mainly monkeys)
687,901 wild birds, not including canaries, parrots, and parakeets, millions of which are annually imported
572,670 amphibians (frogs, toads, salamanders)
2,109,571 reptiles

In research,[6] in the United States alone, for the year 1978 alone, some 400,000 dogs, 200,000 cats, 33,000 apes and monkeys, thousands of calves, sheep, ponies, horses; rabbits in the millions; hamsters, guinea pigs, rats, birds, mice, all also in the millions; an estimated 64 million animals used in research, in the United States alone, in the year 1978 alone. Worldwide, educated estimates place the figure at about 200 million.

As for slaughter,[7] incomplete data suggest that, worldwide, the number of animals killed for food easily exceeds hundreds of millions for individual species: upwards of a hundred million cattle, one hundred million sheep, two hundred million pigs, twenty-six million calves, and, in the United States alone, four billion chickens.

But numbers prove nothing. They neither establish that something is right nor that it is wrong. In this case they merely confirm what we know already: that the use of animals as a primary food source and as subjects in scientific research knows no geographical boundaries; that the extinction of unique species is not the private stomping ground for one or a few political ideologies; that, in general, aside from economic considerations, and excluding the privileged status in some cultures that pet animals have as honorary members of the family, few things are regarded so cheaply as

[6]These figures are suggested by Ryder (1975). See also his "Experiments on Animals" in Godlovitch and Harris (1971) reprinted in Regan and Singer (1976).

[7]These figures are based on estimates furnished for the year 1968 by the United Nations Food and Agriculture Organization and quoted in Altman (1973), p. 59.

an animal's life and few things now call forth less of the collective compassion of humankind than the great suffering animals are made to endure
in the name of human interests.

But these numbers serve another function. They serve as an index of
the complexity of questioning the morality of those activities which
routinely involve the pain or destruction of animals—whaling, say, or
farming, scientific research or the maintenance of zoos. Questions about
the morality of these activities are not likley to have simple answers. To
expect otherwise would be like thinking that knots can easily be untied
simply by taking hold of one end and pulling very hard. Knots resist such
callous treatment. They require patient examination, due attention to
competing forces, an eye for subtlety and nuance, before they yield. Difficult questions about difficult problems are like this, too. They are not
unraveled just by pulling hard on one idea. There is no reason to believe
that there are simple answers to what, if anything, ought to be done in the
case of whaling or animal experimentation. For example, there is no reason to believe that the ideas that all life is sacred, or that we ought to have
reverence for all life, or that all forms of life are good will dissolve the
complexity before us. For these ideas, like many others, will leave
unanswered, if left as stated, the crucial questions that arise when the
needs, desires, or, in general, the interests of one form of life come into
conflict with those of another. Then what is needed is not the simple declaration that all life is sacred, that all life ought to be revered, and so on.
Then what is wanted is some rational way to think through and resolve the
conflict. More particularly, what is wanted are the moral principles that
ought to be applied if the conflict is to be resolved equitably.

Not that these principles will be enough by themselves. They will
not. By themselves, these principles will not tell us, for example, how the
population of whales will be affected if whaling is stopped; how stopping
whaling will affect the population of, say dolphins; how the population of
dolphins will affect the population of tuna; and so on. By themselves, that
is, the moral principles that ought to be applied to questions about whaling or the use of animals in research will not tell us *the facts* about these
practices, nor will they give us a means of predicting what the facts will
be if various alternative courses of action are taken. What these principles
will tell us is what sort of facts to look for, when we attempt to decide the
morality of these practices; in other words, they will tell us what facts are
morally relevant to reaching a rational decision about these matters. The
task of identifying what these principles are, in the present context, as
well as in others, is one of the distinctive tasks of moral philosophy. It is
this task that shall occupy most of my attention in what follows. But I
shall also want to bring the results of my inquiry back to the place from
which we have begun—back to the rabbit, the calf, and the others.

It is a commonplace that morality places limits on how animals may
be treated. We are not to kick dogs, set fire to cats' tails, torment ham-

sters, or parakeets. Philosophically, the starting point is not so much whether, but why these acts are wrong.

An answer favored by many philosophers, including St. Thomas Aquinas and Immanuel Kant,[8] is that people who treat animals in these ways develop a habit that, in time, inclines them to treat humans similarly; in general, that is, people who torment animals will, or are likely to, torment people. This is the crucial morally relevant fact, according to these thinkers. It is this spillover effect that makes mistreating animals wrong. It's not the ill-treatment that the animals themselves receive—not that they are made to suffer; not even that they are brought to an untimely end. Rather, it is the ill this bodes for humankind. So, on this account (henceforth referred to as "the Kantian account"), the moral principle underlying those constraints placed on how animals may be treated goes something like this: Don't treat animals in ways that will lead you to mistreat human beings.

Now, one need have no quarrel with this principle itself. Where the quarrel lies is with the grounds on which, according to the Kantian account, this principle is allegedly based. Peter Singer (1976) argues that there is a close parallel between this view and those of the racist and sexist, a view that, following Richard Ryder (1975), he denominates 'speciesism.' The racist, for example, believes that the interests of others matter only if they happen to be members of his own race; the speciesist, that the interests of others matter only if they happen to be members of his own species. Racism has been unmasked and denounced for the prejudice that it is. The color of one's skin cannot be used to determine the relevance of an individual's interests. Following Bentham,[9] Singer and Ryder both argue that neither can the number of one's legs, whether one

---

[8]Relevant selections from both St. Thomas and Kant are included in Regan and Singer (1976). What I call the Kantian account is criticized further in Regan (1979a). Kant's views are criticized at length by Elizabeth Pybus and Alexander Broadie (1974). I defend Kant against their objections in Regan (1976a), and Broadie and Pybus reply (1978).

[9]The famous passage from Bentham reads as follows (Bentham, 1789, Chapter XVII, Section 1; reprinted in Regan and Singer, 1976):

The day has been, I grieve to say in many places it is not yet past, in which the greater part of the species, under the denomination of slaves, have been treated by the law exactly upon the same footing as, in England for example, the inferior races of animals are still. The day may come, when the rest of the animal creation may acquire those rights which never could have been withholden from them but by the hand of tyranny. The French have already discovered that the blackness of the skin is no reason why a human being should be abandoned without redress to the caprice of a tormentor. It may come one day to be recognized, that the number of the legs, the villosity of the skin, or the termination of the os sacrum, are reasons equally insufficient for abandoning a sensitive being to the same fate. What else is it that should trace the insuperable line? Is it the faculty of reason, or, perhaps, the faculty of discourse? But a full-grown horse or dog is beyond comparison a more rational, as well as a more conversable animal, than an infant of a day, or a week, or even a month, old. But suppose the case were otherwise, what would it avail? the question is not, Can they reason? nor, Can they talk? but, Can they suffer?

walks upright or on all fours, lives in the trees, the sea or the suburbs. There is, they argue forcefully, no rational, nonprejudicial way to exclude the interests of nonhuman animals just because they are not the interests of human beings, and it is because the Kantian account would have us think otherwise that we are right to reject it.

A second view about the constraints morality places on how animals may be treated involves the idea of cruelty. The reason we are not to, say, kick dogs, according to this way of thinking, is that we are not to be cruel to animals and kicking dogs is cruel. It is the prohibition against cruelty, then, that this second view covers and conveniently sums up our negative duties to animals (i.e., those duties concerning how animals are *not* to be treated.)

The prohibition against cruelty can be and sometimes is given a distinctively Kantian twist. This happens when the grounds given for prohibiting cruelty to animals are that this leads people to be cruel to other people. The philosopher John Locke suggests, but does not clearly endorse, this view in the following passage from his *Thoughts on Education* (1905, Sec. 116, pp. 225–226):

> One thing I have frequently observed in Children, that when they have got possession of any poor Creature, they are apt to use it ill: They often *torment,* and treat very roughly, young Birds, Butterflies, and such other poor Animals, which fall into their Hands, and that with a seeming kind of Pleasure. This I think should be watched in them, and if they incline to any such *Cruelty,* they should be taught the contrary Usage. For the Custom of Tormenting and Killing of Beasts, will, by Degrees, harden their Minds even towards Men; and they who delight in the Suffering and Destruction of Inferior Creatures, will not be apt to be very compassionate, or benign to those of their own kind. . . .

Locke's position strongly suggests the speciesism that characterizes the Kantian account, and, to the extent that the Kantian account is unsatisfactory, it will not do to attempt to ground the prohibition against cruelty to animals on *just* the adverse consequences this has for humankind. However, Locke's understanding of what cruelty is—tormenting a sentient creature or, more generally, causing it to suffer, "with a seeming kind of Pleasure"—seems to be correct and has important implications for those who want to accept the prohibition against cruelty as a summation of our negative duties to animals and who endeavor to give this prohibition a nonspeciesist foundation. These are those thinkers who, like many persons active in the humane movement today, champion the prohibition against cruelty to animals, not *just* because this will lead people to treat human beings in similar ways (which would be a speciesist basis); rather, we are not to be cruel to animals, these thinkers submit, because it is wrong to be cruel to the animals themselves, independently of how human beings will fare as a consequence. This way of grounding the prohibition against cruelty, which I shall refer to as "the cruelty account," deserves our critical attention.

It is difficult to overestimate the importance the idea of preventing cruelty has played, and continues to play, in the movement to secure better treatment for animals. Whole societies take names that champion this cause, the Society for the Prevention of Cruelty to Animals (SPCA) in the United States and the Royal Society for the Prevention of Cruelty to Animals (RSPCA) in Great Britain being perhaps the two best known examples. But while not wishing to deny the importance of preventing cruelty nor to deprecate the crusading work done by these and similar organizations, I think it must be concluded that to stake so much on the prevention of cruelty not only beclouds the fundamental moral issues, it actually runs the serious risk of being counterproductive to the cause for which these organizations labor.

'Cruel' is a term of moral appraisal that we use to refer either to the character of a person or to an individual action. People are cruel if they are inclined to delight in or, in Locke's phrase, take "a seeming kind of Pleasure" in causing another individual pain. Individual actions are cruel if they manifest one's taking pleasure in making others suffer. Conceptually it is clear that someone's being cruel is distinct from someone's causing pain. Surgeons cause pain. Dentists cause pain. Wrestlers, boxers, football players cause pain. But it does not follow that persons engaged in these activities are cruel people or that any of their individual actions are cruel acts. Clearly, to establish cruelty in any case we would need to know more than that someone caused pain; we would also need to know the state of mind of the agent and whether, in particular, he/she took "a seeming kind of Pleasure" in the pain inflicted. He reasons poorly, therefore, who reasons thus:

> Those who are cruel cause pain. Surgeons (football players, etc.)
> cause pain. Therefore, surgeons (football players, etc.) are cruel.

But just as clearly, he reasons poorly who reasons in the following way:

> Those who cause pain are cruel. Those who experiment on animals
> (or kill whales, or raise veal calves in isolation, etc.) cause pain.
> Therefore, those who treat animals in these ways are cruel.

Recognizing the speciousness of this line of reasoning is one point those who are inclined to march under the banner of anti-cruelty must soon realize, if their thought, however well intentioned, is not to becloud the issues.

A second point is this: Once cruelty is understood in the way Locke saw that it should be, we can understand why more than the prohibition against cruelty is needed by those who object morally to the way in which animals frequently are treated. Take the case of the use of animals in the Draize test. Increasingly people want to object morally to this; increasingly people want to sy it is wrong. However, if establishing that this is wrong required establishing cruelty, the weight of the evidence would be on the side of the experimenters and against the objectors. For there is no adequate evidence for believing that people who administer the Draize test are cruel people or that, at the very least, they are cruel when they administer this test—i.e., that they take "a seeming kind of Pleasure" in causing the animals pain. That they cause *pain* to the animals is certain. But causing pain, once again, does not establish cruelty and, except for some few sadists in the scientific community, there is good reason to believe that researchers are no more cruel than the general lot of humankind.

Does this mean that using animals in the Draize test is all right? Precisely not. What it means is that to ask whether this is right or not is logically distinct from, and should not be confused with, asking whether or not someone is cruel. Cruelty has to do with a person's character or state of mind—whether someone takes pleasure in causing another pain. The moral rightness or wrongness of a person's actions is different. People can do what is wrong (or right) whatever their state of mind. In particular, researchers can be doing what is wrong, when they use animals in the Draize test, whether or not they enjoy causing the animals to suffer. If they enjoy this, we shall certainly think less of them as people; but even if they enjoy the pain they cause it will not follow that the pain is unjustified (wrong), any more than it will follow that the pain is justified (right) if they feel sorry for the animals or feel nothing at all. The more we are able to keep this in view—the better we recall that the morality of what a person does is distinct from his/her state of mind in general and from the presence or absence of taking pleasure in causing pain in particular—the better the chances will be for significant dialog between, say, those who experiment on animals and those who oppose this. For to charge experimenters with cruelty *can* only serve the purpose of calling forth all their defenses, and this because the charge will be taken as a denunciation of *what they are* (evil people) rather than of *what they do*. It will also, as I mentioned earlier, give them an easy way out: After all, *they* are in a rather privileged position to know their own mental states; *they* can take a sober moment's view and see whether in fact they do take a "seeming kind of Pleasure" in causing animals' pain; and if, as will be the case in the vast percentage of cases, they find that they honestly do not, then their reply to the charge of cruelty is clear: They are not cruel (evil) people, and—(and this is the crucial point, the point where the well-intentioned efforts of those "on the side of the animals" can be and, I think, often are most counterproductive)—those charged with cruelty can come away with a feeling that their hands are clean, that their work (their acts) are quite all right. If it is cruelty they are charged with, and if cruel they are not, then they win, and the litany of accusations about cruelty will be so much water off their back, as, indeed, it should be. It is no good trying to improve the lot of animals by trying to convince people that they are cruel when they are not.

I can imagine members of some anticruelty organization complaining that the preceding is "too picky." They might say that 'cruelty' has been interpreted too narrowly, that what is meant by being against cruelty is being against treating animals badly, treating them in ways they do not deserve, harming or wronging them. In practice perhaps this is what anticruelty charges often come to. But then this is the way the charges should be made since, as I have tried to explain, persisting in making them in terms of cruelty not only beclouds the issues (to do what is wrong to an animal is not the same as, and does not presuppose or entail, cruelty) and actually can be and, I believe, frequently is counterproductive. To ask for more care in the charges leveled, therefore, is not to strain at gnats. It is to take an important step in the direction of making the charges more difficult to answer. If something like "Society for the Prevention of Maltreatment of Animals" is not as euphonious as, say "Anti-Cruelty Society," a lack of euphony is a price those laboring for animal welfare should gladly pay.

Utilitarians give a different account of the constraints regarding how animals ought to be treated. The utilitarian account—or, to speak more precisely, one version of the utilitarian account[10]—involves the acceptance of two principles. The first is a principle of equality. This principle declares that the desires, needs, hopes, and so on of different individuals, when these desires, needs, and the like are of equal importance *to* these individuals, *are* of equal importance or value, no matter who the individuals are, be they prince or pauper, genius or moron, white or black, male or female, or—and this is the most significant case for our purposes—human or animal. Thus, this equality of interests principle, as I shall call it, would seem to provide a philosophical basis for avoiding the grossest forms of prejudice: racism, sexism, and, following Ryder and Singer, speciesism. Whether it succeeds in doing this is an issue we shall turn to in what follows.

The second principle that figures in the utilitarian account is the principle of utility itself. According to this principle, roughly speaking, we are to act so as to bring about the greatest possible balance of good over evil (e.g., the greatest balance of the satisfaction of interests over the frustration of interests), taking the interests of everyone affected by the outcome of the action into account *and* counting equal interests equally. Now, since animals have interests (desires, needs, etc.), *their* interests must be taken into account; and because their interests are frequently as

---

[10]The utilitarian position I consider is the one associated with Bentham and forcefully presented by Peter Singer. That Singer is a utilitarian is made unmistakably clear by him in Singer (1978).

important to them as comparable interests are to human beings, *their* interests must be given the same weight, the same importance, as these comparable human interests. In acting so as to forward the utilitarian objective of maximizing good over evil for all concerned, therefore, we must do this in a way that does not violate the equality of interests principle, as this applies to any being with interests, including any animal who qualifies. It is because kicking dogs, setting fire to cats' tails, and the like run counter to the principles of equality and utility that, on the utilitarian account, they are wrong.

Granted this is a very rough sketch of the (or a) utilitarian position; nonetheless it enables us to understand the main features of the utilitarian account of the constraints morality places on our treatment of animals and to sketch the main points of resemblance and contrast between it and the other accounts described so far. Briefly, these points are as follows. Like the Kantian account, but unlike the cruelty account, the utilitarian account emphasizes the importance of results or consequences for determining what is right or wrong; but unlike the Kantian account, and in this respect resembling the cruelty account, the utilitarian account recognizes the moral status of animals in their own right: For the utilitarian, we are not to measure morality by the speciesist yardstick of human interest alone. Finally, unlike the cruelty account, but now in concert with the Kantian, the utilitarian does not conflate the morality of an act with the mental state of the agent, though the utilitarian can certainly be as opposed to cruelty as anyone else; it is just that, within the utilitarian theory, right and wrong are determined by consequences, not feelings, motives, intentions, and the like. Within the utilitarian position, therefore, the ordinary moral constraints placed on how we may treat animals can be accounted for by arguing that these constraints are necessary if we are not to violate the equality of interests principle *or* if we are to succeed in bringing about the greatest possible balance of good over bad.

The utilitarian account has much to recommend it, and it is interesting to ask how far it can take us in challenging the way in which animals are routinely treated—for example, as subjects in scientific research. Peter Singer is a utilitarian whose work has had, and continues to have, considerable influence. It is his position that utilitarianism leads to far-reaching consequences, when applied to the way animals are treated. In particular, Singer argues that utilitarianism requires that we become vegetarians *and* that we oppose much (even if not quite all) research involving animal subjects. Singer's main argument for these conclusions is that the intensive rearing of animals as well as their routine use in experimentation violate the equality of interests principle. Briefly, his argument appears to be this. The animals involved, we have reason to believe, have an interest in not being made to suffer, and this is an interest that we have further reason to believe is as important to them as is the comparable interest in the case of human beings; thus, the equality of interests principle entails that the relevant interests are equal in value or importance. This being so, Singer contends, it is wrong to do to animals what we would condemn being done to humans; in particular, it cannot be right to raise animals intensively or use them in research if we would morally oppose these things to human beings. Since we do oppose this in these latter cases—we do condemn cannibalism and the coerced use of humans in research—we must, Singer argues, morally condemn and oppose the comparable treatment of animals. And this means that we have a moral obligation to become vegetarians and oppose much, if not quite all, experimentation on animals.

As clear and powerful as this argument is, I do not believe that Singer quite succeeds in making a convincing case for his call to action. What he shows is that animals are treated differently from human beings. What he does not show is that this differential treatment violates either the equality of interests principle or the principle of utility. I shall consider the equality of interests principle first. That principle, it will be recalled, tells us to count equal interests equally, no matter whose interests they are. Now, this is a requirement we can respect and still treat the individuals whose interests are involved quite differently. For example, I might correctly regard my son's and my neighbor's son's interests in receiving a medical education as being equal and yet help my son and not help my neighbor's. Thus, I *do* treat them differently, but I do *not* necessarily count their equal interests differently, and neither do I thereby do anything that is in any obvious sense morally reprehensible. As a father, I have duties to my son that I do not have to other people's children.

The general point, then, is this: The differential treatment of individuals with equal interests does not by itself prove a violation of the equality of interests principle. To show that this principle is violated, given the different way that animals and humans are treated, therefore, Singer has to give an *argument* that shows *more* than that they are treated differently. What argument does he give and how adequate is it? Characteristically, the way Singer proceeds (1975, especially pp. 78–83) is by asking whether we would do to humans what we allow to be done to animals—for example, whether a researcher would use an orphaned, profoundly retarded human baby in a painful experiment in which he is willing to use a more intellectually and emotionally developed animal. If the researcher says no, then Singer charges him with speciesism—that is, charges him with violating the equality of interests principles—on the grounds that the animal's interest in avoiding pain is just as important to it as is the infant's interest to him/her.

This argument begs the question. It assumes that, by treating the individuals involved differently, we count their equal interests differently. As I have explained, however, this is not always true; so, whether it is true in any particular case is something that must be established, not simply assumed on the basis of the differential treatment of the individuals involved. Singer, I believe, assumes just this, which is why his argument is question-begging.

But Singer's argument has a further deficiency, I believe. This involves the other principle that is central to his position—namely, the principle of utility. Regarding this principle two things at least must be seen. The first is that Singer does not show that the differential treatment of animals to which he objects runs counter to the utilitarian objective of bringing about the greatest possible balance of good over evil. To show this Singer would have to give a very elaborate, detailed description, not only of how animals are treated, a part of the task which he completes with great skill, but of what, all considered, are the consequences for everyone involved—e.g., how the world's economy depends on the present level of productivity in the animal industry, how many people's lives are directly and indirectly involved with the maintenance or growth of this industry, and so on. More even than this, he would have to show what would be or, if certainty is impossible, what would most likely be the consequences of a sudden (or gradual) collapse (or slow-down) of the animal industry's productivity; and this, too, would need to be very elaborate and detailed. But more even than this, Singer would need to make a compelling case for the view that *not* raising animals intensively or *not* using them routinely in research would lead to better consequences, all considered, than those that now result from treating animals in these ways. It is important to emphasize that Singer is required to show that better consequences *would* result, or, if certainty is impossible, at least that it is *very probable* that they would, not merely that they might (that is, that it is possible or conceivable that they might occur), if he is to provide a compelling utilitarian basis for, say, vegetarianism. It must come as a disappointment, therefore, that we do not find anything approaching the necessary empirical data Singer's utilitarian basis requires. What we find, instead, are passages where, for example, he bemoans (rightly, I believe) the fact that animals are fed protein-rich grains that might instead be fed to malnourished human beings (Singer, 1975, Ch. 4). The point, however, is not whether these grains *could* be fed to the malnourished; it is whether we have solid empirical grounds for believing that they *would* be made available to and eaten by these people, if they were not fed to animals, *and* that the consequences resulting from this shift would be better, all considered. I hope it neither seems to be, nor that it is, unfair to Singer to observe that these necessary calculations are missing, not only here, but, to my knowledge, throughout the body of his published writings.

This, then, is the first thing to note regarding Singer's position vis-a-vis the principle of utility: He fails to show, with reference to this principle, that it is wrong to treat animals as they are at present being treated in modern farming and scientific research. The second thing to note is this: That, for all we know, and so long as we rely on the principle of utility, the present treatment of animals might actually be justified (might actually be right). The grounds for thinking this are as follows.

On the face of it utilitarianism looks to be the fairest, least prejudicial view around. Everyone's interests count, and no one's counts for more or less than the equal interest of anyone else. The trouble is, as we have seen, there is no necessary connection, no pre-established harmony between respect for the equality of interests principle *and* promoting the utilitarian objective of maximizing the balance of good over bad. On the contrary, the principle of utility might be used to justify the most radical kinds of differential treatment between individuals or groups of individuals, and thus might justify recognizable forms of racism and sexism. For these prejudices can take different forms, can find expression in different ways. One form consists in not even taking the interests of a given race or sex into account at all; another takes these interests into account, but does not count them equally with those of the equal interests of the favored group; but another does take their interests into account, does count equal interests equally, but adopts laws and policies, engages in practices and customs, that give greater opportunities to the members of the favored group because doing so promotes the greatest balance of good over evil, all considered. Thus, recognizable forms of racism or sexism, prejudices that seem to be eliminated by the utilitarian's use of the principle of the equality of interests, could well be resurrected and justified by the principle of utility. If the utilitarian replies that denying certain humans an equal opportunity to satisfy or promote their equal interests on racial or sexual grounds must violate the equality of interests principle and so, on his position, is wrong, we must remind him that differential treatment is not the same as, and does not entail, violating the equality of interests principle. It is quite possible, for example, to count the equal interests of blacks and whites the same (and thus honor the equality principle) and still discriminate between races when it comes to what members of each race are permitted to do to pursue those interests, on the grounds that such discrimination promotes the utilitarian objective. So, utilitarianism, despite initial appearances to the contrary, does not after all provide us with solid grounds on which to exclude all recognizable forms of racism or sexism.

Or speciesism. For exactly the same kind of argument can be given to show the possible utilitarian justification of an analogous form of speciesism: We count the equal interests of animals and humans equally; it just so happens that the consequences of treating animals in ways that humans are not treated (e.g., raising animals, but not humans, intensively) are better, all considered, than other arrangements. Utilitarianism, therefore, appears to be so far from providing a satisfactory basis against all forms of speciesism that it might itself actually provide a basis for speciesist practices. Whether it *actually* does or not will depend on whether the consequences would be better, all considered, if animals continue to be treated as they are, than they would be if these practices were stopped or altered. And since, as noted earlier, Singer fails to provide us with the necessary empirical data to show that the consequences would be better, if we changed how animals are treated in farming and research, it follows—and this, once again, is the second main objection we raise against his use of the principle of utility—it follows that, for all we know, and so long as Singer continues to rely on utility, the present speciesist way of treating animals might actually be justified (i.e., might actually be right).

The results to this point are mainly negative. In our search for the principles to use, when we ask about the morality of burning out a rabbit's eye to test toothpaste, what we have thus far argued is that (1) these principles cannot make reference to the agent's state of mind (in particular, to whether the agent takes a "seeming kind of Pleasure" in causing animal suffering); (2) these principles cannot refer only to the consequences that harm or benefit human beings, since this prejudicially leaves out of account the harms and benefits to the animals themselves; and (3) these principles cannot refer only to the utilitarian objective of maximizing the balance of good over evil, even if animal harms and benefits are taken into account, because this (utilitarian) way of thinking could justify morally reprehensible prejudicies (e.g., recognizable forms of racism or sexism, not to mention speciesism). What is wanted, then, is an account of our obligations to animals that avoids each of these shortcomings. This account is to be found, I believe, by postulating the existence of animal rights. Indeed, for reasons that I hope will be clear as they unfold, I believe it is only if we postulate human rights that we can provide a theory that adequately guards humans against the abuses utilitarianism is heir to. I hope I can make this clear in what follows, even if, because of the limits of space if for no other, I cannot establish it conclusively.[11]

Various analyses of the concept of a right have been proposed. For present purposes the nooks and crannies of these competing analyses will have to be bypassed and attention focused on the role that moral rights play in our thinking about the status of the individual, relative to the interests of the group. On this matter the truth seems to lie where Ronald

---

[11]The case for animal rights cannot be made briefly. The complete case involves first making a case for human rights and then arguing that the case for human rights implies that (some) animals have rights too. Thus, the complete case can be challenged at two vital points: (1) in posting rights in the case of humans and (2) in arguing that positing human rights implies that animals have rights. Understandably I am not here able even to attempt the complete case. That is a project that commands my present attention. I have examined some of the issues at greater length in Regan (1979b).

Dworkin (1977) sees it: The rights of the individual trump the goals of the group.

What does this mean? It means that the moral rights of the individual place a justifiable moral limit on what the group can do to the individual in the group's pursuit of what is beneficial to it. For example, suppose a group of people stand to gain a lot of enjoyment by arranging for another person to be harmed—imagine the Romans looking on while Christians go up against lions. In such a case the group does what is wrong because they allow their interests for enjoyment to override the individual's moral rights—e.g., the right not to be forced to do something against one's will, or the right not to be harmed. This does not mean that there are no circumstances, real or otherwise, in which an individual's rights must give way to the collective interest. Imagine that Bert has inadvertently swallowed the microfilmed code that we must have in order to prevent a massive nuclear explosion in, say, New Zealand. We (and Bert) sit safely in, say, Tucson, Arizona. We explain the situation to Bert but he refuses to consent to our request that we operate on him, retrieve the code, and prevent the explosion. He says, sorry, but he'll have none of it, and does he not, after all, have a right to determine what is to be done to his body, a right that, if we force him to undergo the operation, we will be overriding unjustifiably? Well, in such a case it is not implausible to say that Bert's right must give way to the collective interests of others, whether Bert likes this or not.

Individual rights, then, normally, but not always, trump collective interests. To give a precise or complete statement of the conditions that determine which ought to prevail, the rights of the individual or the interests of the group, is very difficult indeed, but each of the following three conditions, which deal only with the right not be be harmed, at least seems to incorporate a sufficient condition, and the satisfaction of any one of them, a necessary condition for justifiably overriding this right.[12]

An individual's right not to be harmed can be justifiably overriden if:

(a) We have very good reason to believe that overriding the individual's right by itself will prevent, and is the only realistic way to prevent, vastly greater harm to other innocent individuals.

(b) Or if we have very good reason to believe that allowing the individual to be harmed is a necessary link in a chain of events that collectively will prevent vastly greater harm to innocent indivdiuals

---

[12]The present statement of these conditions deviates somewhat from my earlier effort in Regan (1975). I believe the present statement's inclusion of conditions (b) and (c) marks an improvement over the earlier formulation. However, a fuller statement would have to include more than simply the idea of *preventing* vastly greater harm; for example, *reducing* already existing harm would also have to find a place.

and at the same time we have very good reason to believe that this chain of events is the only realistic way to prevent this vastly greater harm.

(c) Or if we have very good reason to believe that it is only if we override the individual's right that we can have a reasonable hope of preventing vastly greater harm to other innocent individuals.

There is much that is vague in these conditions—e.g., "vastly greater 'innocent individuals,'" "reasonable hope." On another occasion it might be possible to make the meaning of these concepts more exact. On the present occasion, however, we will have to make do with them as they stand and take what we can get from them in their present state. Even in this state we can understand that these conditions attempt to do justice to the complexity of conflicts of interest. In particular, they attempt to explain how, in a principled way, we might justify overriding an individual's right not to be harmed even though *just* doing this will not guarantee the prevention of vastly greater harm, either because [as (b) brings out] harming an individual is only one part of a more complex series of events that we have very good reason to believe will prevent vastly greater harm, or because [as (c) brings out] we simply do not know how things will turn out, but do know, we do have very good reason to believe, that we can have no reasonable hope of preventing some catastrophe unless we allow an individual to be harmed. It is possible that some might find these conditions too liberal; (c) in particular might seem too lenient to some; and even (b) might go too far in the view of others. I am not certain what to say in either case, and shall beg leave of this issue on this occasion, except to say that the case for the unjustifiability of harming animals is proportionately greater the more one is inclined to restrict the justifiability of harming individuals just to condition (a). For reasons that will become clearer as we proceed, however, even the more liberal view, that harm can be justified if any one of the three conditions is met, is sufficient to make a strong case against the routine abuse animals are made to endure.

These conditions share one extremely important feature. Each specifies what we must know (what we must have very good reason to believe) if we are to be justified in overriding an individual's right not to be harmed. As such, each places the burden of justification squarely on the shoulders of anyone who would harm an individual to show how this does not involve violating the individual's right. It is not those who treat individuals in ways that appear to conform with the treatment due them as possessors of rights who must justify doing so; rather, it is those who would treat them otherwise.

Part of the importance of the question, Do animals have rights and, in particular, do they have the right not to be harmed?, now comes clearly into focus. For *if* they have this right, then it will be violated whenever animals are harmed and condition (a), (b), or (c) is not satisfied. Moreover, given what was just said about these conditions, the onus of justification will always be on those who harm animals to explain how it is that they are not violating the right of animals not to be harmed, if animals have this right. So, the question presses itself upon us: Do animals have the right not to be harmed?

This is not an easy question to answer. One is immediately reminded of Bentham's observation that the idea of moral rights is "nonsense upon stilts." And Bentham meant this in the case of *human* moral rights. One can only speculate regarding what he might have thought concerning the moral rights of *animals*! So, how is one to proceed? If I am right, the circuitous path we must travel, albeit cautiously, in broad outline, is as follows (see Regan, 1979b).

We begin by asking about our reasons for thinking that human beings have the moral right not to be harmed; then we go on to ask whether, given these reasons, a case can be made for saying that animals have this right as well. So let us go back to the idea that individual human beings have this right and recall that, except in extreme cases, this right trumps collective interest. Why? What is there, in other words, about being a human being, to which we might point, as it were, and say, *That* is why you must not harm the individual just so that the group might benefit?

The heart of the answer lies, I believe, in thinking of human beings as having a certain kind of value, what I shall term inherent value. By this I mean that each human being[13] has value logically independently of whether he/she is valued by anyone else (or, what perhaps comes to the same thing, whether he/she is the object of anyone else's interest). Put another way, the view that human beings have inherent value implies that the kind of value properly attributable to them is not exclusively instrumental. Humans have value not just because, and not just so long as, they are good for something. They have a different, an inherent type of value, distinct from their utility and, say, their skill.

Now, *if* this is true, we can explain, in very general terms, reminiscent of Kant certainly, what is involved in mistreating human beings. Humans are mistreated when they are treated as if they have value only if (and only so long as) they forward the interests of other beings (for instance, other human beings). To treat a human being thus is to be guilty of showing a lack of proper respect for the sort of value humans have. In Kant's terms, what has value in itself must always be treated as an end, never merely as a means. However, this is precisely what we would be doing if we were to harm an individual so that others might gain pleasure, profit, or analogous benefits; we would be treating the individual merely

---

[13]Whether sense can be made of including irreversibly comatose human beings in the class of beings having inherent value is a troublesome question indeed. I consider this issue, but perhaps not very adequately, in the essay referred to in the previous footnote.

as a means, as if the individual had value only to the extent that he/she contributed to the realization of the collective interest.

Again, *if* we accept the view sketched to this point—in particular, the postulate that human beings have inherent value—we can press on and ask how rights enter the picture. They enter, I believe, in being grounded in inherent value. In other words, it is individuals who have inherent value, individuals who have moral rights, and it is *because* they have value of this kind that they have a moral right not to be treated in ways that deny their having value of this kind. Thus, rather than rights being connected with *the value of consequences* that affect individuals for good or ill, rather than rights being justified by the utility of recognizing them (a point I shall return to below), rights are based on *the value of individuals*. In the particular case of the right not to be harmed, then, what we can say is that individuals who have inherent value have the right not to be harmed, a right that precludes the justifiability of treating them merely as a means to the pleasure or profit, say, of the group, since this would fail to treat these individuals in ways that, because of the kind of value they have, they are entitled.

Now, certainly the foregoing is not a definitive account of the view that individuals having inherent value have basic moral rights, in particular the right not to be harmed. But there is one omission that is more conspicuous than others. This is some answer to the question, What is there about being a human being that underlies the inherent value attributed to them? Like the other questions encountered along the way, this one, too, is very controversial, and a sustained defense of the answer proposed here is not possible. But the answer I would give is as follows.[14]

Human beings not only are alive; *they have a life*.[15] What is more, we are the subjects of a life that is better or worse for us, logically independently of anyone else's valuing us (e.g., finding us useful). By this I do not mean that others cannot contribute to or detract from the quality of our lives. On the contrary, both the great goods of life (love, friendship, and, in general, fellow feeling) and life's great evils (hatred, enmity, loneliness, alienation) involve our relationships with other people or our failure to relate to them. What I mean, rather, is that our being *the subject* of a life that is better or worse for us does not depend logically on what others do or do not do for us or to us. And it is this fact, I believe, that

---

[14]I do not believe it is absurd (unintelligible) to think of natural objects that lack consciousness, or collections of such objects, as having inherent value, in the sense in which I use this expression—i.e., *X* has inherent value if it has value logically independently of anyone's valuing *X*. I do not say this is an easy view to clarify or defend, and it may be plain wrongheaded. At present, however, I believe it is a view that must be held, if we are to be able to develop an environmental ethic, as distinct from an ethic for the use of the environment. On this, see Regan (1981).

[15]The distinction between being alive and having a life is one James Rachels frequently makes. See, for example, Rachels (1980). Rachels does not, so far as I am aware, relate this distinction to the idea of inherent value.

provides the illumination we seek, when we ask why human beings have inherent value. Humans have inherent value because we are ourselves the subjects of a life that fares well or ill for us. Thus, in broad outline, the position I am advocating comes to this:

> Human beings have inherent value because, logically independently of the interests of others, each individual is the subject of a life that is better or worse for that individual. Because of the type of value human beings have, it is wrong (a sign of disrespect and a violation of rights) to treat humans as if they had value merely as a means (e.g., to use humans merely to advance the pleasures of the group). In particular, to harm human beings for the sake of the profit, or pleasures or curiosity of the group is to violate their right not to be harmed.

That, very roughly, is the view I am forwarding in the case of human beings. The question that now arises is, Can this same line of argument be developed in the case of animals? The answer is, It can, at least in the case of those numerous species of animals whose memebers are, like humans, the subjects of a life that is better or worse for them, logically independently of whether they are valued by anyone else. And there can be no rational doubt, I believe, that there *are* numerous species of animals of which this is true, including, not incidentally, those species whose members include the great blue whale, gibbons, and the other individuals with whom we began and to whom we shall eventually return. In their case, if in ours, they have a distinctive kind of value in their own right; in their case, if in ours, therefore, they have a right not to be treated in ways that fail to respect this value—in particular, they, like us, have the right not to be harmed; and in their case, if in ours, this right of theirs will be unjustifiably overridden if they are harmed merely to advance the profits, pleasures, or to satisfy the curiosity of others (e.g., human beings).

Two final philosophical points are in order, before I attempt to bring the results of my argument to bear on how animals are treated in the world at large. First, it is important to realize that I have not, and that I do not suppose that I have, *proven* that animals have rights, including the right not to be harmed. This I have not proven because I have not proven that *human* beings have rights. Rather, what I have argued, to put it crudely, is that *if* humans have rights, *then* so do many animals; more particularly, I have argued that what appears to me to be the most promising line of argument for accounting for human rights—namely, that we have inherent value, in the sense explained—can rationally be extended to the members of numerous species of animals. So, while I freely admit that I have not "proven" that animals (or humans) have rights, what I hope I have at least made clear is the direction in which I think future argument ought to procede. Erecting pointers, to be sure, is not the same as constructing proofs, but pointers are the best I can do on the present occasion.

Second, if the history of moral philosophy teaches us anything, it teaches us that utilitarianism dies hard. Just when one thinks it has been forced off-stage for good, one finds it loitering in the wings, awaiting yet another curtain call. So it is in the present case. The utilitarian can be counted on to say that there is nothing introduced by the idea of "rights" that he cannot account for.[16] All that one has to see is that the utilitarian objective is promoted if we recognize a strict obligation not to harm individuals except in extreme cases, *and* that, furthermore, utility is promoted by saying that individuals have the right not to be harmed, the invocation of a right functioning as an especially forceful way of conveying the idea that we ought not to harm individuals.

I am not persuaded by this attempt to resurrect utilitarianism. I do not think the utilitarian is in a position to say that he knows that the utilitarian objective is promoted by talk of individuals having rights. But— and here I raise my final and, if I am correct, what constitutes the most fundamental objection to utilitarianism—even if it were true that talk of rights helped promote the utilitarian objective and thus, given utilitarian principles, ought to be encouraged and honored, there could only be a *contingent* connection between the right not to be harmed, say, and the fact that respecting this right forwarded the utilitarian objective. At the very most, that is, the utilitarian can say that recognizing and respecting the right not to be harmed *as a matter of fact* fits in with forwarding his goal of maximizing the balance of good over evil. However, since the most he can say is that this is true as a matter of fact, the utilitarian must also accept the idea that things could have been (and could become) otherwise; he must accept the possibility, in other words, that it could have been or might become perfectly all right to harm individuals if this happened to (or if in the future it happens to) forward the utilitarian objective. But the wrongness of harming individuals, and their right not be harmed, cannot change in the ways utilitarian theory implies they could. The wrongness of harming an individual, and the individual's right not be harmed, are not contingent upon the *utility* of not harming individuals. Neither depends on the value of consequences. Instead, each depends on *the value of individuals,* if the view outlined in the above is correct.[17]

Let us put the preceding argument in perspective before applying it to the issues at hand. Earlier we said that, before we can make an informed judgment about the morality of, say, whaling or the use of rabbits

[16]It is possible that Mill meant to give rights a utilitarian basis. On this see Lyons (1977). The principal objection to this enterprise, I believe, is the third objection I raise against utilitarianism in the body of the present essay.

[17]I do not believe utilitarianism is alone in implying that the duty not to harm an individual (or the individual's right not to be harmed) are *continent* moral truths, which *might* have been otherwise (or *might* become otherwise). Certain aspects of Kant's theory as well as ethical egoism arguably imply this as well. I believe this is absolutely fatal to these theories, a point I argue in Regan (1980).

in the Draize test, we must not only know what the facts are, we must also know what are the moral principles that we ought to apply to these practices. Otherwise, we cannot know which of these facts are the morally relevant ones; and without this preliminary knowledge, we will not know what moral judgments to make. To determine what these principles are, we said, is one of the distinctive tasks of moral philosophy, and with that much said, we set out to see what these principles might be. Three positions were examined and found wanting, the Kantian account because it failed to acknowledge the moral status of animals; the cruelty account because it confused questions about the morality of actions with those dealing with the mental state of the agent; and the utilitarian account for the variety of reasons most recently adduced. The failure of these accounts then led us to consider one that involves ascribing rights to animals (which I shall call "the rights account"), a position that, it bears emphasis, meets the objections that prove fatal to the three views examined earlier. Unlike the Kantian account, the rights account concedes and, indeed, insists upon the moral status of animals in their own right; unlike the cruelty account, the rights account does not confuse the morality of acts with the mental states of agents; and unlike utilitarianism, this account closes the door to the institution and justification of prejudices that just so happen to bring about the best consequences by placing emphasis on the value of individuals as distinct from the value of consequences. This emphasis on the value of individuals will occupy a place of prominence when, as now we shall, we turn at last to the task of applying the rights account to the creatures whose place in this world gave rise to this inquiry—to the whale, the calf, and the others. It would be grotesque to suggest that the whale, the rabbit, the gibbons, the bobby calf, the millions upon millions of animals whose treatment at the hands of humans adds up to something like light years of pain and death—it would be grotesque to say that they are not harmed, as if harm were the privilege only of human beings. *They* are harmed, and harmed in the literal, not in the metaphorical sense: They *are* made to endure what is detrimental to their welfare, even to the point of death. Those who would harm them, therefore, must justify doing so. It is not up to us, who do not harm them, to show that our not doing so is justified. Thus, members of the whaling industry, the cosmetic industry, the farming industry, the network of hunters–exporters–importers, these people, not us, must justify the harm they bring down upon animals in a way that is consistent with recognizing the animals' right not to be harmed. Now, to produce such a justification, it is not enough to argue that people profit, satisfy their curiosity or derive pleasure from allowing animals to be treated in the ways they are in these industries. These facts, if they are facts, are not the morally relevant ones. Rather, what must be shown is that overriding the right of animals not to be harmed is justified because of other, relevant facts—for example, because we have very good reason to believe that overriding the individual's right by itself will prevent, and is the only realistic way to prevent, vastly greater harm to other innocent individuals.

So, let us ask of the whaling industry: Have they justified their trade by showing that it satisfies one or another of these conditions? Have they made their case in terms of the morally relevant facts? Our answer must be, No! And the cosmetic industry: Have they justified the harm they cause? Again, our answer must be, No! The farmers who raise veal calves? No! The retailer of exotic animals? No! Perhaps a thousand times we must say "No!," when asked whether those who profit from practices that harm animals have justified their actions.

I do not say—we need not say—that they cannot possibly justify what they do. For though the individual's right not to be harmed almost always trumps the interests of the group, it is possible that the right must give way sometimes. Possibly the rights of animals must sometimes give way to human interests. It would be a mistake to rule this out as an impossibility. But while allowing that this might be so, we must insist that, just as in the case of harming human beings, so also in the case of harming animals, the onus of justification must be borne by those who cause the harm to show that they do not violate the rights of the individuals involved, *and* that this justification cannot be carried out by citing morally irrelevant facts, e.g., facts about how much pleasure or profit are derived.

The situation, then, is this. We allow that it is *possible* that harming animals might be justified; but we also insist that those involved in practices that harm animals frequently fail to show that the harm caused is *actually* justified. And the question we must ask of ourselves is, What, morally speaking, ought we to do in such a situation? Reflection on comparable situations involving human beings will help make the answer clear.

Consider racism and sexism again. Imagine that an institution of slavery is the order of the day and that it is built on racist or sexist lines: Blacks or women are assigned the rank of slave. And suppose that, when asked about how this is to be justified, we are told that, since (given extreme circumstances) even slavery might conceivably be justified, it follows that we ought not to object to it, we ought not to try to bring it down, even though no one has shown that it is actually justified in the present case. Well, I do not believe for a moment that we would (or should) accept this attempt to dissuade us from toppling the institution of slavery. Not for a moment would we (or should we) accept the general principle involved here—namely, that an institution or practice is actually justified because it might conceivably be justified. Indeed, I believe we would (and should) accept a quite different principle—namely, that we are morally obligated to oppose any practice that appears to violate the rights of individuals unless or until we are shown that it really does not do so. To be satisfied with anything less, to look past the need to justify what is actually being done, is to cheapen the value properly attributable to the individuals who are the victims of the practice. In the particular case of slavery, therefore, I believe we would (and should) accept the view that we ought to act to bring it down.

Exactly the same line of reasoning applies in the case of how animals are treated by the industries mentioned earlier, where animals are regarded as so many dispensible "commodities," "models," "subjects," and so on. We ought not back away from bringing these industries and kindred practices to a halt just because it is *possible* that the harm caused to the animals *might* be justified. If we do, we fail to mean it, when we say that, like us, animals are not mere things; that, like us, they are subjects of a life that is better or worse for them, logically independent of whether they are of value to anyone else; that, like us, they have inherent value. As in the comparable case involving harm done to human beings, therefore, our duty is to act, to do all that we can to put an end to the harm animals are made to endure. The fact that the animals themselves cannot speak out on their own behalf; the fact that they cannot organize, petition, march or, in general, exert political pressure or raise our level of consciousness; all this does not weaken our obligation to act on their behalf. If anything, the fact of their impotence makes our obligation all the greater.

At this moment, then, we can hear, if we will but listen, the muffled detonation of the explosive harpoon, the sharp crack of the Thai hunter's rifle, the stock clamping shut around the rabbit's neck, the bobby calf's forlorn sigh. At this moment, we can see, if we will but look, the last convulsive gasps of the great blue, the dazed terror of the gibbons' eyes, the frenzied activity of the rabbit's feet, the stark immobility of the bobby calf. But not at this moment only. If we look, if we listen, we can see and hear the future. Tomorrow, other whales, other rabbits will be made to suffer; tomorrow, other gibbons, other calves will be killed. And others the day after. And others. And others. And others stretching into a future that, at this moment, we can see and hear, if we will but look and listen. All this we know of a certainty. All this and more, incalculably more, *will* go on, if we do not act, and act today, as act we must. Respect for the value and rights of the animals, just as respect for the value and rights of humans, cannot be satisfied with anything less.[18]

---

[18]For a more complete list of mainly recent philosophical work relating to the topics discussed in the present essay, see Magel and Regan (1979).

Animal Rights Revisited

Jan Narveson

What do we owe to the animals? What, that is to say, do we owe them *qua* animal, rather than in their various possible roles as pets, watchdogs, potential sources of protein, or potential sources of knowledge on various matters of medical interest? Our usual repertoire of moral ideas does not give us a very clear answer to this question, for those ideas have been framed for dealing with our fellow humans, by and large. When we address ourselves to this nonstandard case, then, we must scrutinize those ideas rather closely. Our habit of relying on our "intuitions" will not get us far, for those intuitions, I believe, push us in different directions when we try to extend them, and extension is necessary under the circumstances.

It may be well to begin by trying to assemble the options, though even to do this is assuredly to begin to do moral theory. Here, then, are the main ones as I see it:

(1) The moral status of animals is simply that of things, potentially useful or dangerous in various ways; the proper way to deal with them is simply whatever way is dictated by our interests in such things.

(2) Animals are in the same moral boat as we are: to wit, they have the capacity to suffer or prosper, to be better or worse off, and we ought to attach the same weight to a given degree of well-or ill-being on their part as we do to our own, endeavoring to do the best we can for all concerned.

(3) Animals are in the same moral boat as we are, but it is a different boat: to wit, they have the right to lead their lives as they choose, without interference from us—but also, without *help* from us, if we do not wish to give it.

This list of options is not exhaustive of the logical possibilities, obviously. I have come to suspect, however, that it exhausts all the *interesting* possibilities. And curiously enough, those are the same possibilities that

45

obtain with respect to our moral dealings with our fellow humans. In effect, the three views are (1) that morality is based on the interests of the agent only, at bottom; (2) that morality is based on the interests of all parties, taken equally; and (3) that morality is based not on interests as such, but rather on respect for the capacity to lead a life as one chooses. These, as I say, seem to me to be the interesting possibilities, and in the present paper I wish to explore each at some length. Eventually, as will be seen, I will be inclining towards view (1). The plan of this paper will simply be to discuss each one in turn. I will, however, discuss them in reverse order, beginning with what is, in a sense, the most committal view, in the sense of the one that would make our duties to animals the most stringent of the three.

1

View (3) is, of course, Libertarianism. What makes it interesting for our purposes is not so much that it is a plausible view about animals' rights, as that it is an interesting and even plausible view about people's rights that prompts us to think about why it should not apply to animals too, if in fact, it does not.

The essence of libertarianism, I believe, is to be found in this passage from Robert Nozick's much-discussed *Anarchy, State and Utopia:* "A line (or hyper-plane) circumscribes an area in moral space around an individual" (1974, p. 57). Some entirely non-arbitrary method for determining that space is supposed to be in principle available, and once determined, justice consists simply in not crossing that line. Moral beings are not to be damaged, harmed, used exclusively as means to the ends of others, and so on, and the prohibitions on doing so are supposed to be absolute rather than *prima facie*. It is essentially a property view of justice. Within the region enclosed by one's moral boundaries, the individual may do whatever he wishes: his will is there absolute, as absolute as that of the most absolute despot. His freedom of action is limited by the boundaries of others. Often, of course, the relevant boundaries will be literal, physical ones delimiting regions of land or whatever, but the idea is the same whether or not the items in the region are physically "in" it.

I believe that Libertarianism can be sufficiently characterized as the view that people have only *negative rights,* and that those rights are absolute. But since there is some confusion about notions of negative and positive rights, it might be well to pause for a moment to fix this idea. The *normative content* of any statement about rights consists in a correlative statement(s) about the obligations or duties of a specified or specifiable class of others: the right of A to x consists (normatively speaking) in the duty of B to do y, for relevant B. Now, the difference between negative and positive rights is this:

One thing that this hard-headed theory does allow is the acknowledgment that there are circumstances in which we morally *could* eat *people*. For on it, I need only have a valid contract with someone that calls
for my eating him/her under certain conditions, and then have those circumstances arise. Someone might find the prospect of living the kind of
life I offer prior to his/her tenure expiring so attractive as to be quite prepared to live on those terms. (Anything less than this would be too paternalistic, one supposes.) Now, the question arises whether Nozick was not
too quick in dismissing the suggestion that animals might find it a good
deal from their point of view to be well-fed and otherwise looked after in
exchange for allowing us to slaughter them and turn them into ham-
burger. I rather think he was; but we gain some insight into moral theory
by considering the matter, anyway.

Nozick's objection to this is that even if I bring someone into existence for the purpose of violating his rights, that still does not justify my
violating them. "An existing person has claims, even against those
whose purpose in creating him was to violate those claims" (1974, p.
38). Yes, but those claims are only those involved in negative rights. So
the proper procedure is as follows: Bring the newborn individual up until
it reaches the age of Libertarian discretion. At that point, you give it its
choice. Either stay in the game, which involves being well treated until
age such-and-such, and then be eaten, or we turn you out next Tuesday.
This seems consistent with Libertarian principles.

Now it seems very likely that human individuals would rarely or
never have to accept such a macabre option. For one thing, the neighbors
would surely put up a great fuss at the idea that such a deal is even being
offered, let alone accepted; but more importantly, the individual will invariably have a better deal if the other option to move out is elected. But
this may well not be true with animals. Few have such sympathy with the
cow that if someone makes the cow that offer and it elects to move out,
someone else will take reasonable care of it. And then we run up against
the other snag, which is that the cow is not about to *address* itself to the
question of which horn of this bovine dilemma to take. We would have to
do its thinking for it. That, of course, is the very reason why cows do not
seem eligible for libertarian constraints anyway: not even the world-class
geniuses in that species are very plausibly thought to be busy formulating
weltanschauungen by the light of which to make their life decisions!

But the interesting question is whether the slightly macabre deal we
envisage above would not be right up the libertarian alley, if we turn up
that alley in the first place. For remember: libertarian rights are not positive rights. They are *not* rights to good treatment, rights that others provide their possessors with a certain minimum level of welfare or whatever. They are merely rights not to be treated badly without one's
consent. So if we toy with the idea of giving such things as animals this

kind of right, we must consider what we can do, in the absence of linguistic ability on the part of those creatures, making it problematic just when we do or do not have consent. Now, Nozick understandably throws very cold water on the idea of *implicit,* that is, tacit consent, which "isn't worth the paper it isn't written on" (ASU 287). But beings who cannot communicate create a problem. There would be no problem if they had welfare rights, but within the libertarian scheme they do not (if they did, libertarianism would be in the curious position of granting more rights to animals than it does to people!). But if we cannot have implicit consent, and say that after all the cow does not seem to object when we sell it arbitrarily to another owner, so *it* must think it is acceptable, we could surely contemplate *hypothetical* consent: we could say that addressing ourselves to the cow's evident utility schedules, it looks as though accepting our offer of decent room and (very filling!) board for 2 years, followed by painless slaughter, would make eminently good sense for it.

There is one hitch in this that I have skirted thus far. The argument Nozick was worrying about has it that it is better to exist in *this* way than not to exist *at all.* But that way of putting it gives me the logical willies. It does not seem to me that in producing a cow, or a person, we are endowing some hitherto merely noumenal bovine or human self with concrete actuality, thus enabling us to make sense of that self's addressing itself to the question of whether it would be better off being real for a few years or decades rather than remaining a mere unrealized quiddity for the duration (better luck next time?). Once it has arrived, however, we can then address ourselves to the options, as indeed we do—and so would they, if they could.

In current circumstances, it is certainly true that there would be very few domestic animals if vegetarianism held sway, and so it is true that from the species point of view, so to speak, The Cow is doing a lot better with slaughter than it would without. But if we reject the "species point of view," as it seems to me we should (*whose* point of view is it, after all?), that is beside the point. What is not beside the point is that the cow would do very badly indeed if turned out into the woods to fend for itself, even if one could find a wood to let it fend in. It may perhaps also be true, though, that given animal utility schedules, the ones that are raised by "methods which as nearly as possible reduce animals to the level of vegetables" (as I put it when first addressing myself to this matter (Narveson, 1977, p. 162) might nevertheless do better on their own. Perhaps from the pig's point of view, it is better to root about half-starved in the forest, dying of frost or at the mouths of hungry wolves, than to be crowded into a penful of fellow pigs and forcefed until one could scarcely move even if there were room to.

It may seem that this is social contract talk rather than libertarian talk. In section 3 below, I will explain why I think that is not so. Meanwhile, my conclusion is that libertarianism either leaves the animals with no rights at all, or with the same rights we have; and if the latter, then it is unclear that vegetarianism issues from it. Humane slaughter, perhaps, and possibly the condemnation of certain particularly unattractive methods of animal husbandry, and quite likely the condemnation of sheer gratuitous torture, but that is about it. Above all, meanwhile, there remains the enormous problem of what sort of foundations might be available for this theory; and even if provided it is problematic at best whether animals get anything out of them. Perhaps if clear foundations are discerned, they will make it excruciatingly clear that animals do not qualify for rights at all. Let us, then, move on to the next theory, utilitarianism.

2

At this stage of the game, it is presumably not necessary to say very much about the general framework of utilitarianism. Its leading principle is statable in two words: maximize utility! Or in five: utility counts; nothing else does. The simplicity ends there, unfortunately. We do not know quite what it means to say that everyone else's utility counts as much as our own, if it is even meaningful at all; nor do we know whether it is possible to act in accordance with that principle. Most unfortunately of all, we do not know *why* we ought to do so, if we ought. Utilitarianism, I think, has the same kind of foundational problem as does libertarianism; perhaps even worse. Standardly, its assumption about human nature is that people are guided, or perhaps rationally guided, by their *own* utility. If so, how do we go about showing that they should really be guided by everyone's? Classical utilitarians seem to make the very assumptions that should make it impossible that the principle of utility would be adhered to by people of whom the assumptions are true (viz., that they try to maximize their own utility). Despite this (to which we will return), let us carry on for the present.

What assumptions is it reasonable to make about the utility of animals? It seems very reasonable indeed to suppose that animals can feel pain and pleasure. It seems reasonable to attribute to them some degree of intelligence (but unclear just what we are attributing to them in doing so, nor whether it is a capacity of the same sort we attribute to humans). Does that matter? Mill thought it did. It is tempting to say that he thought that the utility of intelligent beings counts more than the utility of less intelligent ones, but that surely will not do. What we must say instead, and what Mill really does say (I think), is that intelligent beings have a greater capability of utility than less intelligent ones: the satisfaction of a satisfied Socrates (if that is possible) involves a great deal more utility than the satisfaction of a satisfied pig. For that matter, even the satisfactions of a dissatisfied Socrates outweigh those of the pig. One question to worry about is: Is Mill right, or even believable, about that? Another is: What follows if he is? We tend, I believe, to be too fast on the second question, but let us take up the first anyway.

Utility is to be a quantity of something; but what? Classically, it is pleasure; but it is not clear how helpful it is to say that. Another idea (perhaps) is that utility is really just preference. More precisely, it is that x has more utility for A than y iff A prefers x to y. This is the view employed in recent decision theory. But recent decision theory does not assume cardinal utility, which classical utilitarianism certainly does. (Cardinal utility is utility of which it makes sense to say not only that x has more utility than y, but also how much.) Or rather, recent decision theory does not employ interpersonally comparable cardinal utility, for there are ways of cardinalizing over preferences, at least in a sizable class of cases. Without interpersonally comparable cardinal utility, the pig would properly be able to criticize Mill for anthropocentricity. Even with it, it is not clear how we are to proceed. Mill's point of reference in this matter is the judgment of those who have experienced both; but how much does this include? How do we go about having the pig's experiences? And if we do not and cannot, then how do we know that getting ourselves into piggish-seeming situations is even relevant, let alone decisive? If *we* were pigs, rather than just philosophers attempting to find out what it is presumably like to be them, perhaps we would enjoy wallowing in the mud ever so much more than we now enjoy reading Plato?

We do, certainly, make judgments of the form "people would in general be happier if . . . " Although there is a good deal of disagreement about such judgments, it may also be admitted that we are not entirely out in left field in making them. The problem is to make judgments of the form, "people *and animals* would be happier if . . . ," and that is trickier. It is acutely trickier in just the cases we have to worry about in the present paper—all the cases wherein there is a genuine conflict between the interests of us and the animals: namely, if our main interest in animals is realized, then their interest in *whatever* they may be interested in is thwarted, because they end up on our dinner plates. And that is a loss of utility that, in the case of humans, would certainly not be thought to be outweighed by the gourmet's interest in them, however powerful that interest might be. So we would surely be headed for vegetarianism if there were no reason for downrating the animals' utility quite substantially.

Actually, there are two sorts of "downrating." One way is to claim that the utility of animals, although admittedly quite comparable to ours, simply does not count, or that it counts very little: as if, for instance, we were allowed to multiply the animal's utility score by 0.01. The other way is to claim that animals have very little utility, really, at least by comparison with our own. As we have noted, utilitarianism must surely take the latter tack. It is axiomatic on it, after all, that everyone counts for one and none for more (or less).

What might reasonably (as opposed to just self-interestedly) persuade us that animals *do* have a lesser capacity for utility than we? Many would point to their supposedly lesser intelligence as a justification for treating them as we do. But they may or may not have in mind intelligence as a factor influencing utility. They may instead be thinking of it as an intrinsic good. Can we find a reason for supposing that intellect affects capacity for utility, then?

One thing that has long intrigued me in this connection is the involvement of intelligent beings with their own, or indeed, any futures. We are acutely aware of the future stretching out before us, and of the past in the other direction. We are, indeed, often so involved with time that we might be accused of neglecting our present. And we can at least conjecture that with animals things are different. Perhaps it is still excessively anthropomorphic to think so, but we do seem to think that animal awareness of their own future, indeed of their own identity in general, is rather dim; this despite homing pigeons and whatnot, who certainly seem to have a clear idea where to go next. But we do suppose that they are, as the saying goes, guided by instinct rather than reason. (It is unclear what to think of that. But we will suppose it makes some sense, and is reasonably close to true.)

Still, *why* might this matter? I have suggested that animals might "experience only more or less isolated sensations and uninterpreted feelings. If such beings are killed, all that happens is that a certain series of such feelings which would otherwise have occurred, do not occur . . . When beings having a future are killed, they lose that future; when beings lacking it are killed, they do not. So no interest in continued life is lost in their case" (Narveson, 1977). Well, setting aside the critical question of whether some such thing is true of animals, there remains the question just why it might make the kind of difference I supposed it did *on utilitarian grounds*. If two beings, one of whom has and one of whom lacks a future, each had a nonutilitarian-type right to its future, then we could agree that if we painlessly killed each of them, we would have violated one creature's rights, but not the other's. Unfortunately, utilitarians are not entitled to nonutilitarian rights. So if this difference is to make a difference, it must be because beings with futures experience more utility than beings without. For otherwise, the two creatures in question might be exactly on all fours *qua* utility sources. The universe might have lost exactly the *same* amount of utility by their deaths, instead of vastly more in the one case than the other.

Can we say "Well, so much the worse for the universe, then: let it go its way, and we will go ours!"? Not easily. If we *care* only about sophisticated beings with futures and the ability to master differential equations or to comprehend the late quartets of Beethoven *and* we admit that these favored beings do not actually experience any more utility than the average cow, then we obviously have abandoned utilitarianism.

What we need to think, therefore, if we are to remain utilitarians *and* we think that normal humans are much greater in their capacity for utility than animals, is that at each typical moment in the sentient life of a human, he or she is chalking up a much higher utility score than a beast at any typical moment for it. And what the basis of this judgment would be is, again, unclear. There is certainly the danger of anthropocentric bias here, if anywhere, one would think. Presumably utilitarianism supposes that utility estimates are not simply value judgments: when we compare the utility of experiencer X with that of experiencer Y, we are *not* supposed to be simply reporting our preference for having X or having Y around. Instead, I take it we are supposed to be reporting on the value *to* X of having X around and the value *to Y* of having Y around (so to speak). There is supposed to be an objective quantity, existing independently of the feelings and interests of observers, that simply is the amount of utility that being enjoys at that time. The basis of our report may still be preferences, but if so, they are its preferences rather than ours. The utilitarian, indeed, is someone who prefers to maximize over the preferences of all and sundry sources of utility, whereas nonutilitarians do not, but in principle the quantities at issue are what they are independently of this difference, or so we suppose.

Perhaps this affords some hold on the matter in the following way. If utility is based on preference, then perhaps we could say that if being X is able to have in mind more possible states of affairs over which to exercise preference than Y then X has a greater capacity for utility. As stated, this raises some rather kinky problems about individuating states of affairs so that we can get a fair count; and there is a lingering suspicion that the whole idea is wrong anyway, and utility should not really be thought of as preference at all. But it might offer some explanation of how we manage to account for so great a proportion of the universe's known supply of utility, or at least why we think we do.

If that amount of elitism is accepted, what about vegetarianism? It is axiomatic that some beings may, in principle, be sacrificed for others, on the utilitarian view. But may animals be sacrificed merely in order to enable humans to have a wider variety of gustatory pleasures? In order for it to be so, the marginal increment of such pleasure for humans has to exceed the marginal cost to the animals. Consider, then, the case of Kentucky Fried Chicken. Suppose that one chicken feeds three people for one meal. We might suppose that the cost of this is all the utility that the chicken might have experienced had it been allowed to live to a ripe old age. But wrongly. For that is only the cost to *that* chicken. But it is also reasonable to believe that, under the carnivorous regime we are investigating, this chicken will be replaced by another one which would not have existed at all if its predecessor were not eaten. In fact, the plot is thicker than that, for as Derek Parfit[1] points out, its predecessor would most likely not have existed either, were it not for the prospect of *its* being eaten. Given that we in fact raise animals to be eaten, it is not unreasonable to believe that the total utility of the animal population is enormously higher than it would be if we did not eat them, because so comparatively few would exist at all otherwise. And if we count that way, then the marginal cost to any given animal is the wrong thing to weigh against the marginal benefit to us of eating it. Viewed globally, those costs are very handily outweighed by the total utility increase in question.

That, of course, is to assume that we can apply "total" rather than average or some other sort of utilitarianism here. If we do not, and insist that it is the average utility of animals that should be our sole concern, the prospects for animal rights are much better, perhaps. Or are they? For now we also must reckon the cost of upkeep and care for the animals, which is born by people. Their cost would certainly not be born, in fact, if the animals were not beneficial to people in this way. Chickens would be raised only to lay eggs, cows for milk; but most would have little if any use. It might be argued that the loss in utility to people from having to care for useless beasts would exceed the loss to the beasts if they were (painlessly) killed. So even on average principles, it is far from clear that maximization would preclude the eating of animals.

Of course, if we do use total utilitarianism, then we have another small matter to contend with. Animals are, in fact, quite an inefficient source of food. If humans ate only vegetarian diets, it would be possible for there to be a great many more of *them*. And if, as has been imagined above, each human is so much larger a source of utility than any animal, it might seem that the tables are turned again, since the large animal population is keeping the human population smaller, and yet the human population is so much more efficient a source of utility than the animal one. But that, in turn, is to assume that the marginal utility change associated with the addition of each further human is in fact positive, and it can be argued that *that* is not so. Perhaps a world with two billion humans would have more total utility than one with ten billions. If so, we would have a global justification of carnivorousness from the above arguments.

I am sure that no one will think me excessively conservative if I conclude with the observation that the situation regarding the ethics of our treatment of animals is not entirely clear if we opt for utilitarianism. This is not exactly a surprise, but it is of some importance that it should be so. The vegetarians do not have things all their way on that theory; and it is, I think, the theory that offers the best prospects for animal rights among those I am considering.

[1] But perhaps not in print.

This brings us to the last one, contractarianism. On that theory, I shall argue, the situation is not unclear at all, at least if what is in question is basic animal rights: they don't have any.

3

The tendency in the past few years has been to take John Rawls' well-known theory of justice as the model of contractualist moral theory. I must therefore begin by explaining why that is a mistake.

On the contract view of morality, morality is a sort of agreement among rational, independent, self-interested persons, persons who have something to gain from entering into such an agreement. It is of the very essence, on such a theory, that the parties to the agreement know who they are and what they want—what they in particular want, and not just what a certain general class of beings of which they are members generally tend to want. Now, Rawls' theory has his parties constrained by agreements that they would have made if they *did not* know who they were. But if we can have that constraint, why should we not go just a little further and specify that one is not only not to know *which* person he or she is, but also whether he or she will be a person *at all:* reason on the assumption that you might turn out to be an owl, say, or a vermin, or a cow. We may imagine that *that* possibility would make quite a difference, especially if one were tempted by maximin! (Some proponents of vegetarianism, I believe, are tempted by it, and do extend the veil of ignorance that far.)

The "agreement" of which morality consists is a voluntary undertaking to limit one's behavior in various respects. In a sense, it consists in a renunciation of action on unconstrained self-interest. It is, however, self-interested overall. The idea is to come out ahead in the long run, by refraining, contingently on others' likewise refraining, from certain actions, the general indulgence in which would be worse for all and therefore for oneself. There are well-known problems generated by this characterization, and I do not claim to have solutions for them. I only claim that this is an important and plausible conception of morality, worth investigating in the present context.

A major feature of this view of morality is that it explains why we have it and who is a party to it. We have it for reasons of long-run self-interest, and parties to it include all and only those who have *both* of the following characteristics: (1) they stand to gain by subscribing to it, at least in the long run, compared with not doing so, and (2) they are *capable* of entering into (and keeping) an agreement. Those not capable of it obviously cannot be parties to it, and among those capable of it, there is no reason for them to enter into it if there is nothing to gain for them from it, no matter how much the others might benefit.

Given these requirements, it will be clear why animals do not have rights. For there are evident shortcomings on both scores. On the one hand, humans have nothing generally to gain by voluntarily refraining from (for instance) killing animals or "treating them as mere means." And on the other, animals cannot generally make agreements with us anyway, even if we wanted to have them do so. Both points are worth expanding on briefly.

(1) In saying that humans have "nothing generally to gain" from adopting principled restraints against behavior harmful to animals, I am in one respect certainly overstating the case, for it is possible that animal food, for instance, is bad for us, or that something else about animals, which requires such restraint from us, would be for our long-term benefit. Those are issues I mostly leave on one side here, except to note that some people may think that we gain on the score of purity of soul by treating animals better. But if the purity in question is moral purity, then that would be question-begging on the contractarian conception of morality. In any case, those people are, of course, welcome to treat animals as nicely as they like. The question is whether others may be prevented from treating animals badly, e.g., by eating them, and the "purity of soul" factor cannot be appealed to in that context.

A main motive for morality on the contract view is, of course, diffidence. Humans have excellent reason to be fearful about each other. Our fellows, all and sundry, are quite capable of doing damage to us, and not only capable but often quite interested in doing so; and their rational (or at least, calculative) capacities only make things worse. There is compelling need for mutual restraint. Now, animals can, many of them, be harmful to us. But the danger is rather specialized and limited in most cases, and in those cases we can deal with it by such methods as caging the animals in question, or by shooting them, and so on. There is no general need for moral methods, and there is also the question whether they are available. In any case, we have much to gain from eating them, and if one of the main planks in a moral platform is refraining from killing merely for self-interest, then it is quite clear that such a plank, in the case of animals, would not be worth it from the point of view of most of us. Taking our chances in the state of nature would be preferable.

(2) What about the capability of entering into and keeping such agreements? Animals have been pretty badly maligned on this matter in the past, I gather. Really beastly behavior is a phenomenon pretty nearly unique to the human species. But still, when animals refrain from killing other animals or people just for the fun of it, there is no good reason to think that they do so out of moral principle. Rather, it is just that it is not really their idea of fun!

There remains a genuine question about the eligibility of animals for morality on the score of their abilities. A very few individuals among some animal species have been enabled, after years of highly specialized work, to communicate in fairly simple ways with people. That does not augur well for animals' entering quite generally into something as apparently sophisticated as an agreement. But of course agreements can be tacit and unwritten, even unspoken. Should we postulate, at some such inexplicit level, an "agreement" among humans, it is largely tacit there. People do not enter into agreements to refrain from killing each other, except in fairly specialized cases; the rule against killing that we (virtually) all acknowledge is one we adopt out of common sense and antecedent inculcation by our mentors. Still, it is reasonable to say that when one person does kill another one, he or she is (among other things) taking *unfair advantage* of the restraint that one's fellows have exercised with regard toward one over many years. But can any such thing be reasonably said of animals? I would think not.

On the whole, therefore, it seems clear that contractarianism leaves animals out of it, so far as rights are concerned. They are, by and large, to be dealt with in terms of our self-interest, unconstrained by the terms of hypothetical agreements with them. Just exactly what our interest in them is may, of course, be matter for debate; but that those are the terms on which we may deal with them is, on this view of morality, overwhelmingly indicated.

There is an evident problem about the treatment of what I have called "marginal cases" on this view, of course: infants, the feeble-minded, and the incapacitated are in varying degrees in the position of the animals in relation to us, are they not? True: but the situation is very different in several ways. For one thing, we generally have very little to gain from treating such people badly, and we often have much to gain from treating them well. For another, marginal humans are invariably members of families, or members of other groupings, which makes them the object of love and interest on the part of other members of those groups. Even if there were an interest in treating a particular marginal person badly, there would be others who have an interest in their being treated well and who are themselves clearly members of the moral community on contractarian premises. Finally, it does have to be pointed out that there is genuine question about the morality of, for instance, euthanasia, and that infanticide has been approved of in various human communities at various times. On the whole, it seems to me not an insurmountable objection to the contractarian account that we grant marginal humans fairly strong rights.

It remains that we may think that suffering is a bad thing, no matter whose. But although we think so, we do not think it is so bad as to require us to become vegetarians. Here by 'we,' of course, I mean most of us.

And what most of us think is that, although suffering is too bad and it is unfortunate for animals that they are turned into hamburgers at a tender age, we nevertheless are justified on the whole in eating them. If contractarianism is correct, then these attitudes are not inconsistent. And perhaps it is.

Knowing Our Place in the Animal World

Annette C. Baier

Rawls, in *A Theory of Justice,* says that "a correct account of our relations to animals and to nature would seem to depend upon a theory of the natural order and of our place in it. How far justice as fairness will have to be revised to fit into this larger theory it is impossible to say" (1971, p. 512). Rawls had not felt the need for a theory larger than his Kantian interpretation of fairness when, earlier, he had discussed principles of right for individuals, such as duties not to injure and duties of mutual aid. Such duties, when due to other humans, can be seen as arising from principles that would be chosen in the original position. But since, in that position, the hypothetical choosers are self-interested rational persons, duties to animals, or to anyone or anything not party to the hypothetical agreement, cannot be derived from that starting point. Rawls recognizes this, and recognizes it as a limitation on his theory. He does *not* conclude that since the theory shows nothing wrong, there *is* nothing wrong with testing our drugs at the cost of animal suffering and contrived death, nor with destroying whole species in our determination to convert their bodies into materials that increase our profit, comfort, or pleasure. He acknowledges the need for a larger theory that would accomodate at least some beliefs concerning wrongs to animals.

Rawls' own account of what a moral theory is, and how it is related to the moral beliefs of nontheorizers, allows for the possibility that a good theory may have implications that clash with some moral convictions, both with some convictions the theorists had before working out the theory, and perhaps with some still held after the theory is developed. A theory may lead us to revise our earlier judgments, to change some of our convictions. It should be in "reflective equilibrium" with our considered judgments, and part of the consideration and reflection to which those judgments are submitted is provided by the theory itself, and by alternative competing theories. I agree with Rawls that a good moral theory

61

should not merely systematize existing intuitions on particular matters, but deepen moral insight, and even correct moral error. The trouble comes when there is conflict or disagreement in "considered" beliefs. Some people, like Rawls, noting that a given theory cannot account for the belief that some ways of treating animals are wrong, conclude that the theory therefore fails to maintain reflective equilibrium with the full range of one's considered beliefs, and they therefore look for a "larger theory." But others, enamored of their theory, dig their heels in and insist that since the theory shows nothing wrong in exploiting animals for human ends, there *is* nothing wrong in it; thus, they believe that those who disagree, even they themselves some of the time, are sentimental or unrealistic, and that their beliefs are not properly "considered" ones.

Should we, when faced with an array of moral theories *none* of which, as Narveson has argued, provide a theoretical basis for worrying very much about the fate of animals, give up worrying? Or should we declare the theories all inadequate precisely because none of them *do* accomodate the belief that we are not morally free to torture, hunt, and kill animals as we please? If there were unanimity in moral belief—if everyone agreed with Rawls that there are wrongs to animals—then it would be very clear that the theories, not the intuitions, are what need revision. But there is no such agreement. What we need, in this situation, is a reasonable way of deciding whose intuitions to discount, since *no* theory can hope to accomodate the intuitions of both the animal lovers and their enemies. If contemporary scientific experimenters, heirs of those Cartesians who nailed living cats by their paws to a wall, then slit them open to study the circulation of the blood in a living creature, are strongly convinced that the advance of science demands and justifies similar practices, while those who protest are equally strongly convinced that only moral monsters could act this way, then no theory can hope to gain acceptance by both parties, unless the theory also succeeded in converting one of them. But on this issue, as on some other moral issues—for example abortion— feelings run very high, and the chances that any mere theory will convert anyone are negligible. I am not content, like Rawls (1971, p. 50), to retreat to the aim of finding a theory that accomodates all *my* considered judgments, while resigning myself to the likelihood that yours are different and may require a different theory. I see no point at all in having a moral theory unless it can serve to help us live together more successfully and, therefore, unless it *aims* at general acceptance. Private moral theories, like private moralities, are more likely to divide us further than to harmonize our differences. So I see as unavoidable the issue of how we can with good conscience discount the moral convictions of some of our fellows and adopt a theory that has implications they reject. Even if one did adopt Rawls' modest program of squaring a moral theory with one's own ineradicable moral beliefs, one is likely to find oneself ambivalent on some issues, to find that one's own mind reflects the disagreement in the culture that formed one's intuitions. When am I thinking straight and feeling appropriately: At those times when contrived animal suffering for human ends, whether frivolous ones such as the taste for hunting, or more serious ones such as testing medical drugs, seems clearly wrong, seems as firm a moral given as any could be, and seems finally a decisive objection to any moral theory that would allow such practices? Or at those times when the claims of self-conscious human purpose seem to drown out, if not silence, all others, times when a moral theory that bases duties on mutually acceptable ties between persons seems obviously correct? Which side of my ambivalent mind should I suspect of muddle or ulterior motive, and so reform or control? Who, in the community of moral thinkers, should I heed: those who, like Stephen Clark, suggest that no one has "any standing" in discussions about our duty to animals unless they have taken "the simple minimal step of abandoning flesh-foods" (Clark, 1977, p. 183), that "those who still eat flesh when they could do otherwise have no claim to be serious moralists" (Ibid.), or those who tell me that I indulge in anthropomorphism and sentimentality when I worry about the fate of animals at human hands, that on "lifeboat earth" serious moral thinking is restricted to the question of how humans can save themselves from self-destruction, that "animals are not selfconscious and are merely means to an end" (Kant), so that it is only right and proper that they serve the ends of rational beings like us, thus becoming our food, our clothing, our playthings, our prey, our experimental subjects, our guineapigs, and our sacrificial lambs. Both to cure the instability and incoherence in my own beliefs and to attempt to face up to the real disagreement in the community of moral thinkers, I must raise the issue: When is there good reason to discount moral intuition, however recalcitrant and ineradicable it may be? I shall list some sources of likely prejudice and moral error on *any* issue, and so on this issue, before turning to the question of which moral theory squares best with the intuitions we have least reason to suspect.

One ground for suspicion of apparently sincere moral convictions is their link with some special interest of those who hold them. The questions *cui bono* and *cui malo* are appropriate questions to raise when we are searching for possible contaminants of conscience. Entrenched privilege, and fear of losing it, distorts one's moral sense. Just as, on an issue such as abortion, we should not listen too respectfully to the views of those who derive special profit from the legality or illegality of abortion, so on this issue of human treatment of animals we must be on the watch for special interest. We need to ask who has something special to lose or to gain from any change in our practices. As far as special gain goes, I see no interest group whose views ought to be treated with suspicion. Animals themselves cannot plead their cause, and those who plead it for them have no obvious financial or other selfish interest in the issue, although many may have "vested" their emotions in it. When we turn to special gain from maintaining existing practices, special loss if they were to be changed, we find a large number of groups whose views might be discounted. Butchers, furriers, hunters, cattlemen, chicken farmers, scientific experimenters on animals would, unless compensated, all have to suffer significant personal loss if we were to change our practices. They cannot therefore be expected to see the moral issue without the distortion of special interest. The scientists might claim that in *their* case their own interest coincides with a universal human interest, but I think the butcher and the furrier could make a similar claim—that their private gain from their occupation comes from their success at providing a public benefit— the food the rest of us want to eat, the sort of clothes we wish to wear. Even if some case can be made for giving a special place to the scientific and health professions, for seeing their practitioners as providing what is more obviously a public benefit than butchers and furriers, who simply meet a public demand, and I am not at all sure this case can be so convincingly made, it would not affect the claim I am making, that since private personal gain, and group interest, is also involved, the views of scientists who use animals in research ought to be discounted. The public benefits they supposedly offer can more impartially be weighed by other members of the public, not by those whose individual reputation, privilege, and income are tied to the practices whose morality is in question.

At this point it may be objected that my procedure in eliminating from the serious debate all those who stand to gain or to lose from a change in practices is itself contaminated with prejudice, namely a prejudgment of the relation between morality and interest. To avoid this objection let me emphasize that it is not interest, but *special* interest which I am citing as a disqualification. I am indeed assuming that the point of morality is to harmonize and advance interests and other components of the good of those concerned. But I am also assuming that one is not in the best position to judge overall or long term interest, nor to appreciate the value of some proposed measure that will harmonize interests, when one is threatened with loss of what one has come to count on and depend on. It is not *interest* that interferes with moral vision, it is the threat of loss of special privilege, of short term loss, or the lure of special privilege and of short term gain. I follow Hume in thinking that the moral point of view is "steady and general" and that such a point of view is one from which I "prefer the greater good whether at that time it will be more contiguous or remote" (T, 536),[1] a view that gives both contiguous and remote goods their due weight. But, as Hume noted, the "violent propension to prefer contiguous to remote" makes it difficult for a person whose near interest is involved to adopt this point of view, to "overlook our own interest in those general judgments, and blame not a man for opposing us in

---

[1] References to Hume's works are to the Selby-Bigge editions: T, Treatise of Human Nature; E, Enquiries.

any of our pretensions . . . " (T, 583). A realistic assessment of the difficulty, for any of us, of adopting the moral point of view when our own immediate profit or loss is at issue lies behind commonly accepted rules such as those requiring judges and politicians to disqualify themselves when private and special interest may conflict with the public interest their role requires them to consider.

It still might be objected that this disqualification of all especially interested parties is a hopeless measure when the issue involves a possible clash of "interest" between all humans and their animal victims. Must we disqualify ourselves because our species' advantage is at issue? This would indeed reduce *ad absurdum* my suggested way of discriminating between conflicting intuitions. They cannot *all* be suspect if moral theory must square with some of them. If moral theory is more than a rationalization of superior advantage in a competitive struggle between would-be exploiters of others, then there must be some way, not merely of fairly adjudicating, but of avoiding irreconcilable conflicts of interest. If there is an unavoidable conflict between our human interest and that of other living species, then it is unlikely that, in that battle for survival, we will (of necessity unilaterally) tie our hands with moral bonds. Yet can we really still call a theory a moral theory if it openly avows itself as a self-serving justification of ruthless acquisition and retention of group or species advantage? If morality is to be neither an ineffective guilty gesture in the direction of those who must lose so that we may win, nor a glorification of the advantage of the stronger, there must be some faith in the possibility of a harmonization of interests, in some way of preventing the moral game from being itself a zero-sum one. I find it no more idealistic or utopian to believe that we do not *have* to choose "animal interests or our own interests?" than to believe that we do not *have* to choose between the interests of males and of females. The moral enterprise is built on the faith that interests can to some extent be reconciled, that flourishing need not always be at the expense of others. I am supposing that such a faith in a peaceable kingdom is no more Utopian when extended beyond the human species than it is within the species of us moralizing animals. The fact that we can theorize, moralize, accuse, and condemn does not itself make our different individual self-perceived interests any easier to harmonize than they would be if we did not. Our special human skills give us ways of seeing solutions to the difficult problems of coexistence, but the mere fact that we are talkers, theorizers, and political and religious zealots creates as many conflicts as it solves.

I therefore do not need, absurdly, to conclude that because human interests are involved, humans must disqualify themselves from the moral debate. If the moral discussion aims at a harmony of interests, only those with a vested interest in avoiding harmony, or in what in fact prevents harmony, need to be disqualified. On this issue, as on any other moral issue, we need not listen to the gunrunners.

So far I have suggested only one ground to discount a belief—its contamination by vested special interests. There are, of course, other sources of moral prejudice and error. Besides perceived immediate special interest, there are also tastes and likings that might warp one's moral judgment. One hangs on tenaciously to the pleasant, as well as to the profitable. Those who just *like* eating meat, although they have no special interest in eating it, may disguise from themselves the full costs of catering to their tastes. Should we therefore, as Clark suggests, disqualify all meat-eaters and fur-wearers from the debate on this issue? It seems to me that many of us who eat meat do not even believe our lives would be any the worse if we did not. This may make continuing to eat it all the more inexcusable, but at least it will not be our tastes that, by their strength, distort our moral perceptions. (If distortion is there, it will more likely come from guilt, and the defensive mechanisms that burden brings with it.) So I am inclined to think that our present tastes are a much less important source of contamination of conscience than our present vested interests, some of them interests in catering to tastes that could fairly easily be changed.

Another possible source of prejudice is dogmatic attachment to some set of beliefs that one keeps safe from critical examination. If, as I have suggested, our moral intuitions derive from our moral upbringing, then a particularly rigid one may indoctrinate a person so successfully that no serious re-thinking is possible. I do not think it is easy to detect such inflexible commitments to *inherited* theory in one's own set of beliefs. If a particular moral intuition survives confrontation with a theory one explicitly considers, that might be only because one's loyalty is already irrevocably enlisted by the theory one's parents held, which one received before critical reflection was a possibility for one, and which therefore became an incorrigible belief. It is theoretically possible that any stubborn moral intuition is of this sort, and not really susceptible at all to "consideration." One may hope and believe that all one's beliefs are sensitive to rational criticism and challenge, but this too seems to me part of the faith on which moral philosophy is built. It is not, therefore, easy to tell the dogmatists from the open-minded, in order to discount the intuitions of the dogmatists. Presumably if there were a religious cult fanatically dedicated to animal sacrifice, their beliefs would be suspect, and so equally would those of fanatic Hindus, hopelessly prejudiced on the other side of the issue. We will suspect and discount those recalcitrant moral intuitions that are clearly tied to dogmatically held theories we do not accept. But one possibility that we must face is that *all* recalcitrant moral intuitions are tied, perhaps not obviously, to some older moral theory, so that all the "data" against which we might test a theory are vestiges of the faith of our fathers. If that were so, then the only external constraint on any moral theory would be that of conservatism. To demand that a moral theory square with incorrigible or invincible moral intuition will then be to demand that it not depart too radically from its predecessor theories.

I have tentatively suggested some reasons we might have for discounting some moral intuitions—the taint of special interest, the less powerful poison of special tastes, the contaminant of obvious and dogmatic prior theoretical commitment. The last may be only an obvious version of an unavoidable component in every moral intuition, but I shall nevertheless treat avowed dogmatism as a reason to discount a moral intuition, while not discounting the intuitions whose dogmatic component is hidden and unavowed. If one is to avoid total moral scepticism one must salvage some intuitions as less suspect than others. "Byasses from prejudice, education, passion, part, etc., hang more on one mind than another" (Hume, E. 107 note.)

Where does all this methodological preliminary leave us with respect to intuitions concerning duties to animals? It seems to me to leave us with most of the intuitions of most of the defenders of the cause of animals free from suspicion, or tainted only with the suspicion of religious or prior theoretical prejudice, while most of the intuitions of their opponents are tainted with special interest, over-attachment to their current tastes, or a combination of these. These may also be a large dose of religious prejudice, in the form of a commitment to the view that humans are the crown of creation, uniquely made in the image of God, behind many of the intuitions of those who would subordinate animals to our subjective ends. So the result of my "method of doubt," applied to moral intuitions, leaves me in agreement with Rawls and others who share his intuitions here, that these intuitions are as *nonsuspect* a foundation for a moral theory as any we are likely to find.

What moral theory has any hope of doing justice to these intuitions? If Narveson is right, none of the serious contenders among existent moral theories accomodate even a fairly weak set of beliefs about the scope of our duties to animals. Utilitarians may be able to derive a prima facie duty to avoid making animals suffer, but it is only prima facie, and usually easily balanced by some promised "higher" pleasure for us, the cost of which is animal suffering. Utilitarians can derive none but prudential reasons for not destroying whole species. I suppose the consistent Utilitarian would keep in existence such animal species as serve our needs, or as we find cute, and would have no objection at all to the preferably painless destruction of the rest. Contractarians, as Rawls has pointed out, cannot see any duties as owed *to* animals since they are not and cannot be members of a moral community who formally enter in into agreements, actual or hypothetical, with one another. They may be the *beneficiaries* of obligations accepted by members of a moral community, but if the reason for accepting such obligations is the perceived rational self-interest of the contractors, then only enough seals to stock the zoos and nature reserves

will be the beneficiaries of our hypothetically contractually based obligations, and the rest will be fair game for the fur trade.

Is there then no moral theory on the horizon that might accomodate
the intuitions of those who, like me, believe that it is primarily for the
seals' sake, not mine or yours, that one should desist from bashing baby
seals, however much profit it might bring, for the cat's sake, not mine or
yours, that one should not slit it open while keeping it alive and
unanesthetized, however much one might learn from the vivisection? I
think there is at least one moral theory of respectable lineage and good
independent credentials that can accommodate such fairly minimal intuitions about us and animals.

This is the theory Hume offers us. I do *not* consider Hume a forerun-
ner of utilitarianism, and therefore what I shall go on to say in defense of
Hume is not intended as a defense of any version of utilitarianism. I see
Hume to be much closer to Aristotle than to Mill, to be offering us a
theory about human virtues, not a theory about utility maximization and
the duties that might involve.

Let me first say something about the basis on which Hume rests his
specifically ethical views. Rawls says that any moral theory that generates duties to animals will have to be based on a metaphysical theory of
the "natural order and our place in it" (1971, p. 512). Hume cannot be
said to give us a metaphysical theory of the natural order, but he does give
us a psychological theory, which with a bit of exaggeration can be said to
be a theory of the animal order. In his section "Of the Reason of Animals" in Book One of the *Treatise* and in the first *Enquiry,* he relates
human cognitive powers to animal intelligence, and in Book Two of the
Treatise, he applies his account of both pride and love to animals. Indeed
one might say of Hume's version of human nature, in all its aspects, that
it presents us as not radically different from other animals. Hume emphasizes the continuities, and these continuities go well beyond shared sensitivity to pleasure and pain. Both in our cognitive habits and in our emotional range, human nature as Hume sees it is a special case of animal
nature. Not merely the direct passions of desire, contentment, and fear,
but the idea-mediated indirect passions of pride and love, and the spread
of all these passions to others by sympathy (T, 363) are attributed by
Hume to animals. Since the moral sentiment, on Hume's account of it,
depends both on the capacity for pride and love,[2] and on the capacity for
sympathetic sharing of another's feelings, it is quite significant that Hume
attributes to other animals all the basic emotional prerequisites of a moral
sense, including some, such as pride, that require a sort of self-
consciousness. Hume says that "all the internal principles that are neces-

---

[2]I have discussed the relation between Hume's version of the moral sentiment and
Humean pride in "Master Passions" (Baier, 1979a), and in "Hume on Resentment"
(Baier, 1980), *Hume Studies,* 6, 133–149, Nov '80.

sary in us to produce either pride or humility are common to all creatures; and since the causes which excite these passions are likewise the same, we may justly conclude that these causes operate after the same manner thro' the whole animal creation'' (T, 327–328). He had previously allowed that some causes for human pride cannot be causes for canine pride, that we must make ''a just allowance for our superior knowledge and understanding. Thus animals have little or no sense of virtue or vice; they quickly lose sight of the relations of blood, and are incapable of right and property. For which reason the causes of their pride and humility must lie solely in the body, and can never be plac'd either in the mind or external objects . . . '' (T, 326). Animals, Hume says, have no sense of property or right, and so there can be no question of any obligations, based on convention or agreement, either on their part or owed to them. Hume would reject any attempt to give sense to the concept of rights of animals, since all rights arise from artifice. But since the artificial virtues are only a small subset of the virtues, the fact that animals have no obligations or rights would not mean that no moral wrongs can be done to them, nor even that they themselves can have no ''duties.'' Hume significantly says not that animals have no sense of virtue or vice, but that they have ''little or none.'' He is, of course, both in these sections in Book Two, and in the earlier section on ''Reason of Animals,'' deliberately debunking the inflated rationalist conception of a mental substance uniquely capable of truth-seeking and a moral life. He demotes human truth-seeking to a version of instinct, and likens the philosopher's love of truth to the passion for hunting (T, 451), the desire to *collect,* and to get prize catches. Similarly he is intent on de-intellectualizing and de-sanctifying the moral endeavor, in presenting it as the human equivalent of various social controls in animal or insect populations. So he may be merely teasing us when he tells us that animals have little or no sense of virtue and vice. But he is in earnest in presenting human capacities, including moral ones, as special cases of animal capacities.

One curious feature of Hume's discussion of the indirect ''passions of pride and love is that he both says that their ''object'' must be a person like ourselves, with a sense of self (T, 329), and also says that animals feel both pride and love. What is more, in the case of love, it is not only that they, in their fashion, feel, towards something like them—a member of the same species—a version of what we feel towards creatures like us, Hume goes out of his way to say that ''love in animals has not for its only object animals of the same species but extends itself farther and comprehends almost every sensible and thinking being. A dog naturally loves a man above his own species and very commonly meets with a return of affection'' (T, 397). Hume might be said to imply, by these claims, that animals *are* persons, but it would be more reasonable to interpret him as pointing to a very strong analogy between human emotions, of the most self-conscious and personal kind, and their animal equivalents.

At this stage of my exposition of the "theory of nature" on which
Hume's ethics are founded I need to say something about his treatment of
self-consciousness. As is known, if anything too well, Hume in Book
One gave a skeptical account of our consciousness of ourselves as enduring persons. But that was an account carefully restricted to intellectual
self-consciousness, or "personal identity as it concerns our thought and
imagination," as distinct from "our passions or the concern we take in
ourselves" (T, 253). When it comes to the latter he shows no skepticism
at all—pride involves the occurrence of an "idea or rather impression of
ourself" that he says is "constantly present to us." "The immediate object of pride and humility is that identical person of whose thoughts, actions and sensations we are intimately conscious" (T, 329). Because this
self-reference is part of a *passion,* not the culmination of a merely intellectual search for self, Hume treats it as escaping the incoherences of attempted intellectual self-survey. Indeed part of his case for claiming that
animals can feel pride is that his account of it "supposes so little reflexion
and judgment that 'tis applicable to every sensible creature" (T, 328).
These words, however, should not be taken to deny what Hume has previously insisted on, that all passions are "reflexive impressions," and
that pride in particular involves a special sort of reflection, namely reinforcement of one's own self-assessment from the respect others pay one.
Hume does not believe that pride requires any *intellectual* reflexion, any
reflective or thoughtful ideas, but, like all passions, it does involve a reaction, or reflex, to a given pleasure, and for its own prolongation it requires reinforcement from other passionate beings, it calls out to be "se-
conded by the opinions and sentiments of others" (T, 316). So Hume in
fact has quite an intricate account of the varieties or levels both of "reflexion" and of the self consciousness reflexion may involve.[3] The pride
he attributes to animals involves both non-intellectual sorts of reflexion—
both the capacity for passion, or reactive sentiment—and also the need
for reinforcement from others of the passion of pride, which involves
non-intellectual self-consciousness. He has therefore provided us with an
account of human and animal psychology that gives us a basis for an ethical theory that can say something much more interesting about our relation to animals that Bentham's simple-minded trichotomy: "The question
is not can they *reason,* nor can they *talk,* but can they *suffer?*" It is the
range of feeling and suffering, the presence of self-conscious and social
feeling, which is important for deciding the moral issues. I believe that
Hume gives us a plausible and discriminating account both of what we
share and of what we do not share with other animals (and, of course,
how much we share will vary from species to species).

[3]I have discussed the intricacies of Humean self-consciousness in "Hume on Heaps
and Bundles" (Baier, 1979b).

To turn at last to the moral issue itself. What sort of account can a Humean moral theory give of the place of animals in a moral order? Hume's version of a moral order is of a community of persons who assess one another's virtue by shared standards. The virtues that are prized include both "natural" ones, which presuppose no convention, and "artificial" ones, which are displayed in respect for conventions. The qualities of persons that count as virtues are those that are agreeable or useful to oneself or to others. Agreeable or useful to which others? To those who recognize the virtues, but they are of necessity persons who can sympathize with a wider circle of others, who can appreciate what is agreeable or disagreeable to others, whether or not those others are approvers and disapprovers. The virtue of kindness, for example, is displayed in one's treatment of children, long before they are themselves aspirants to and judges of virtue. It seems to me a virtue of Hume's account that there is absolutely no need to say that the only reason we should not be cruel to infants is because they are potential full members of a moral community. That fact may point us to some specific harms that we should avoid doing to them—for example exclusion from the circle of those who are mutual approvers and disapprovers—but it is not itself the reason for not harming them. The harm *itself* is the reason for that, and our sense of its disagreeability. So although Hume himself does not address the question of whether vices and virtues are shown in our treatment of animals, it seems quite evident that his answers would be "Of course." The duties of justice will not be owed to them, but all the natural virtues will cover our treatment of animals. Because we can recognize what constitutes harm to them, because they, like us, are potential victims of human vices, we have both sympathetic and self-interested reasons to condemn humanly inflicted harm to them. Hume makes only a half-hearted attempt to separate out the self-interested from the altruistic component in his account of the natural virtues, and I find this a strength not a weakness. Sympathy with others is, for Hume, both altruistic and also an indirect way of getting what one needs oneself, insight into how others react to one's own character, and their approval. Because Hume believes not only that "the minds of men are mirrors to one another," but also that each person needs such a mirror to sustain self-confidence, as well as to make confident cooperation possible, it is impossible to separate the self-concerned from the other-concerned reasons for regarding a particular character trait as a virtue.[4]

[4]Both in the *Treatise,* and more thoroughly in the *Enquiry Concerning the Principles of Morals,* Hume does attempt to classify virtues by asking whether their beneficial effect falls mainly on the virtue-possessor or on that person's associates. A virtue is a character trait that, by corrected sympathy, we see to give pleasure or to advance the interest of "the person himself whose character is examined; or that of persons who have a connection with him," (T, 591) where the "or" is definitely not exclusive. As Hume says, a virtue

Only when we turn to Hume's account of justice, that is to the group of artificial virtues comprised of respect for property, promises, and governmental authority, do we find a rationale that is based only on utility or interest, and even there it takes artifices to *create* this interest that is a share in a public interest. Clearly animals cannot have either the obligations or the rights that the artifices of property, promise, and government give rise to, but when one looks at Hume's general definition of a convention, rather than at the specific conventions he discusses, I think one finds that, in his sense of convention, we could have conventional obligations to animals. A convention involves both a mutually expressed sense of common interest, and mutually referential intentions. "The actions of each of us have a reference to those of the other, and are performed on the supposition that something is to be performed on the other part" (T, 490). A horse and its rider, or a man and his dog, seem as good an example of this as Hume's boatload of rowers. For simple "conventions," involving no symbolic expression such as money or promise, the higher animals seem not disqualified from inclusion in the conventional agreement. As long as the expression of the mutual interest and interdependent intentions can be natural, surely Hume's conditions can be fulfilled, are fulfilled, in every case when humans and animals can be said to cooperate for common advantage. The fact, which even Kant recognized, that a man who can shoot his trustworthy and trusting dog shows a bad character, can be given a quite straightforward interpretation on a Humean theory. It is *not* because such a man might well also shoot a trusty and trusting human companion, it is simply because he betrays the animal's trust, breaks the "agreement" with the animal. I see nothing at all anthropomorphic or in any other way absurd in saying that one may "break faith with" an animal, exploit its trust, disappoint expectations one has encouraged it to have. I see no reason at all to treat such a case as discontinuous with breaking faith with humans. Hume half recognizes that the virtues of honesty and fidelity and loyalty extend more widely than respect for property, for promises, for governments, since he on one occasion includes fidelity on a list of natural virtues (T, 603). The distinction between artificial and natural virtues may not be as sharp as Hume sometimes makes it appear, especially when one realizes the broad scope of his official definition of a convention. (After all there surely is some link between the agreeable and what has been agreed on.)

I have suggested that all the Humean natural virtues and vices, and a primitive informal precursor to the artificial virtues and vices, are as

---

may have "complicated sources" (E, 328) so that, although the main reason for approving of persons who are kind and generous is the good they do others, still these virtues are also "sweet, smooth, tender and agreeable" (E, 282) to the kind and generous person, and "keep us in humor with ourselves" (ibid.) as well as others. In the *Enquiry* Hume is at pains both to show that we do not need a self-interested motive to cultivate the virtues, and also to show that we need not lose by cultivating them.

much displayed in our treatment of animals as in our treatment of fellow humans. Indeed, if we are to believe Hume, some animals may themselves display a rudimentary sense of virtue. It would I think be absurd to suggest either that we morally assess an animal's character, or that we should be sensitive to the 'disapproval' of animals. Animals can not disapprove, but they can complain and protest, at least until their vocal chords are cut to spare experimenters their protests. Children too have complaints heeded before they are capable of disapproval.

One question that still remains is why, granting that we *might* treat a virtue's scope as extending to our treatment of animals, we should not narrow it to exclude them, if that appears to be easier or more convenient for us.

To that question I answer: for almost the same reason as we do not exclude future generations of humans from the scope of our moral concern.[5] We gain nothing from considering them, and, if we harden our hearts, we *need* not consider them. Rawls makes his self-interested contractors heads of families, in part to ensure that there will be some self-anchored reason for each hypothetical contractor to care about some of those in the next generation, to let concern spread. To make the contractors heads of families, caring about the members of their families, is to modify the claim that it is *self*-interest, in the right conditions (the veil of ignorance), that generates agreement on moral principles. We could perhaps extend the same tactic and make the hypothetical contractors heads of households that include domestic and other animals tolerated in the household, a group of Noahs each with an ark of humans and animals in his or her care. The principles then agreed on would include ones that would accommodate Rawls' intuitions about duties to animals.

Would anything be gained by approaching the matter this way? Hume, as I have already argued, could have let the agreement-based artificial virtues include more than he did, by a strict application of his weak definition of agreement or convention. Had his parties to the agreement been taken to be, not individuals, but heads of households, or, to get the effects of Rawls's veil of ignorance, spokesmen for "the party of humankind," all the virtues could in a vacuous sense be seen as artificial virtues, ones founded on agreement. There would still remain the important differences both between nonformal agreements and symbol-involving formal agreements, and the even more important distinction between the cases where the agreement is possible only because of the artifice that agreement creates, and the cases where the parties agree about what was and is agreeable independently of their agreement. Would anything be gained, is anything gained, by forcing all moral demands into the mold of agreement-based demands?

[5] I have discussed obligations to future generations in "The Rights of Past and Future Persons" (Baier, 1981). See also my "Frankena and Hume on Points of View" (Baier, 1981). *Monist* Vol. 64,3 342–358.

The gain, as Rawls' work shows, is the availability to the moral philosopher of the formal techniques of game theory and the theory of rational choice. One can *calculate* what would be agreed on by one's hypothetical contractors, given their hypothetical preference-based interests. One thereby gets at least the appearance of greater precision in one's moral theory than is present in, say, Hume's theory as it stands. But we should not aim at a greater precision than our subject matter admits of, and I am not at all convinced that the precision gained by any moral calculus, utilitarian or contractarian, is more than a surface one. What is more, it is gained at the cost of what might be seen as a false pretence that moral reasoning is just a special case of individualistic self-interested reasoning,[6] a special case of a competitive game. It may be a hard historical fact that the developed formal techniques in decision theory are those developed to meet the needs of capitalists and heads of armies. But that is no good reason to make ethics the moral equivalent of war, or of business competition. The hard questions about the relation between moral reasoning and the calculations of individual self-seekers are pushed back into the description of the contractors, their primary goods, and the circumstances of the contract. It would seem to me better if they were directly addressed. A moral theory that makes the admittedly fuzzy concepts of virtue and vice the central ones at least openly avows both the imprecision concerning the relation between individual interest and the interest of others that is at the heart of the theory,[7] and also the imprecise guidance the theory yields both concerning the details of moral duties and also concerning the proper limits of state coercion. On the last question, Hume speaks as if magistrates are invented to clarify, formulate, and enforce already existent conventions and the conventions creating government—to prevent theft and fraud and treason. But clearly he expected them not only to do this, as well as to organize cooperative enterprises of "complicated design," to "build bridges, open harbours, raise ramparts, form canals, equip fleets and discipline armies" (T, 539), but also to prevent assault and murder. The "care of the governments" includes care for the persons as well as the property of their subjects, so they must treat some manifestations of some natural vices as criminal, in addition to making the artificial vices punishable by law.

[6]As Hume put it, "It is not conceivable how a *real* sentiment or passion can ever arise from a known *imaginary* interest" (E, 217; see also, E, 300).

[7]Hume sees a need to distinguish *mine* from *thine* with definiteness only for those scarce alienable goods that are covered by the institution or artifice or property, not for such benefits as accrue from the cultivation of natural virtues. Those who insist on sharply distinguishing benefits to oneself from those to others, and on making narrowly self-interested calculations, may find themselves the "greatest dupes" (E, 283). Because of this view about convergence of interests, Hume can afford to say what he does about motivation to virtuous action: "It seems a happiness in the present theory that it enters not into that vulgar dispute concerning the degrees of benevolence or self love which prevail in human nature" (E, 270).

What guidance would a theory of virtues and vices give us in deciding what treatment of animals is wrong, and which wrongs ought to be made into criminal offenses? Hume's lists of virtues includes, besides the artificial or convention-dependent virtues, the natural virtues of generosity, humanity, kindness, friendship, good nature, considerateness, prudence, benevolence, good sense, economy, all of which would be involved in appropriate treatment of animals. Lists of virtues and vices in themselves do not also embody decisions on what degree of vice is sufficiently evil to render an action criminal, but we surely could and should forbid the obvious manifestations of extreme cruelty as well as of extreme lack of ecological good sense. And the less formal sanctions of disapproval and withdrawal will be directed at those who display less than criminal degrees of that cold insensibility, and that foolishness, that is involved in treating animals as mere things for our use. Hume describes (E, 235) as a "fancied monster" a man who has "no manner of concern to his fellow-creatures but to regard the happiness and misery of all sensible beings with greater indifference than even two contiguous shades of the same color" (ibid.). To limit one's concern to those sensible beings who are of one's own species is to be part-monster, but such monsters, alas, are not merely fancied ones.

Hume's moral philosophy, then, gives no very precise ruling on wrongs to animals, either to individuals or to legislators. But I think that the serious guidance given by other theories in real life situations is equally vague. So the wish for definite guidance provides no reason to switch from a theory of virtues to a theory of agreement-based obligations. When we are trying to find a theory that holds out some promise of accommodating widely shared nonsuspect moral intuitions about how animals may and may not be treated, a theory such as Hume's seems at least to contain no theoretical obstacles, even if it gives us no very precise guidance. The more difficult question that I have still not answered is whether the theory not merely fails to present obstacles to extending moral concern to animals, but gives us positive reason to do so. Hume's version of human nature is a good basis from which to start, since it encourages us to respect, not to downgrade, the capacities we share with other animals, to recognize that what we respect in fellow humans has its basis in what we share with other animals. When we refrain from hurting an animal, it will not be because by hurting it we would "damage in ourselves that humanity which it is our duty to show towards mankind," (Kant) it will rather be because important elements of the humanity we respect in our fellow humans are also really present in the animal, and so demand respect in their own right. It is also intrinsic to any theory of virtues to include some, such as Aristotle's equity and practical wisdom, that discriminate among the relevant *differences* between cases, and are displayed in good judgment concerning how they should be treated. A theory of virtues must give an important place to such discrimination and judgement, and a Humean theory that extends our moral concern from those humans (future generations, idiots, madmen) who cannot have a return concern for us, to animals who must fail to reciprocate that moral concern, will value as a special virtue that "delicacy" and judgment that is shown by a thoughtful appreciation of the difference between proper treatment of a child, an adult, an idiot, a mad person, an ape, a pet lamb, a spider (and the proper appreciation of the differences between eating human and other flesh).

Does a moral theory that, like Hume's, bases discernment of approved character traits on sentiment build on a nonsuspect foundation? If tastes and perceived interests can poison our moral intuitions with prejudice in our own favor, may not sentiments do the same? Why should human sentiment, even when it is moral sentiment, be a reliable guide to the good of a larger than human community of morality-affected beings? As I have already acknowledged, emotions *can* be invested in a cause, good or bad, and so distort perception of issues. And one sometimes suspects that some of those who plead the cause of animals are as much human haters as they are animal lovers. Sentiment, as much as taste or perceived interest, can contaminate one's conscience. But if Hume is right, certain sentiments have both a built-in potential for enlargement by sympathetic spread, and also a built-in need for reinforcement from others, so that there can be a genuinely moral sentiment that results from the correction and adjustment of more partial sentiments. For this moral sentiment to arise, "much reasoning should precede" to "pave the way" (E, 173). Hume does not believe reason alone can discern what is right, since he sees reason as always serving some sentiment or passion. But no sentiment can count as the moral sentiment unless it is "steady and general," unless it does arise from a reflective reasoned consideration of the good of all concerned, since "I am uneasy to think that I approve of one object, and disapprove of another . . . without knowing upon what principles I proceed" (T, 271). The moral sentiment, to *be* moral, must be one "in which every man, in some degree, concurs" (E, 273). Hume speaks of this shareable sentiment as the sentiment of *humanity,* not of animality, since it is only humans who approve, or are themselves objects of approval. It does take some reconstruction of Hume's theory to ensure that the scope of moral concern extends to animals, but I believe this can be done. As long as we distinguish the community of moral judges and judged, restricted to humans, from the wider community of those with whom some sympathetic concern must be felt in order to do the judging, then we can find in Hume's moral theory at least the basis for an adequate account of proper attitudes to animals.

Nevertheless I want to end by admitting that I do not think that a Humean moral theory can show anything positively incoherent in refusing to consider the fate of animals or animal species,[8] that is, in restricting the scope of a virtue's reach to humans. To get any strong reason why we should extend our concern to animals, I think the best theory is one that is Kantian to this limited extent, that the moral order is seen as continuous with a natural order, that the question to ask is "Can I will this as a law in a system of nature?" The reference to system, the insistence on seeing human good as the good of a being who is part of a wider system that is a system of *nature,* a nature containing a great variety of beings of different kinds, was only one component in Kant's theory, one that owes much to Leibniz and Aquinas. If Rawls looked to that element in Kant's ethics, rather than to Kant's dualistic Cartesian refusal to respect anything we share not with God, but with animals, then the larger theory he needs may be surprisingly close at hand.

[8]The fate of the human species may be more bound up than is immediately obvious with that of other species. Although *this* generation would probably live well enough if it exterminates species as it pleased, the long-term effects on the human environment of human policies of "improving" their natural environment have almost all been bad. So if we *are* considering future generations of humans, we might do best for them by not attempting to pit their interests against those of other species. We know too little about our own species' interests to rest our policies on fallible calculations aimed to advance only them. Consideration of other species may be the best policy, whoever and whatever we care about. At least we do not know Hume to be wrong in thinking that those who take a universal moral point of view will not be the greatest dupes.

The Clouded Mirror

Animal Stereotypes and Human Cruelty

Thomas L. Benson

Misdeeds and moral illusions keep close company. Acts of cruelty may pass for something better where the victims can be seen as either undeserving or of no moral account. In turn, low estimations of another's moral worth do little to discourage hostile conduct. The long history of injustice to women and the sordid traditions of racial and ethnic discrimination amply illustrate this pattern. In each case a system of degrading stereotypes has served both to legitimate and to stimulate immoral conduct. Somewhat less recognized, but no less potent are the stereotypes that shape and are shaped by the traditions of human cruelty to animals.

Walter Lippmann referred to stereotypes as pictures we "carry about in our heads,"—*a priori* representations all too often at odds with complex reality (1922, p. 59). Their function, according to Gordon Allport, is to justify categorical acceptance or rejection of select groups and to screen out unwelcome details that threaten simplicity in perception and thinking (1958, p. 187). Not surprisingly, stereotypes have their greatest currency with respect to groups that are at best vaguely understood and, more often than not, thoroughly mistrusted. The stereotype defines the status of such groups as being either second class or beyond the pale of moral accounting altogether. As such, these groups may be denied the full protection of the society's legal and moral sanctions. In the case of animals, a cluster of tenacious stereotypes has served to maintain widespread indifference to their suffering at the hands of humans. Having been hunted, roped, and penned-in by our stereotypes, animals face such treatment all over again in nonmetaphorical terms.

The stereotype enjoys a shadowy existence, influencing perceptions and shaping ideas from behind the scenes. The controlling picture is, typically, quite unflattering. Its grip on the imagination is registered in a wide range of hostile behaviors, from acts of physical cruelty to psychological abuse and derisive speech. Among the group of dominant animal stereotypes, however, there are some interesting exceptions to this pattern. As we shall see, not all of the stereotypes associated with animals are obviously pejorative. Indeed, some of them involve apparently innocuous or even idealized versions of animal nature. Unflattering or not, however, each of the stereotypes responds more to human needs than to the realities of animal nature. Such stereotypes are, quite simply, lies told at the expense of animals; and here, as elsewhere, deception acquires its own momentum. One lie tends to beget another. The seemingly benign distortion is all too easily replaced by an unmistakably vicious one. There is, thus, an essential instability among the animal stereotypes. Cut adrift from the demands of discovering and responding to animals as they present themselves to us, we are free to invent their natures, floating at the impulse of need and fantasy from one false image to another.

The full range and variety of animal stereotypes defies neat and rapid review. We can do no more than note some of the more popular and influential ones. In examining each of these stereotypes, it should be kept in mind that none of them dominates the field nor functions in isolation. The attitude that many people have toward animals appears to be a composite of several of the stereotypes, with some animals being viewed according to one, and others according to another. Moreover, the stereotypes associated with a particular species may vary with changing circumstances. The five stereotypes I shall discuss in the following pages are *alien, child, moral paragon, demon,* and *machine.* Although the first of these is, perhaps, the most formidable and far-reaching stereotype and the last the most recently developed, no significance should be assigned to the order of presentation.

THE ANIMAL AS ALIEN

There is between the human and the nonhuman animal what John Berger has called an "abyss of noncomprehension" (1979, p. 3). Beyond the distractions of scent, sound, form, and movement, there is a lack of focus in the eyes, an inability to slice through the fog of species separation. It is not simply that the animals cannot talk with us; rather, it is the deeper, more sobering realization that, as Wittgenstein noted, "If a lion could talk, we could not understand him" (1958, p. 223). The encounter of the human and the nonhuman animal is always a meeting of radically different forms of life. It is not surprising, then, that humans have been tempted to regard animals as aliens. An animal is, after all, like many aliens, on leave from a strange region and possessed of unfamiliar habits. And, as with human aliens, there are always ugly flashes of the sentiment: "He doesn't really belong here." The alien comes to us an intruder or a guest—in either case as one without widely acknowledged rights to freedom or the resources of the land. Moreover, like the human alien, the animal inspires both fascination and suspicion. Contrary to expectations, the fascination appears to derive from the margin of dissimilarity between the animal and the human, while the suspicion owes to the perception of crucial similarities. We are attracted by the novelties of appearance and behavior, and, at the same time, unnerved by the possibility that these features mask evils we have discovered in ourselves.

As aliens, animals are dimly thought capable of the same greed and treachery humans too often visit on one another. Our repertory of metaphorical epithets reflects this distrust of animals. A human may be censured as a "snake," "vulture," "rat," "pig," "turkey," "shark," "leech," and much more. It is suspicion, then, that dominates the response to the animal as alien. This suspicion, in turn, stimuates the desire to control the animal's behavior. Such control tends to fall into one or the other of three broad categories: assimilation, confinement, or banishment. In the case of diverse immigrant groups and the Native American "alien," it is not difficult to trace these categories—and the pattern is no more mysterious with respect to animals. The direct or indirect extermination of hundreds of species of wild animals, the domestication of others, and the confinement of a wide variety of animals in game preserves, zoos, and, more benignly, in animal sanctuaries, all conform to the categories of control.

In zoos and safari parks, there remains, of course, a fascination with the peculiar appearance and behavior of the animal, but this interest is indulged only within a framework of strict controls. It is these controls and the larger process of subjugation that leads to disappointment at such places of amusement, for what we encounter at the zoo or the safari park are, at best, imitation animals, possessing the genes and broad forms of their kinds, but behaviorally transformed. John Berger refers to zoos as monuments to an historic loss. Through the centuries of accelerating control of animals, humans have succeeded in marginalizing animal life—to the point of nearly total dependence upon favorable human behavior. Berger writes: "Looking at each animal, the unaccompanied zoo visitor is alone. As for the crowds, they belong to a species which has at last been isolated" (1979, p. 26).

THE ANIMAL AS CHILD

The tendency to view the animal as a child—cute, cuddly, and dependent—is, like the institution of human childhood, a relatively modern development (see Aries, 1962). Although we can find some isolated anticipations of this attitude in earlier periods, e.g., the paternalistic attitude shared by St. Godric of Finchale and St. Francis of Assisi, the roots of the child/animal stereotype are to be found in the romantic ideals of the late eighteenth and early nineteenth centuries. At a time of unprecedented growth in the cities and in the experimental sciences and technology, animal life was being exploited and sacrificed at an alarming rate. In reaction to these depredations, the humane movement arose and began to exercise a growing influence on public policy and private habits, advocating a wide range of reforms, from anti-cruelty statutes to vegetarian diets. Much of the thought and literature associated with the early humane movement is intensely anthropomorphic and sentimental. Appeals for reform were based, more often than not, on the plight of the defenseless and innocent animal, engulfed in a gathering tide of human abuses (see Turner, 1964). In the ensuing years and struggles, many wings of the humane movement have retained this paternalistic, somewhat sentimentalized outlook. The remarks of Hans Ruesch, in his recently published attack on vivisection, reflect the persistence of this attitude: "The desire to protect animals derives from better acquaintance with them, from the realization that they are sensitive and intelligent creatures, affectionate and seeking affection, powerless in a cruel and incomprehensible world, exposed to all the whims of the master species" (1978, p. 45).

In our time, the notion of the animal as a dependent child has acquired considerable popularity, with the media of film, television, and photography providing animal images of great sentimental appeal. A culture of cuteness now surrounds many animal species. Every nursery is equipped with a menagerie of stuffed animals; and children's literature, films, and television programs are nearly monopolized by winsome child/animals. This partnership of children and anthropomorphized animals may well be desirable from a psychological point of view. The vulnerable, dependent, and frequently incompetent child/animal may approximate the child's own experience of himself and may invite the exploration of attitudes of trust and responsibility toward others. The unthreatening companionship of a puppy or of a soft teddy bear may be just what Dr. Spock ordered. Nevertheless, some cautionary notes are in order. It should be observed, for example, that the child/animal stereotype is selectively imposed. Not all animals are cute and cuddly. And many of those who are cute soon become less so, thereby losing their appeal. The cute duckling/ugly duck pattern is tragically familiar to animal welfare workers who must cope each spring with the suddenly matured and no longer wanted Easter presents. The same story can be told even more dramatically with respect to puppies and kittens. Whatever its psychological value, the commercially driven emphasis on the cute and cuddly animal predisposes the human child to relate to animals on an unrealistic and anthropocentric basis. The value of the animal consists in its docility, playfulness, and charm as a human companion. Animals that fail to meet such standards may be written off as of no account.

The sentimental paternalism associated with the child/animal stereotype may be, in some respects, unwholesome for the favored species as well. In his book on social values and human childhood, *Escape from*

Childhood, John Holt argues that cuteness is the handmaiden of subjugation. Children are usually held to be most cute when they are exhibiting ignorance and incompetence. The intelligent and thoroughly competent child seldom strikes us as cute (1975, p. 85). Moreover, Holt argues that there is a direct connection between sentimental perceptions of children as cute and cruelty towards them (p. 81):

> The trouble with sentimentality, and the reason it always leads to callousness and cruelty, is that it is abstract and unreal. We look at the lives and concerns and troubles of children as we might look at actors on a stage, a comedy as long as it does not become a nuisance. And so, since their feelings and their pain are neither serious nor real, any pain we may cause them is not real either. In any conflict of interest with us, they must give way; only our needs are real.

The ease with which some people are able to move from Mickey Mouse cartoons to the setting of spring action mouse traps, from a chorus of "Old MacDonald Had a Farm" to a round of hamburgers at the local fast food outlet, suggests the applicability of Holt's argument to the child/animal stereotype. At bottom, as Holt observes, it is the abstractness, the unreality of the cute child image that creates the difficulties. So also with the child/animal stereotype. The sentimentalized representations of animals provide a distorted picture of their natures and needs. Children reared in an environment that portrays animals exclusively as cute, docile, and innocent are ill-prepared to appreciate the complex behavior and very real suffering of mature and, from a human standpoint, not always attractive animals.

In registering these points of caution about the child/animal stereotype, there need be no concern about the imminent exile of Winnie the Pooh or Kermit the Frog from the crib or nursery cupboard. It would be foolish to deny or attempt to suppress the charm and appeal that many animals and animal-like toys hold for us. Similarly, abandonment of the sentimentalizing excesses associated with the child/animal stereotype need not and should not cause us to discount our custodial responsibilities toward domestic animals and endangered and threatened species. Indeed, in outgrowing our tendency to regard animals as docile and dependent children, we may discover fresh possibilities for more responsible and satisfying companionship with them.

THE ANIMAL AS MORAL PARAGON

If the child is frequently confronted with images of the animal as helpless and cuddly, he also becomes well-acquainted with images of the animal as protective older brother or surrogate parent. Lassie, Rin Tin Tin, Rikki Tikki Tavi, Flipper, Flicka, Lobo, the parade of gallant and plucky animal heroes goes on and on, trailing back into the childhood library and the neighborhood moviehouse. In each case, the animal hero, while retaining much of the charm and appeal of the child/animal, models some virtue or cluster of moral ideals. In the preface to *Lives of the Hunted*. Ernest Thompson Seton, a storyteller with few rivals in the representation of animals as moral heroes, claims that his aim has always been "to emphasize our kinship with animals by showing that in them we can find the virtues most admired in Man" (1901, p. 9). Seton provides an inventory of the virtues associated with his animal heroes (p. 9): "Lobo stands for Dignity and Love-constancy; Silverspot, for Sagacity; Redruff, for Obedience; Bingo, for Fidelity; Vixen and Molly Cottontail, for Mother-Love; Wahb, for Physical Force; and the Pacing Mustang, for the Love of Liberty."

There is, of course, nothing new in the latter-day representations of animals as moral exemplars. Kenneth Clark suggests that the ancient cave-paintings at sites such as Lascaux in Southwest France were motivated primarily by admiration for animals and by a humility before their strength and speed (1977, p. 72). Indeed, reverence for certain animals as supremely gifted and/or virtuous appears to underlie both totemistic worship practices and the ample use of animal symbols in many primitive religious traditions. In the Psalms of David and the preaching of Jesus, as well as in the Buddhist Sutras and the teachings of Lao Tzu and Chuang Tzu, animals appear as models of diverse moral and spiritual virtues. It is also an idealized view of animal nature that inspired the early animal fables and the development of the *Physiologus,* the most significant of the early bestiaries.

In all of these traditions, old and new, there is boundless anthropomorphism. The virtues manifested by the animals are distinctly human ideals. What the animals do effortlessly and consistently, humans achieve only sporadically and with much effort. On this account, the exemplary animals are seen as innocent and, in some accounts, beatific. Catherine Roberts advances such a view in her recently published book, *Science, Animals, and Evolution.* Animals are uniquely virtuous, Roberts claims, and incapable of immoral conduct. "Nature thinks no evil and does no evil. How blessed is man to live in an environment of purity" (1980, p. 182). The difficulties posed by crediting animals with both virtue and the inability to do wrong appear to trouble Roberts no more than all the others before her who extol animals as moral paragons.

The moral paragon stereotype presupposes that animals are moral agents, capable of understanding, however dimly, the principles of right conduct and equally capable of pursuing such principles. To this extent, it is misleadingly anthropomorphic and inaccurate. Although animals may possess moral rights, there is no good reason to include them in the class of moral agents. There is, nevertheless, something to be said for one of the other, perhaps more fundamental, presuppositions of the moral paragon stereotype—the notion that we can draw moral lessons from animal behavior. The restraint shown by most animals in intraspecies conflict and the economies practiced by animals in manipulating the means of survival have clear instructional value for us, given our long history of internecine conflict and our reckless wasting of precious natural resources. Our greed and destructiveness cannot be written off to biology. It is within our power to adapt more harmoniously to each other and to the environment.

In evaluating the moral paragon stereotype, it should also be noted that there is a dark side to the picture. A pattern of species favoritism is built into the stereotype. Not all animals are regarded as virtuous. The animal kingdom includes both the virtuous and the derelict. From Aesop to Disney, the sheep and the wolf, the ant and the grasshopper divisions persist. With only rare exceptions, the virtuous animals are docile and attractive, while the ranks of the morally errant are filled by animals perceived as ugly or menacing. As long as lines are drawn in this arbitrary and anthropomorphic manner, it will remain difficult to win broad support for efforts to improve the lot of the laboratory mouse, the pork farm pig, or the endangered red wolf—among scores of maligned species— none of whom is conventionally thought of as morally exemplary.

The moral paragon stereotype poses still another problem. In the sometimes romantic representations of animals as heroes (and villans), there is a tendency to deprive them of their natural identities. The objections that feminists and Native American activists have leveled at efforts to place them on pedestals, as guileless and innately wise, are relevant here. The latter groups have recognized the potential for mischief in such romantic portraits. To be set apart as morally exemplary by means of patently distorted characterizations is to run the risk of being set apart and, indeed, left out—on the basis of other distorted characterizations—when it is a matter of distributing rights and opportunities. As noted earlier, one lie begets another; one false characterization, even if inoffensive on its face, invites another. The distorted picture of animals conveyed by the moral paragon stereotype may well hold comparable risks for animals. Animals have suffered enormously at the hands of humans who have insisted on viewing animals according to one nature-disorting stereotype or another. The sooner animals are recognized as animals, nothing less and nothing more, the better their fortunes will be.

THE ANIMAL AS DEMON

Beyond the traditions that regard some animals as morally derelict, there is a still grimmer view, one that represents some animals as demonic beings, utterly outside the pale of decency. Rats, snakes, sharks, spiders, vultures, bats, and crocodiles, among other species, are sometimes viewed in this manner. They are seen as treacherous predators who compound their crimes of greed and destruction by resorting to methods of stealth and cruel surprise. The representation of such animals as demonic serves to legitimate programs of annihilation. There is, after all, only one proper disposition for demons, devils, and witches. Such was the logic used in the persecution of dissident and eccentric women in seventeenth century New England. Still another application of this logic, and of the underlying demon stereotype, is discussed by J. Glenn Gray in his classic study of men in battle, *The Warriors*. Gray writes (1967, p. 163):

> There is another image of the enemy . . . even more abstract and deadly in its psychological effects. In this the enemy is conceived to be not merely a loathsome animal, below the human level, but also above it in being a devil or at least demon-possessed and, as such, an enemy of God.

Gray adds that this image of the enemy leads to one policy—annihilation (p. 154): "Killing them becomes a kind of sacrament; after enough of it, the killers come to feel like high priests."

Although the demon/animal stereotype is often associated with widespread fears concerning such animals as snakes and rats, it may derive from deeper currents of distrust in all animals. In the mythic traditions of many cultures, the animal is a symbol of chaos and the irrational. A number of primitive cosmogonies, such as the Babylonian Epic of Creation, the *Enuma elish,* represent the world as having been fashioned out of the remains of a slain chaos-monster. The ritual sacrifices of animals in some ancient societies can be understood as commemorating the primordial conquest of the chaos-monster (Eliade, 1961, pp. 54–58). In contemporary life, there are grim resonances of such sacrifice in the spectacles of the bullfight and the rodeo, as well as in large game hunting and deep sea fishing.

Secular philosophical traditions in both the East and the West have also contributed to the notion of the animal as an irrational being. In the West, Aristotelian thought excluded animals from the class of rational beings, while in the East, the Confucian ethic regarded "birds and beasts" as proceeding in the opposite direction from the ideal man, the *chün tzu*. Additional sources of suspicion concerning animal nature may be found in the Gnostic and Manichaean traditions, with their radical rejection of things earthly and carnal. Here the animals, devoid of *nous,* are seen as minions of darkness and evil. These and other intellectual currents have created a legacy of profound distrust toward animals and have inclined some to view them as inherently defective and evil.

The terms "animal" and "beast" are widely used to refer to people who are utterly lacking in decency and respect for others. Not infrequently, we will hear an advocate of the death penalty speak of criminals as "animals who should be taken out and shot." Lost in the racing flood of metaphor is the fact that, for the most part, the *beasts* are simply not *beastly*. The human animal, on the other hand, too often is—especially in his conduct toward animals. Rats, sharks, snakes, bats, and spiders—however much they may frighten or appear to menace us—are not models of lechery, greed, treachery, and the like. We must look to the annals of human behavior for proper examples of such conduct. Mary Midgley has exposed the distortions and covert anthropomorphism underlying what I have called the demon stereotype. She points out that " . . . man has always been unwilling to admit his own ferocity, and has tried to deflect attention from it by making animals out more ferocious than they are"(Midgley, 1976, p. 99).

Humans have an abiding fascination with the terrible and the macabre. It is no accident that monster movies and horror stories featuring giant spiders, armies of rats, and beachcombing sharks are perennial favorites. At the roots of this taste for terror may lie apprehensions, not about dread monsters and demonic animals, but about our own, distinctly human capacities to inflict and to experience evil. Perhaps, it is the beast within ourselves who parades and prowls through these suspenseful stories. And that inner beast is *not* our "animal nature," that which we share with the other species, but rather the uniquely human potential for irrationality and moral corruption. In still broader terms, it may be argued that the appeal of the demon/animal stereotype owes substantially to our unwillingness to face our own demonic tendencies. And thus, the animals are mustered once again; this time, as Richard Lewinsohn observes, " . . . to pay by being made to mirror man's depravity" (1954, p. 191).

THE ANIMAL AS MACHINE

Although Rene Descartes' notion of the animal automaton stimulated a great deal of heated discussion, its influence on the rise of what Konrad Lorenz has called the "mechanomorphic" view of animals is negligible. It was the imperatives of industrialization and urbanization, rather than metaphysics, that gave currency to the image of the animal as a machine. The new era brought a fresh emphasis on the rationalization of productive processess—a striving for greater control over the elements in such processes and increasing efficiency in manipulating them. As elements in a large number of productive processes, animals came to be seen more and more in mechanistic terms.

Throughout much of human history, animals have been used for food, clothing, labor, transportation, companionship, religious ritual, and entertainment. In most of these associations—however cruel and onerous—some trace of recognition of the animals' "creatureliness" remained. Such rudimentary respect for animal life derived, in part, from a sober realization of the vulnerability and value of animals and, in part, from a spiritual sense, here faint and there distinct, of the intrinsic worth of animals. The new atmosphere in the age of the machine was much less hospitable to such considerations. Nicholas Berdyaev accounts for this important shift in attitudes (1934, p. 39):

> The supremacy of technique and the machine is primarily a transition from organic to organized life, from growth to construction. From the viewpoint of organic life, technique spells disincarnation . . . . Technique destroys ancient bodies and the new ones it creates do not resemble organic bodies; they are organized bodies.

The animal/machine stereotype is associated almost exclusively with farm and laboratory animals. The breeding, maintenance, and feeding of these groups is usually controlled according to strict standards. Most farm and laboratory animals are bred to fulfill quite specific roles in diverse productive processes. In the case of the farm animals, it is usually the production of foodstuffs and a wide range of organic byproducts. The laboratory animal, on the other hand, is involved in quite different manufacturing process: the production of experimental data. Just as machines are designed, tested, and maintained according to precisely defined production standards, so also the fortunes of farm and laboratory animals depend increasingly on their levels of performance, judged in terms of strict reliability and cost-effectivenss standards. Irregularities in performance, stress reactions, diseases, and other difficulties are understood and treated as engineering problems.

In *Animal Factories,* a survey of recent developments in agricultural technology, Jim Mason and Peter Singer reveal some of the neglected ''moral costs'' associated with the treatment of animals as machines. Noting the expanding application of such Brave New Farm concepts as the mechanized total-confinement system, synthetic environmental control, reproductive engineering, and chemical manipulation of animal growth and behavior, Mason and Singer conclude (1980, p. 125):

> Productivity for and catering to the whims of the market may be all right in the plastics or automobile industries, but it can be cruel and abusive when the factory method is applied to animals . . . . Animal factories are one more sign of the extent to which our technological capacities have advanced faster than our ethics.

The animal/machine stereotype represents the last stop on the way to the total marginalization of animal life. However unhappy and defeating the other stereotypes may be, they allow for some measure of feeling, some organic kinship, faint or not, between animal and human life. Seen in mechanistic terms, the animal becomes merely a thing, an object to be manipulated according to the designs of the human technician. The advantages in this for the laboratory researcher and the factory farm operator seem to be considerable. Freed of the sometimes inconvenient moral constraints that regulate our behavior toward sentient beings, he can concentrate on the efficiency and productivity of his enterprise.

The tendency to reduce animals to the level of cogs in a productive system is not without parallels in relations among human beings. The development of modern industry, with its assembly lines, engineered environments, and systems management approaches to personnel issues, has stimulated much social criticism and moral concern. Alarm over the mechanization of human life is also to be found in recent controversies involving such matters as behavior modification strategies in the schools, health care technology, city planning concepts, and the expanding role of the computer in everyday affairs. Although some of the concern may be unwarranted, there can be no doubt that an earnest struggle is underway between the human and his muscle-flexing machines. At stake is the domestication of the machine or the subjugation of man. In the face of such difficulties, the animal/machine stereotype is even less welcome. To disregard the animal's manifest capacity for suffering, treating it, at best, as so much noise in the machine, may facilitate a comparable attitude toward human beings. We would do well to remember Kant's cautionary words (1963, p. 240): "If (a man) is not to stifle his human feelings, he must practice kindness toward animals, for he who is cruel to animals becomes hard also in dealing with men. We can judge the heart of a man by his treatment of animals."

BEYOND THE STEREOTYPES: THE ANIMAL AS ANIMAL

How animals are seen and how they are treated are interrelated matters. As long as the stereotypes we have examined persist, there is little prospect for meaningful changes in human conduct toward animal life. What, then, is the ideal attitude toward animals? How are they to be seen? Any answer will be inadequate that fails to acknowledge the mystery of animal nature. As noted earlier, there is an "abyss of noncomprehension" dividing us, and it is unlikely to yield to any amount of zoological and ethological research. We cannot know the animals, however much we may know *about* them. Much of the clatter and dither of anthropomorphic stereotyping derives from an inability to accept the essential inaccessibility of the myriad forms of animal life. Rather than contriving false identities for animals, identities that serve human interests, we must make peace with the mystery of animal life. At the same time, we must appreciate the importance of what we do know about the animals—that they are active, complex, sentient, and valuable neighbors. In short, we must come to accept animals as animals. Such an attitude will do justice both to what we know and what we cannot know.

Although we may succeed in adopting this new attitude toward animals, we will, inescapably, continue to employ many anthropomorphic categories. Some animal behavior will appear admirable, some ignoble.

Some animals will strike us as lovable, others as aloof and unlikable. We will be frightened by some animals, and moved by paternalistic impulses toward others. Movement beyond the animal stereotypes requires some vigilance in our indulgence of such sentiments, but it does not mean the eradication of all vestiges of anthropomorphism. It does mean, however, the renunciation of anthropocentrism, those modes of viewing animals that serve only to justify patterns of exploitation by humans. To perceive animals nonanthropocentrically is, above all else, to regard them as possessing worth independently of any uses we may have for them.

Given the extraordinary involvement of the diverse animal stereotypes in popular attitudes and contemporary institutions, the task of transforming attitudes toward animals will not be easy. The continuing struggles of many minority and feminist groups in this area provide only moderate encouragement. Nevertheless, if we are serious about achieving substantial reforms in human conduct toward animals, we cannot avoid such work. As the effort is begun, it may be useful to attempt to rekindle that awareness of a common cause that motivated some of the early advocates of abolitionism, women's rights, and animal welfare reforms. The traditions of bigotry and injustice are joined in complex and mutually reinforcing patterns. As we have seen, the stereotypes that promote indifference to animal suffering are closely related to modes of racist and sexist thought. If genuine and lasting changes are to be made, we must recognize anew that the degradation of life anywhere threatens the dignity of life everywhere.

SECTION II

Morality, Legality, and Animals

Introduction

James M. Buchanan draws a distinction between 'moral order' and 'moral community.' An individual's moral community (in Buchanan's terms) is the set of other individuals for whom the individual feels significant moral concern. For almost all humans, as a characteristic of our species, this moral community is quite limited. Many (perhaps most) of these limited moral communities include some nonhuman animal individuals but exclude the great majority of humanity. A moral order, on the other hand, is a system of rules that govern behavior independently of an agent's felt concern. The structure of law is the paradigm of a moral order. Since moral order requires abiding by abstract rules, and nonhuman animals cannot abide by such rules, they cannot be members of a moral order. (The argument here is similar to Narveson's concerning contractarianism.) But some animals are usually within human moral community.

Bernard Rollin claims that our legal structures ('moral orders') rely on moral principles, and that some of the most central such principles are those assigning moral rights to individuals. Since no morally relevant difference between humans and other animals can be established, he argues, animals must be recognized as moral patients. Like Tom Regan, and for some of the same reasons, Rollin concludes that animals should be granted rights. His proposal is that animals become bearers of legal rights in themselves, not as valuable property or for human good. This might be achieved either by *extending* the range of holders of rights (usually a gradual process, predominantly judicial) or by *conferring* rights upon animals (a matter for legislation). Just what rights should be granted to what species remains a complicated question that cannot be usefully addressed without considerable empirical information.

Even though perceptions of right and wrong (moral intuitions) underlie our laws, Rollin points out that one need not wait for intuitive consensus before changing laws. For our intuitions evolve over time, and legal changes influence this evolution. A change in 'moral order' can lead to a change in perceived 'moral community.'

Moral Community
and Moral Order

The Intensive and Extensive
Limits of Interaction

James M. Buchanan

In a paper presented at a 1978 conference under the same sponsorship at this one, I argued that human moral capacity, defined as the potential outreach of one's moralistic behavior toward others of the species, are stretched beyond tolerance limits in modern political settings (Buchanan, 1978). Humans are called upon to "care about" unknown persons with whom they have no means of identification, and with whom they share no common loyalty to external symbolic entities capable of stirring emotions. Thrust involuntarily into such settings, modern people necessarily behave so as to further their own narrowly defined self-interest. I inferred from this essentially empirical hypothesis that the moral potential of humankind could only be exploited if the institutions of political interaction were to be reconstructed so as to correspond more closely with the limits of the human moral "community."

In this paper, I want to build upon the earlier analysis and, in particular, to examine the position of animals in that human moral community. I now realize, however, that I failed in my earlier paper, to make a necessary and categorical distinction between moral "community" and what I shall here call a "moral order." I shall argue that animals may be treated as members of the human moral community, but that they have no place in the human "moral order," as the latter is distinguished from "community."

MEMBERSHIP IN MORAL COMMUNITY

I want to discuss the question of membership in a moral community empirically. I want to discuss the behavior of persons precisely as we may, conceptually or actually, observe it, rather than how such behavior ought or ought not to be. In this context, I hypothesize that persons do include animals (more generally, nonhuman animals) in their moral communities, defined as relevant for behavior, and, further, that persons exclude *some* humans from such communities. In saying this, I am not, of course, advancing anything at all new or novel. Nonetheless, it does continue to surprise me to find how scarce are discussions of the implications of this simple hypothesis, or at least as I have found them. Discussions in morals and ethics concentrate on how, why, and when a person does or should "love my neighbor," without paying much attention to the more important question: Who is to count as "my neighbor"?

Human beings have within them instincts and drives that were developed, that evolved, from thousands of years of living in essentially tribal groupings. These instincts and drives presumably are related to those behavior patterns that enhance the survival characteristics of the tribal group. Such groups probably numbered between fifty and one-hundred persons, and the human instinctual structure is presumably related to behavioral traits consistent with the survival of the tribe as a unit or entity. I am not concerned here with the problem of distinguishing that portion of these instinctive behavior patterns that may be genetic in origin from those that may have evolved culturally as a set of rules or codes of conduct. I want to postulate here only that human beings have within them such instincts that they do not, indeed cannot, understand and explain rationally in any scientific sense.[1]

In the tribal setting, the rules of conduct, or behavior patterns, would have led a modern external observer to attribute "morality" to persons, and to define readily the limits to the moral community of a particular person. Toward members of the tribe, behavior would appear to be "moral." Toward persons outside the tribal membership, no such behavior would be observed.

In such a conjectural setting, what can we postulate about human behavior toward animals? Clearly, to the extent that animals are essential elements in the tribal life-chain, behavior toward them would have attributes of "morality" little if any different from those directed toward human members. For animals beyond the needs of the tribal life-chain, be-

---

[1]In this postulate, and in much of the discussion of this whole paper, I should acknowledge my indebtedness to Professor F. A. Hayek, who has, especially in his most recent, and unpublished works, largely in the form of lectures, stressed the importance of the potential conflicts between human tribal instincts and the social settings that humans confront.

havior directed toward them would be akin to that directed toward human "outsiders."

Human beings have, of course, moved far beyond their tribal heritage, but carry with themselves the basic patterns evolved within the tribal setting. Civilization as such is simply too brief a span of history to have exerted important effects on basic genetic or culturally evolved human traits or proclivities.

As humans shifted from a tribal setting, where they were instinctively "at home," they faced challenging problems of behavioral adjustment. Basic instincts failed as soon as they found that they could not identify tribally with persons they confronted in any sort of personal interaction. There are essentially two separate and distinct socio-institutional responses to humankind's "dilemma of civil order," to the behavioral dilemma that humans in a nontribal setting have confronted, and continued to confront, since they emerged as civilized beings. Perhaps I should not refer to these as responses since, instead, they may be considered to be preconditions that allowed humans to supersede tribal organization in the first place. I shall discuss the two "attempts" or "responses" or "institutions" in the two following sections. At this point, however, I want to stress that the innate *moral* behavior of humans, that which is motivated by genuine "fellow-feeling" in an unthinking, unrationalized sense, remains tribal in its extent. Human beings react instinctively toward members of their "tribe" in a way that we might classify as "moral"; they do not extend such "morality" to those outside the tribal membership, whether the units encountered be human or nonhuman.

EXTENDING THE INTENSIVE LIMITS OF MORAL COMMUNITY

In order to behave toward persons (and/or animals) beyond tribal boundaries in a manner that might be considered moral; that is, in order to behave as if motivated by genuine concern for the well-being and the survival of what we might call nonnatural members, *artificial* or *artifactual* criteria had to be established that would allow persons to extend the insider group beyond its natural limits. And rules of conduct based on these essentially arbitrary (noninstinctive) criteria had to be established and generally acknowledged.

The great religions can be interpreted as having been born in response to such a set of requirements, and the treatment of persons beyond the pale reflects the straightforward extension (and aberration) of the categorical moral differentiation between tribal members and outsiders. As tribal interbreeding took place, but within narrowly defined groupings, something akin to race emerged to offer yet a further correlate criterion for arbitrary moral differentiation. As communication was developed, and politics established, geographical differentiation reflected in manners, in speech, in dress, provided still further grounds for moral classification. In each case, however, the basically artificial nature of the rules for morality must be stressed. And those humans who remained outside the new moral community were "heathens," to whom no moral rules of behavior were to be applied, and to whom no sense of deservingness was owed. In this setting, then and now, animals (nonhuman) and "heathens" are not treated differently in any moral sense. Some of the great religions included the treatment of animals within their artifically derived rules of conduct; some did not. In any case, however, the membership of some animals within the instinctively based moral community of the tribe remained as a behavioral force.

Some of the religions were "open" in the sense that all persons might qualify for membership, and hence for a status of moral equality, but the moral differentiation between members and nonmembers carries over throughout almost all religious history. Humanism, considered as a great religion, may offer an exception to this generalization, but it never seems to have exerted an influence comparable to the more discriminatory cults that do, indeed, promise members better treatment on earth and/or elsewhere.

Have the great religions, inclusively defined to embody racial mythologies, nationalisms, and other symbolic bases for classification, been successful in pushing human moral capacities beyond tribal limits? A dual response is indicated here. As human interaction extended beyond the tribal limits, "the tribe," in any determinate sense, disappeared. Individual family units were left, more or less as floating islands, without a "tribal home." They found it almost necessary to adopt, in one form or another, the arbitrary schemata suggested to them by the religions. At one and the same time, however, individual human beings retained tribal instincts that did not allow them to behave morally toward arbitrarily defined and nonidentified humans. There remain distinct limits to human "moral capacities," the point that I discussed in my earlier paper. Humankind is thus caught up in the moral dilemma; it can and does have genuine fellow-feeling for a potentially limited set of others of its species, and for some animals. But it cannot, *naturally,* possess comparable fellow-feeling for a potentially unlimited set extending to the whole of humankind, and, beyond, to all of life itself. Tribal constraints prohibit this extension of moral community. Those persons who exhibit such generality of moral concern are those who have most successfully sublimated or superseded the tribal constraints.

The conjectural "history" that I have sketched here may be open to criticism on many grounds. I have played fast and loose with subject matter from many disciplines: anthropology, history, psychology, genetics,

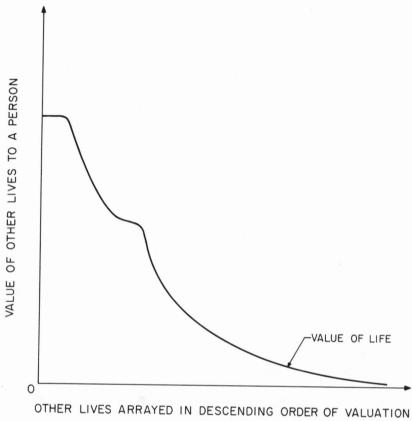

OTHER LIVES ARRAYED IN DESCENDING ORDER OF VALUATION

Figure 1

theology, ethics, and others, in none of which do I claim any expertise. But I emphasize again that my central hypothesis is subject to empirical test and to refutation. I present it as such, and I use conjectural history in the 18th century sense of offering an interesting story that you may or may not take as plausible. In a more formal variation, I should suggest only that human moral attitudes, as reflected in potentially observable behavior, may be depicted in the shape of a curve such as that drawn in Fig. 1, where the ordinate reflects the value an individual places on a life other than his own, and the abscissa represents an array of other lives (human and animal) in descending order of such evaluation. My hypothesis is summarized in the relatively sharp inflection points reached beyond members of the individual's family and beyond plausible limits of his localized community. I am not suggesting, of course, that *all* persons are identical in their moral communities, far from it. There may well exist saints whose appropriate curve would be horizontal out to the limits of existence. As regards ordinary people, however, I am hypothesizing the generalized shape of the curve in Fig. 1, and, further, I am suggesting that for some people some animals fall in the array well above the limits where the human species is exhausted.[2]

THE EXTENSIVE LIMITS OF "MORAL ORDER" AS A MEANS OF EFFECTIVELY SUPERSEDING MORAL COMMUNITY

I now want to discuss the second response to humankind's challenge when it moved beyond tribal limits, a response that does not require an extension of its moral community, as this term might be ordinarily interpreted. That is to say, this second response does not require that humans extend, at least directly, their sense of concern for the well-being of others. The second response involves the generation of a "moral order," as opposed to "moral community," an order in which behavior may be constrained by rules that, in themselves, seem basically *amoral*. Indeed, I should argue that it is through the development of the rules of such a "moral order" that humans have successfully confronted the challenge of civilization. The moral order that allows humankind to supersede the effective limits of its expressly revealed moral community is best described as the order of *law,* in which persons abide by abstract rules of law, that lay down rights of separate persons and that provide bases for mutual respect, tolerance, and, most importantly, for trade and exchange that, in turn, greatly enhances the level of personal well-being for all participants.

It is important to clarify the distinction between the explicitly *moral* behavior extended to members of a person's appropriately defined moral community, and the sort of behavior toward others, which is basically *amoral*, that characterizes relationships in the market. An example may be helpful. A beggar will starve without bread; a person who recognizes the beggar as a member of his or her moral community and gives that beggar bread is behaving *morally* in the standard usage of this term. The donor carries out a moral act, motivated by a sense of concern for the beggar's well-being, directly or indirectly experienced. Let us now modify the example only slightly. A second beggar is observed to possess a loaf of bread, without which the indigent will starve. A person who re-

---

[2]Economists have discussed in some detail the interesting problems that emerge in placing values on human lives. Almost exclusively, however, the focus of their discussion has been on the question of valuation of his *own* life by the individual, or at least on the question of how such conceptual evaluation should be entered into a social policy benefit-cost comparison. There has been little explicit discussion of the equally interesting set of questions raised here. How do individuals value other lives than their own? A probabilistic approach to behavior similar to that used in the other analysis should yield clearly testable hypotheses. Casual empiricism suggests that, in some cases, e.g., children, individuals may well place a *higher* value on lives other than their own. That is, a person may pay more for an extra margin of safety for his children than for himself.

frains from stealing the bread from that beggar is not carrying out a moral act; such behavior is, in my terminology, or may be, *amoral*. To steal the bread would be *immoral,* but to refrain from stealing implies neither concern or lack of concern for the beggar's well-being. Similarly, in the first example, to refrain from giving bread to the starving beggar cannot, in itself, be classified as immoral action. To refrain from giving bread in this case and to refrain from stealing bread in the second case both reflect amoral behavior, but behavior that is still in accordance with abstract rules consistent with a moral order of law, characterized by mutual tolerance of other persons and mutual respect for other persons' rights.

I hope that this example clarifies the distinction that I am trying to develop here. If there exists no moral order of law, the treatment of the beggars would, in each case, depend strictly on whether or not the other person, the potential donor or potential thief, did or did not accept the beggar as a member of his or her own moral community. There would, in this setting, be no distinction to be made between giving bread on the one hand and refraining from stealing on the other, or, conversely, between refusing to give bread and stealing in the second case. The relatively narrow limits of human interaction that could take place under such restrictions should be clear. General adherence to the moral order of law, however, and to the distinct rules of behavior that such an order embodies, allows persons to interact one with another independently of membership in expressly defined moral community one with another. This extremely simple point should never be overlooked, nor its importance neglected. A personal example may illustrate.

I wrote the initial draft of this paper during a period of several weeks in Vienna in early 1979. My conversational German is almost nonexistent, and my knowledge of Vienna's history and culture is limited. Yet I was able to function, very satisfactorily, in this community of "foreigners," or "aliens," to me, who were not, I am sure, in the least concerned for my own private welfare. I did not qualify for membership in the Viennese or Austrian moral community at all. But I was able to survive well by a knowledge of and adherence to a system of rules that involved a mutuality of respect for the rights of property, that of my own and those persons with whom I had dealings! It is easy to imagine the difficulties I might have encountered in a genuinely "foreign" land that was not characterized by such agreed-on rules of behavior and in which, quite literally, I should have been required to depend on the genuine "morality" of others to survive. And is not this precisely the distinguishing difference between "civil order" and "anarchy"?

The example is helpful also in suggesting that human behavior in the moral order of law is and must be unique to humans, or to that subset of the human species that abides by the appropriate abstract rules. In a "society" of animals outside such a subset, whether the animals be human or nonhuman, there is no way that a person might know what rules of conduct to follow if any at all were appropriate. My professor, Frank Knight, used to say that humans are, or may be, law-abiding animals, which is about all I am really saying here.

CONCLUSIONS

To bring my discussion back somewhat more closely to the subject matter of this conference, my discussion implies that the "moral treatment of animals" simply does not arise in an evaluation of human behavior toward fellow humans in the moral order of law. The moral treatment of animals does arise when the whole issue of limits to humankind's expressly moral community is raised. My emphasis on the necessary distinction between this moral community and the moral order that describes the potential for extending relationships far beyond those that may otherwise be possible reflects the disciplinary perspective or bias from which I commenced my inquiries. As stated at the outset, my hypothesis is that some animals are normally within the human moral community. I am relatively uninterested in the normative question whether any, some, or all life *should* be contained within such moral limits. This absence of interest in the normative question does not imply that I claim to be a "positive" scientist who remains aloof from the "should" or "ought" questions. Far from it. The normative issues that excite my interest are, however, the different ones that involve the proven values of maintaining and extending the moral order of law, within which humans become less dependent on the presence or absence of personal morality in the community sense. Modern politics, with its overreaching and apparently insatiable appetite for further and further intrusion into human behavior, tends to confuse the distinction that I have tried to stress here.

The Legal and Moral Bases of Animal Rights

Bernard E. Rollin

I

It is impossible to discuss the legal status of animals without discussing their moral status, that is, without discussing the ethical position that we hold towards them, because the laws of a society are ultimately dependent on its morality. In one sense, this is very obvious. Our laws concerning prostitution, pornography, obscenity, drugs, and so forth clearly *grow out* of what the public thinks is right and wrong. That is, as these attitudes change, the law changes. And laws, in turn, influence public morality; witness the 55 mile per hour speed limit. But the connection between law and morality is often said to be even deeper and more basic than that: In fact, according to this view, moral notions and ideals are *an inseparable part* of the law, that is, we cannot explain what law is without making reference to morality. This view that law in society is based on unchanging and absolute notions of right and wrong is very old—it goes back to the ancient Greeks (Plato, Aristotle, and the Stoics), and is called *natural law* theory. On this view, if a statute violates these basic rights it is not law. This view reappeared recently after the Nazi era, when discussions of Nazi law and morality were rife. Connected to this view is the idea of *natural rights*—that is, the idea that people have by nature certain rights that no laws can violate. This theory is of course the basis of the American constitution. We need only recall the Declaration of Independence (cf. the UN Declaration of Human Rights for the same notion):

> We hold these truths to be self-evident: that all men are created equal, that they are endowed by their creator with certain inalienable rights, that among these rights are life, liberty, and the pursuit of happiness.

As can be seen in this passage, this theory of natural rights was tied to God, as it was in the Middle Ages. As belief in God grew less important in political theory, some thinkers began to deny the existence of natural rights, to think of them as mystical, to call them—as Bentham did— "nonsense on stilts," i.e., elevated nonsense. They tried to argue that the only rights that exist are those specifically adopted by legislatures and found in statutes. These thinkers are called *legal positivists* and have dominated British and American legal thinking for over 100 years.

Recently, however, some thinkers have begun to revive the notion of natural rights and law—the idea that basic moral principles are the skeleton of all proper law and that these ought not change even as public attitudes change—but this idea has been revived without referring to God. One of the most important of these thinkers is Ronald Dworkin, at Oxford, who has developed these ideas in a recent book called *Taking Rights Seriously* (Dworkin, 1977). Dworkin argues that, in addition to explicit rules adopted by legislatures, the law contains *moral or ethical principles* that judges regularly use to decide cases for which no explicit legislative rule or statute exists. We all know, of course, that much of the law is made by judges when they interpret the statutes and decide their application to hard cases (cf. the Bakke case). Contrary to what legal positivists claim, Dworkin argues that these principles are not chosen arbitrarily by judges, but are based on permanent standards of right and wrong. According to Dworkin (p. 22),

> A principle is a requirement of justice or fairness or some other dimension of morality.

So if one reads any interesting decision—Bakke, Brown vs. Board, Miranda—the judges will be found appealing to such principles regularly, even though the principles were never adopted by legislatures. They are implicit elements of our legal system. Examples of these principles can be found in almost any case. Such principles include: "no one may benefit from one's own wrongdoing," "courts will not permit themselves to be used as instruments of iniquity," "separate is not equal," the principle that "unconscionable profit" should be disallowed, "the state ought not treat a conscientious act as a crime," "the law abhors a forfeiture," "one who acts through another acts through one's self," the principle that a law may be broken to forestall a greater wrong. In fact, judges often use such moral principles to overturn certain rules that were explicitly decided by law-making bodies or government agencies. In fact, it is these unspoken moral principles that prevent judges from saying anything they like, however crazy, in unprecedented cases.

Supreme Court cases are an especially rich source of principles, since by their very nature such cases often involve controversial issues undecidable by precedent. The important point is this, in Dworkin's words (p. 46):

> It is wrong to suppose . . . that in every legal system there will be some commonly recognized fundamental test for determining which standards count as law and which do not. No such fundamental test can be found in complicated legal systems like those in force in the U.S. and Great Britain. In these countries no ultimate distinction can be made between legal and moral standards.

The point, then, is this: Our legal system is tied into a set of moral principles that guide and constrain and limit the explicit laws that are adopted.

Probably the most important of these moral principles is the notion of *persons having moral rights,* stemming from our recognition of them as legitimate objects of moral concern, for from these moral rights flow claims by individuals against the state that ought and do enter into judicial judgment (cf. the civil rights cases). According to Dworkin (p. 139):

> . . . a man has a moral right against the state if for some reason the state would do wrong to treat him a certain way, even if it would be in the general interest to do so.

So a right is a protection of the individual against the common good in some cases. The Constitution, of course, lists some of these rights in the Bill of Rights, but these are not meant to be a complete list of all rights, or even to fully state the content of the rights listed. Remember, the 9th Amendment explicitly says this:

> The enumeration in the Constitution of certain rights shall not be construed to deny or disparage others retained by the people.

In order to establish a right in legal cases, as was done in the civil rights cases, one must use moral arguments. Says Dworkin (p. 147):

> The difficult clauses of the Bill of Rights, like the due process and equal protection clauses, must be understood as appealing to moral concepts rather than laying down particular conceptions; therefore, a court that undertakes the burden of applying these clauses fully as law must be prepared to frame and answer questions of political morality.

What this means, as our legal history clearly shows, is that one cannot separate questions of law from questions of right and wrong, that is, from morality.

So the notion of rights is a moral notion. But it serves a major political and social function. In a democracy such as ours, based on individualism and competition, most public policy decisions are made by *utilitarian considerations,* that is by considering what course of action will produce the greatest amount of happiness or benefit for the greatest number of people. Utilitarianism goes hand in hand with democracy and the free market. If all people are equal, no particular individual or group should have its interests favored, and decisions should be made by considering the total or average benefit across the society. Each person's interests count equally. The test of a course of action is whether it provides for the general or majority welfare. In a sense, this is very fair. But in another sense, it can be very dangerous, because the interests of *any given individual* often become submerged under the weight of the common good. For example, if the majority feels it benefits by preventing an unpopular speaker from expressing certain views, strict utilitarian considerations make it easy to allow his being silenced. Thus for example, consider the Nazis who wished to rally in Skokie, Illinois. From the point of view of the general welfare (and the majority wishes), no good comes from permitting them to speak. An inordinate amount of money must be spent on police protection, many anti-Nazi people and Holocaust survivors will be made to suffer, no new ideas will be aired—in short, considerations of utility clearly militate in such a case against letting such people speak. Nonetheless, in our legal system, such considerations do not outweigh the basic right to free speech.

Rights are moral notions that grow out of respect for the individual. They build protective fences around the individual. They establish areas where the individual is entitled to be protected against the state and the majority *even where a price is paid by the general welfare*. Such a right is freedom of speech, to continue our previous example. Our system protects the right of the holder of unpopular views to air them even if he offends, antagonizes, and upsets *everyone else*.

The same holds of religious belief. This is not of course to suggest that rights are unassailably absolute, but they are abridgeable only for the gravest of reasons. For example, freedom of speech can be trumped if its exercise puts in jeopardy the very survival of the society in question. We would not consider it a violation of freedom of speech if we stopped a speaker from telling grade-school children how to make an H-bomb out of breakfast cereal! The notion of rights is based on the basic moral idea that the individual is the fundamental object of moral concern and attention, that the individual has intrinsic value, has dignity, and has equality, and that these are inseparable from one's being.

It is further predicated on the assumption that each individual is at the center of the moral arena and that certain activities or functions are an irreducible part of one's nature as a human being—free speech, free belief, free communion with other persons, and so on. Suppression of these encroaches upon characteristics essential to a person's humanity. The notion of rights spotlights the individual and protects that person from the "general welfare," and from being used or abused for the convenience of the majority. To summarize: we have argued that law is tied up with morality not only in the sense that public morality determines the passage of many laws, but also in that legal decisions are made by reference to moral principles never decided on as explicit laws. The most basic of these principles are individual rights, which protect individuals from being suppressed for the general benefit.

II

So far, everything we have said deals only with people. Rights grow out of our recognition of individual people as objects of moral concern, that count in moral deliberations, that are worthy of being talked about in the "moral tone of voice."

A few years ago, I became puzzled about how this "ethical anthropocentrism" had encouraged so few thinkers to raise the question of creatures other than humans having rights or being worthy of moral concern. Philosophy, after all, has been alive and well for over 2500 years, and has always dealt with at least some strange issues and concerns. Philosophers have tried to show that motion is impossible, that time is unreal, that tables do not exist when they are not being perceived, that the mind exists in the brain, that the brain exists in the mind, that we do not know how we know where our ears are, and so forth. Yet most philosophers had simply assumed that humankind was the only entity to which moral categories could be directly applied, and gave no justifications for such a claim. Perhaps, I began to wonder, I could find so few justifications just because there were none—perhaps the emperor wore no clothes! Perhaps, for once, philosophers were no clearer in their concepts than other persons. After all, our social history reveals some bizarre behavior towards animals. Although people in the Middle Ages did not recognize animals as objects of moral concern or as rational agents, they sometimes put them on trial for crimes, and even executed them, a practice that continued into the 18th century and beyond, as Pierre Bayle so wittily recounts in his *Dictionary*. Even more remarkably, animals were sometimes excommunicated, even though they were said to lack souls and not to be part of the Christian community in the first place. (The most famous of such cases involved a herd of locusts that had laid extensive waste to French farms.)

I finally found a number of explanations put forward as to why animals were not the sort of beings to which moral notions could be applied and towards which people could have obligations. The discussions usually focused on some difference between humans and animals. But it is not difficult to show that none of these are tenable. Very common, I suppose, even today, is the idea that humans have immortal souls and animals do not. Aside from the fact that this is not the sort of thing one can prove, it is open to a more serious criticism, developed first by a Catholic theologian, namely this: Even if it is true that animals do not have souls and humans do, what does this have to do with morality? As Cardinal Bellarmine pointed out, in fact, the absence of a soul can well be used as an argument *in favor* of moral consideration of animals, indeed in favor of *better* treatment of animals than of people, since people would be rewarded in the afterlife, whereas animals had only one crack at existence. We are not interested in theology, but there is an important philosophical point here, it is not enough to specify some difference between humans and animals—there are countless such differences; humans are the only beings that shave, wear panty-hose, fry their food, and so on. To bear on our problem, the difference in question must be *morally relevant,* must have bearing on moral questions.

Armed with this tool, I looked at some other alleged differences between men and beasts. It has been argued in the past on the basis of the Bible and more recently on the basis of evolutionary theory, that humans have dominion over the other animals because they are at the apex of the evolutionary scale. But again, we may ask, even if this is the case, why is this morally relevant? To say that being on top allows you to treat those beneath you any way you choose is to say that might makes right, and to destroy all morality altogether. Anyone taking this position will be forced to admit that the government has the right (not just the power!) to do whatever it chooses to individuals, for after all, it is more powerful than any individuals. Such a person would also be forced to assert that if creatures from another planet, far more intelligent and powerful than we, landed on earth, they would have the right to use us as they saw fit!

I found one position on this question that was and is quite popular; it was held by St. Thomas Aquinas and by Kant, and is the official Catholic doctrine. This view says that animals are not worthy of being objects of moral attention in themselves, but ought not be cruelly treated because of the brutalizing effect of such treatment on people. In other words, I should not torture kittens not because there is something basically wrong with torturing kittens, but because I might start torturing people. The trouble with this position is obvious. Once again, it is people, not animals, who are moral objects. According to this position, if one could torture kittens without becoming brutalized, it would be morally acceptable to do so. If wrecking cars made one violent, that would be forbidden, yet cars would not of course be objects of moral concern. Worst of all, if a human could calm down and release violent urges that might cause him or her to harm people by torturing kittens, it seems that this position would not only allow, but actually require it. (It is interesting to note that historically this "brutalizing" argument was a major argument used by abolitionists against slavery—many abolitionists did not think of slaves as direct objects of moral concern!)

The most frequent and popular way of distinguishing humans from animals from the point of view of morality has to do with the ability to reason and to use language. This way of distinguishing goes back to Plato and Aristotle, and is still often advanced today. The argument goes like this:

Reason is required for moral consideration
Language is required for reason
∴ Language is required for moral consideration.

Only human beings have language
∴ Only human beings are objects of moral concern

Unfortunately for these theorists, many problems can be raised about this manner of distinguishing humans from animals with regard to moral concern.

1. What is the connection between being able to reason and being an object of moral concern. That is, why is being able to reason *morally* relevant? Perhaps being able to reason is necessary before something can be a responsible moral *agent* or *actor* (after all, we do not hold children or the insane responsible for their actions), but why is reason necessary to be considered an *object* of moral concern? After all, we consider many *humans* who lack the ability to reason to be objects of moral concern: small children and infants, the senescent, people in comas, the insane, the retarded, and so on.

2. Another question that may be asked is this: What is the special connection between reason and language? Why can only beings with language be said to be able to reason? This is far from obvious, though there are interesting arguments in support of this claim.

3. Finally, we may ask whether human language is really all that different from animal communication? Again, this is far from obvious. Though the Western tradition in philosophy, psychology, and linguistics has taken this for granted, I think that strong arguments can be mustered against a hard and fast line between language and other forms of communication. In my book, *Natural and Conventional Meaning: An Examination of the Distinction,* I have tried to develop such arguments at length. Many writers see recent research with teaching "language" to chimpanzees, orangutans and gorillas as a further challenge to this split, though many theoretical linguists, the most noteworthy being Noam Chomsky, are disinclined to view the "speech" behavior of these creatures as genuinely linguistic.

In any case, I have discussed these issues in detail in other papers and books and cannot go into them here. But I can summarize the results as they affect our basic question: *does being able to reason provide grounds for distinguishing humans from animals as moral objects?*

We must, I think, conclude that the possession of reason is not morally relevant to being an object of moral concern. First of all, as we just stated, human beings who cannot reason are still treated as moral objects. That is enough to throw the argument out. Secondly, if being able to reason is the crucial feature in being a moral object, it is hard to see why aspects of human beings that are not connected with reason ought to enter into moral consideration. For example, why is it wrong for me to kick you in the crotch if that kick has no effect on your rational nature? The key point is that our moral regard for human beings extends far beyond aspects of human nature which are connected with reason; our ability to feel pleasure and pain is an obvious example. In fact, what makes human beings objects of moral attention is not simply that they can reason (for we have seen that humans who cannot reason are also such objects), but the fact that they can suffer, feel pain, have interests, experience pleasure, and can be helped or hindered in the pursuits of needs, wants, goals, and desires. And language enters into being an object of moral concern not because it is tied to reason, but because it is the most sophisticated way for a being to *inform* another being of needs, wants, likes, dislikes, and so on. If reason is relevant to moral treatment at all, it is because it is one among many interests that human beings have that can be helped or hindered by other human beings. (Again, this is not, of course, to say that reason is not relevant to being a moral agent.)

I am suggesting that moral concern arises out of recognition that there are certain things we can do to help or hurt another person and the feeling that some things are right and others wrong. If a being could not be helped or hurt, it could not be a moral object—we could do neither right nor wrong relative to that being. (This represented a serious theological problem in the Middle Ages—if God was indeed totally self-contained, how could our actions matter to Him!) The key to a person being an object of moral concern to us then is this:

A. We must have a sense of the person's needs and interests.

B. We must be aware that things we do can help or hinder that person in his or her pursuits.

C. We must have a sense that certain things are right or wrong (e.g., that killing a person for no reason is wrong).

So it is this that renders people objects of moral concern to us. But a moment's reflection makes us realize that the conditions just described are fulfilled for animals just as well as for people, perhaps equally for animals!

Take (A)—we certainly have a sense of animals' needs. We know they suffer and experience pleasure. And they are certainly capable of communicating needs to us through signs. Even if they lack language, it does not follow from this that they cannot communicate! And it certainly does not follow, as Descartes tried to say, that they have no feelings or mental life and are just machines because they have no language.

Descartes argued that even as mathematical (geometrical) physics encapsulated the essence of the material world, language was the essence of mind. Lacking language, animals were compared by many Cartesians to the ingenious mechanical and clockwork devices beginning to appear in Western Europe. Such a hypothesis made it quite easy for the embryonic science of physiology to proceed with a clear conscience, and thus served as a spur to the development of the biological sciences. As Leonora C. Rosenfield has brilliantly shown in her classic study, *From*

Beast Machine to Man Machine (Rosenfield, 1968), only when people followed this hypothesis to its inevitable conclusion, *Man is a Machine,* in de laMettrie's words, did it cease to seem so attractive.

If one finds it hard to believe that anyone ever took that position seriously, we must note that people are still doing so—the National Society for Medical Research, the biomedical research pressure group, has often said in print that animals do not really feel, and an eminent zoologist at Rockefeller University recently felt compelled to write a whole book on the *Question of Animal Awareness* (Griffin, 1976). Some of my colleagues teach their students that fish do not feel pain, without giving any justification for this claim.

In any case, it *is* clear that I can know an animal's needs, as well as I can know a person's. Much human communication takes place nonverbally. The burgeoning field of nonverbal communication, the enormous popularity of recent works on "body language," ever-increasing research in subtle communication between mother and infant all serve to underscore this obvious point. All lovers are well aware that at the most crucial moments linguistic communication breaks down, and feelings are best expressed by a glance, a gesture, or a touch. And in fact, animal communication is often a more reliable indicator of need than human language! After all, it is very rare for an animal to lie. When my dog whimpers at the door, I can be pretty sure she has to relieve herself. On the other hand, when a student raises his hands and says he has to go to the bathroom, it is at least even money that he is simply bored to tears by my lecture! Obviously, then, we can know animal needs.

With respect to (B), similarly, we are certainly aware that we can help or hinder the animal's needs once we are aware of them.

And for (C), finally, we must conclude that if certain things are wrong when done to men, they are wrong when done to animals *for we have not been able to draw a morally relevant distinction between men and animals*. Thus, if it is wrong to hurt a man for no reason, one cannot deny that it is equally wrong to hurt an animal.

Thus, I believe I have shown that if men are objects of moral concern, reason requires that animals also be seen as direct objects of moral concern. Nothing I have said tells us how to weigh the interests of one animal relative to another, or even the interests of man relative to animals. But moral theories that deal exclusively with humans do not answer these questions either. They do not tell us for example, what to do if our car is out of control and we must swerve to the left and kill Mr. X or swerve to the right and kill Mr. Y. Nor do they tell us what to do in a lifeboat situation, as in that marvelous old film where there are 12 survivors of a shipwreck and only 10 places in the lifeboat.

In summary, then, I have tried to show that since no good grounds can be given for drawing a morally relevant distinction between humans and animals, we must conclude that animals are direct objects of moral concern. And thus if people have moral rights, animals do. And so too, if legal rights flow from moral rights in the case of human beings, as we said in Section I, they ought to apply to animals as well. And further, if everything I said about legal rights for humans in the first section applies to animals, then arguments of human convenience or utility do not count against animal rights!

III

What would it mean, from a legal point of view, to grant rights to animals? More specifically, how would this differ from the laws that currently regulate what may be done to animals? Why not simply argue for greater enforcement and wider extension of anti-cruelty or animal protection laws?

We need to be aware that *all* legislation that has so far been enacted has evaded the question of the status of animals as direct objects of moral concern. However well-intentioned, all legislation has seen animals as *property*. Animals are either the *property of individuals* or, in the case of so-called "wild" or stray animals, are *community property*. In the eyes of the law, animals are human possessions, which the owner can more or less handle as he sees fit. It is no wonder, then, that all such laws have been based on utilitarian considerations, or considerations of the greatest good for the greatest number, but with the *human community* serving as the reference group for the computation of utility or benefit. Let us recall Dworkin's point discussed earlier: If laws based exclusively on utilitarianism are morally inadequate even though the interests of all members of a human society have been counted in setting them up, what can one say of laws that govern our treatment of animals, but where the calculation of benefits and interests has been restricted to the effects on *human* happiness and welfare? One of the founders of utilitarian ethical and political theory, Jeremy Bentham, pointed out almost 200 years ago that consistent application of utilitarian principles would weigh the interests of *all* creatures capable of happiness and suffering. Yet even a superficial look at so-called animal welfare legislation makes manifest its anthropocentric base.

Such laws, in essence, are designed to consider human interests and circumstances as primary. A few simple examples will illustrate this point with indelible clarity. Consider for example, the legislative declaration introducing the Colorado Nongame and Endangered Species Conservation Act in my home state:

> The general assembly finds and declares that it is the policy of this state to manage all nongame wildlife for *human enjoyment and welfare*, for scientific purposes, and to insure their perpetuation as members of ecosystems. [emphasis mine]

If there is a smattering of moral concern in this declaration, it is surely towards humans alone. Or consider the Colorado Cruelty to Animals Act, which is typical of such legislation. At first glance, the wording may seem to address itself to animals as moral entities:

> A person commits cruelty to animals if, except as authorized by law, he overdrives, overloads, overworks, tortures, torments, deprives of necessary sustenance, unnecessarily or cruelly beats, needlessly mutilates, needlessly kills, carries in or upon any vehicles in a cruel manner, or otherwise mistreats or neglects any animal . . . .

The loophole is of course the use of words like "needlessly" or "unnecessarily" and, through this loophole, the interests of humans pour and submerge the moral status of animals. Subsequent cases testing this law underscore this point, and result in rulings asserting that:

> Not every act that causes pain and suffering to animals is prohibited. . . . Where the end or object is reasonable and adequate the act resulting in pain is necessary or justifiable, [as where] the act is done to protect life or property, or to minister to some of the necessities of man.

Note that "ministering to the necessities of man" easily trumps the moral status of the animals as far as the law is concerned. This in turn makes prosecution highly problematic, and in 1978 more cases of "unnatural intercourse" were prosecuted in Denver than cases of animal cruelty. In fact, *no* cruelty cases were prosecuted!

It is also worth noting that the extensive catalog of prohibitions cited in the act takes no cognizance of behavioral or psychological cruelty—as if suffering is only physical. This is a vestige of the Cartesian view that animals are simply machines. (Ironically, scientists are now becoming aware that disregard for the psychological state of laboratory animals results in research—even cancer research—that is invalid because it has not taken account of the role of stress variables on the physiology of the animal. Those of us who neurotically monitor our blood pressure at regular intervals know well what stress can do.)

In Waters vs the People, the fundamentally Thomistic view of animals embodied in this law is made clear:

> The aim of this section is not only to protect these animals, but to conserve public morals.

As was the case in some slave protection rulings, the object of the law is to keep *humans* from being brutalized, to forestall a potential danger to the general welfare, rather than to focus on the interests of the animal or slave. In many states, the cruelty laws require that "malicious intent" be proved, again evidence that the law is more concerned about brutalization of people than about welfare of animals.

Or we may consider the pioneering Federal Animal Welfare Acts of 1966 and 1970. This act excludes from its protection rats and mice, creatures that many humans find unaesthetic and uncuddly, as well as farm animals—horses, cows, pigs, goats, sheep, fowl, and so on, and all nonwarm-blooded creatures, although all evidence indicates that reptiles can feel. A primary purpose of the Act was again human utility—it licensed dealers in animals sold to research laboratories in order to allay the fears of pet owners who were concerned that their dogs and cats might be kidnapped and sold to experimenters. No constraints are placed by this Act on any sort of experimentation on animals—in fact, the Act specifically disavows any attempt at controlling the actual conduct of research— nor are individual experimenters licensed. In point of fact, virtually any atrocity can be perpetrated on any kind of animal in the name of scientific research in the United States. Anyone of us can perform "research" in our basements, and at worst run the risk of mild prosecution under the ineffectual anti-cruelty statutes. If we are part of a university, or even a high school or junior high school, if we are spaying dogs or surgically implanting cobalt-60 in guinea pigs in a freshman high school class (this has actually taken place), we run no risk at all, for most anti-cruelty statutes exempt educational institutions. Correlatively, there are no legal constraints on any of the excesses of factory farming.

That some enforceable rules governing research are needed is admitted by an ever-increasing number of scientists. The National Institute of Health Guidelines for Laboratory Animal Care are a sane and reasonable step forward, but they are merely guidelines, and an adequate enforcement structure is lacking, Virtually every scientist I deal with admits to the existence of numerous absurdities masquerading as research, atheoretical empirical dabblings. Many such examples have been cited by Richard Ryder and Peter Singer, and there is no need to chronicle atrocities here. To a large measure, good science and good research go hand in hand. For example, as we mentioned earlier, elimination of psychological stress on laboratory animals controls for a highly relevant set of variables that can skew results badly. As another example, I shall cite a recent case with which I am familiar. An experimenter was starving a group of mule deer to determine changes in the rumen, with an eye towards understanding what happens to deer in the wild. As a control group, the experimenter kept another group of deer that he fed. The second group was separated from the first only by a bit of wire mesh. As a result, the starving deer were subjected to the exquisite agony of watching the other deer eat. Barring the moral outrage, this is clearly absurd as an experiment, for obviously the olfactory and visual food stimuli that the starving deer received could (and very likely did) engender major metabolic changes with profound physiological effect, totally vitiating the experiment!

Legislation alone, however, will not be enough. Along with legislation must come education of scientists. As many writers have pointed out, science education is essentially training in techniques and puzzle solving, and is virtually free of conceptual reflection. Science students must learn that thought and morality cannot and ought not be buried under an avalanche of "professional detachment." Such education is both possible and effective—I myself teach philosophy and ethics to veterinary students as part of the professional curriculum, with great success. I also team-teach a basic biology class in which philosophical and moral issues are taught side-by-side with the hard science. I regularly engage in dialog with the biologist who teaches with me, and the students are encouraged to question, not memorize. Such education is required so that scientists will not see efforts at legislation as just one more set of mindlessly imposed hoops they must jump through and, even worse, circumvent.

Obviously, then, we have demonstrated that the current legal status of animals is woefully inadequate in theory and practice. *Hence, the need for the granting of legal rights to animals, direct legal standing, not as property, that is, the need for the legal recognition of animals as legitimate objects of moral concern, who can institute legal action or, rather, have legal action instituted on their behalf, have injuries to them legally considered, and have legal relief run directly to their benefit.* This, it seems to me, can be accomplished in two ways. In the first place, following Professor Christopher Stone's brilliant analysis of extending legal rights to natural objects, in his book, *Should Trees Have Standing* (Stone, 1974), we can talk of *extending* such rights to animals, even as constitutional rights were extended from native-born white adult male property owners to women, blacks, children, naturalized citizens, and corporations. Like children, animals could be accorded standing, with guardians able to protect their rights before the law.

Ironically, the guardians who would speak for natural objects, the Sierra Club and other environmental groups, are far more readily apparent than the potential spokesmen for animals. The most plausible candidates are of course veterinarians and humane organizations. Yet, as currently constituted, there are problems with both of these candidates. Despite the fact that the essential *raison d'être* of the veterinary profession is the health of animals, the profession has been remiss in its duty to spearhead causes of animal welfare. The discussion of ethics and animal rights has been virtually absent from organized veterinary medicine, as I have discussed (Rollin, 1978).

The AVMA has repeatedly refused to condemn the steel-jawed trap, or cockfighting, and many veterinarians are intimately connected with activities that are inimical to animal health and welfare—rodeo, horse-racing, and laboratories engaged in questionable activities of research and testing. Yet veterinarians *are* concerned with animals, and the potential for constructive effort by the veterinary profession is enormous. Correlatively, the humane organizations have often been duped into doing dirty police work for society, so that for many such organizations, the vast majority of their efforts and energies are expended in animal control and in the "euthanasia" of healthy, but "homeless" animals, a demoralizing activity that ill becomes people whose fundamental moral commitment is to the welfare of animals.

If anything, we have shown that a more plausible case exists for extending rights to animals than to trees, and certainly more so than to corporations. Pragmatically speaking, a test case is easy to imagine and plausible to defend. Though I have argued that rationality and language are strictly irrelevant to moral concern, it is nonetheless a fact that ordinary intuitions and historical precedent militate against ready acceptance of my position. People continue to feel that language somehow makes a moral difference. Recent work on teaching "language" to the higher primates, however, provides a dramatic instance of a borderline case. Though many theoretical linguists, most noteworthy Noam Chomsky, would decline to speak of these animals as linguistic beings, our intuitions push in the other direction. Now consider such an ape who has learned to communicate with men in an enriched environment using some conventional, "linguistic" system. The experiment is terminated, and the animal is of no further use, and is to be turned over to a zoo. (Such a case has actually been reported.) Pressing the analogies with human, rational beings, could one not make a plausible case for the violation of the animal's civil rights? Is this not cruel and unusual punishment? Is it not violation of due process? Such a new "monkey trial," comparable to the Scopes Trial, would be salubrious, for it would force us to confront the legal, philosophical, moral and scientific basis of our metaphysical view of animals. Eventually, and incrementally, by a process stressing the absence of relevant differences, one can envision the judicial *extension* of some rights to all animals.

The second way of establishing legal rights for animals involves not judicial extension, but legislative *conferral*. Laws governing the treatment of animals must be written not in human utilitarian terms, but in the language of rights, with animals seen as objects of moral concern, and with human utilitarian interests relegated to the background. This, as we have seen, is the force of all talk of "rights." Such laws must address themselves to the most fundamental rights that accrue to any object of moral concern—the right to life, the right to be protected from suffering, the right to actualize its nature. Just as human rights are derived from our knowledge of human nature, or *telos* (for example, as a thinking, social being), so too should animal rights be derived. The most obvious targets for such laws are experimentation and factory farming, but there are many others—pet ownership, zoos, and so forth. Such laws must confront the hard question of animals as objects of moral concern, and must further provide meaningful penalties for their violation. We cannot pursue details here, but excellent examples are readily apparent. Laws governing experimentation that take as fundamental the animal's right to be protected from suffering, that license experimenters, and that punish violations by forfeiture of licensure already exist in parts of Europe. Such a law is far from utopian, and I have been privileged to help draft a related law in Colorado. Correlatively, laws governing farming that take as fundamental the animal's right to live its life in accordance with its nature would go a long way toward eliminating major atrocities such as the raising of veal calves in tiny boxes where they cannot turn around and are fed on diets that keep them anemic and in constant distress from diarrhea. Such legislation was recommended by the Brambell Committee in Britain, and in Germany, animals must be housed in ways "which take account of [their] natural behavior." If this is economically impossible, perhaps, as Michael Fox has suggested, we need to breed microcephalic food animals who are incapable of suffering, who are essentially incapable of behavior and feeling, or perhaps work towards cloning sides of beef! Happily, Fox has shown in a report soon to be published, that humane treatment of food animals is often *economically* feasible and profitable. If this is true, the passage of such laws is not a pipe dream.

It is easy to caricature both the judicial and legislative approaches to legal rights for animals—to imagine a case for "freedom of bark," or for giving turtles the vote. (I recall seeing a cartoon in which a dog-catcher is reading from a card to a forlorn dog in a cage in the back of his truck: "You have the right to remain silent . . . . ") But we must remember that the idea of extending rights to slaves, Indians, Orientals, women, and children was once similarly ridiculed, even by the courts. One need only look at newspapers and magazines of these periods, and even today, women's rights are regularly lampooned. Recall, for example, the Dred Scott case, or the lesser known but even more dramatic decision in Bailey vs Poindexter:

> So far as civil rights and relations are concerned, the slave is not a person, but a thing. The investiture of chattel with civil rights or a legal capacity is indeed a legal . . . absurdity. The attribution of legal personality to a chattel-slave—legal conscience, legal intellect, legal freedom or liberty and power of free choice and action . . . implies a palpable contradiction in terms.

Moral arguments cannot be ignored on the grounds that they have always been ignored—"we never did it, why do it now"—anymore than they can be ignored on the grounds of convenience. Each step in moral progress exacts a cost in convenience and utility. Seizing Jewish property was quite useful to the German state and to the majority. Slavery was economically useful. If animals are objects of moral concern, then they have moral rights, and correlatively, they must have legal rights. To be sure, they cannot be the rights of human beings; the rights of a child and a corporation are not those of an adult. *That* animals have rights can be established by reason, as we have done. But *what* these rights are is only partly a matter of reason. Answering this question requires clear factual understanding of what a given animal is, what its nature or essence or *telos* is, to use an outmoded but valid Aristotelian category. This in turn requires that we carefully study animal behavior and animal biology to establish clearly the needs and interests of other creatures, though it does not take a Konrad Lorenz to tell us that our treatment of veal calves is an obscene perversion of the natural. Ethology, biology, and ethics thus go hand in hand.

It will be said that attempting such legal changes is like spitting into the wind without a prior change in people's hearts and minds. A moral gestalt shift, it is said, must precede legal changes, else they are meaningless. There is some validity to this claim; for example, we could not outlaw smoking tomorrow, but it also ignores the fundamental dialectical relation between law and morality. Law does reflect public morality, but also molds it—one *can* legislate morality, as racial progress in the South has shown in the twenty or so years since the momumental civil rights legal decision. We as a people *do* respect law. To grant legal rights to animals is to institutionalize their claim to moral concern, to recognize this status in a way which is writ large, to force us to pause and look at what we take for granted, and to confront the inexpedient and bothersome implications of our moral commitments. Certainly the utilitarian costs are enormous, but so too were the opportunity costs of abolishing slavery and child labor. Our treatment of animals is the last moral frontier, the ultimate test of our humanity, the mirror by which we can see most deeply into our own souls.

SECTION III

Humans and Other Animals— Killing

Introduction

Do the differences between humans and other animals suffice to justify the killing of nonhumans by humans? Edward Johnson examines a number of arguments that humans are more important, ethically, than nonhumans. He rejects the argument (of Mill and others) based on the supposedly broader experience of humans. The argument on the grounds that humans are called upon to justify themselves only to humans is also rebutted, as is the group of justifications relying on the claimed nonreflexiveness of nonhuman consciousness. An animal with no concept of the future may well still have an interest in continuing to live, deriving from desires for future enjoyments. Johnson can find no acceptable ground on which to ascribe moral importance to reflexiveness. It is not obvious that mental complexity is per se ethically significant. (James Rachels comments on this point in Section VI.)

Dale Jamieson presents and defends the claim that simple (nonreflexive) consciousness suffices to ground an obligation not to kill the possessor of such consciousness. Contrary to what he calls the 'HTS' (Hegel, Tooley, Singer—an unlikely trio) view that reflexiveness is of primary moral importance, Jamieson argues that consciousness itself, of the simplest sort, is a good, no matter what its content and that thus we are (at least prima facie) obliged not to kill the conscious. But Jamieson holds, contrary to Johnson, that mental complexity does count for something, if not for everything.

Donald VanDeVeer joins in this discussion by attempting to sketch out an interspecific theory of justice. His model is John Rawls's famous *A Theory of Justice* (Rawls, 1971). First he considers, and refutes, claims that life per se is 'sacrosanct' or of 'infinite' value. There are lives so painful, constricted, and hopeless as not to be worth living. Suicide is sometimes rational. That some lives are worse than no life at all is an important part in VanDeVeer's theory of justice.

The best-known feature of Rawls's theory of justice is the derivation of the principles of a just society from the hypothetical agreement of rational, self-interested, well-informed, but impartial contractors. The incompatibility of the requirements of self-interest and impartiality is removed by placing the contractors behind a 'veil of ignorance' such that, though they know that they will be members of the society, they know nothing of their prospective social, sexual, financial, intellectual, or physical status. VanDeVeer's modification of this theoretical device is to make the veil of ignorance more opaque. The contractors know that they will be sentient animals, but do not know of what species.

In such a position, VanDeVeer argues, the contractors would agree to accept the results of the genetic lottery (because of the enormous costs of any attempt to equalize the abilities of species). But further, the contractors would establish a principle that no sentient being should be so treated as to have a life not worth living, nor should such a being be created if it will have such a life. It does not follow from these principles that it is always wrong to kill animals not capable of the highest sorts of satisfactions. We have no right to breed animals for a life not worth living; but there are often gains, for the animal, in domestication (see Gross's paper in Section VIII). So it need not be wrong to breed some sorts of animals in some sorts of ways in order to kill and eat them. VanDeVeer's conclusion is that while many things that humans do to other animals are indeed unjust, much human use of animals is morally acceptable and in fact beneficial. (He thus agrees, in outline if not in detail, with Cargile's position—see Section V.)

Life, Death, and Animals

Edward Johnson

I

Mill says that it is "better to be a human being dissatisfied than a pig satisfied" (Mill, 1863). Is it? How does Mill know that? MacIver says this (1948, p. 65):

> If I tread wantonly on a woodlouse, I do wrong . . . But it is only a very small wrong, and to exaggerate its wrongfulness is sentimentality . . . Little wrongs have to be done, in order that greater wrongs may be avoided. If I kill a Colorado beetle, I do wrong by the beetle; but, if I fail to kill it, I do wrong by all the growers and consumers of potatoes, and their interests are vastly more important.

Are they? How does MacIver know that? Like Mill and MacIver, we all believe, or act as though we believe, that human life is, always or usually, somehow more important than animal life. Indeed, mere pleasure or convenience for humans is commonly supposed to be more important than the lives of nonhuman animals; as Goodrich says, "it seems generally to be held that human life is infinitely more valuable than animal life: there is no number of animals, however great, that is worth the sacrifice of even one human being" (Goodrich, 1969, p. 128). Are we so important? How do we know?

You could say: *we just know*. But it is best to avoid *ad hoc* intuitionism for as long as we can. Haven't complacent assertions of intuitive superiority turned out wrong often enough before? Haven't we, for example, seen through the intuitions that purported to justify slavery and racism? We have come to see, I suppose, that we don't know (and *never did* know), that the life of a white is more important than that of a black, or that there is no number of slaves, however great, that is worth the sacrifice of even one slave owner. Now, what about the pig?

123

II

"Human beings," Mill tells us, "have faculties more elevated than the animal appetites and, when once made conscious of them, do not regard anything as happiness which does not include their gratification." How can we tell when one faculty is "more elevated" than another? Mill's criterion is well known. "Of two pleasures, if there be one to which all or almost all who have experience of both give a decided preference, irrespective of any feeling of moral obligation to prefer it, that is the more desirable pleasure." Mill no sooner enunciates this criterion than he proceeds to indicate how it can be used, not merely to rank pleasures roughly, but to place the "more elevated" pleasures beyond impeachability.

> If one of the two is, by those who are competently acquainted with both, placed so far above the other that they prefer it, even though knowing it to be attended with a greater amount of discontent, and would not resign it for any quantity of the other pleasure which their nature is capable of, we are justified in ascribing to the preferred enjoyment a superiority in quality so far outweighing quantity as to render it, in comparison, of small account.

Thus, it is "better to be a human being dissatisfied than a pig satisfied; better to be Socrates dissatisfied than a fool satisfied." Note that Mill is not appealing just to the fact that *we* prefer human dissatisfaction to swinish satisfaction. Rather, he thinks that we are specially qualified to judge. The opinions of the fool and the pig are disallowed by Mill. "And if the fool, or the pig, are of a different opinion, it is because they only know their own side of the question. The other party to the comparison knows both sides."

The inadequacy of Mill's discussion here has been much noted. Bertrand Russell makes the drollest comment, remarking that "utilitarians have been strangely anxious to prove that the life of the pig is not happier than that of the philosopher—a most dubious proposition, which, if they had considered the matter frankly, could hardly have been decided in the same way by all of them" (Russell, 1975, p. 162, quoting a letter of 1902). Mill assumes that *we,* unlike the pig, can be "competently acquainted with both" the pig's happiness and our own. That is doubtful. Socrates, in his learned ignorance, may not recognize where ignorance is bliss. How much more difficult is the case where we must deal, across species, with vastly different sensibilities and capacities. What reason is there to suppose that any human really knows anything about what it is like to be a pig, or a bat, or any other animal? As William Blake asks (1966, p. 150):

> How do you know but ev'ry Bird that cuts the airy way,
> Is an immense world of delight, clos'd by your senses five?

In the really difficult cases, it may not even be *possible* to be "competently acquainted with both," since the capacities that make one pleasure possible may be exactly what make another impossible. The hedonist can't go slumming *everywhere*. Socrates cannot know the pleasures that the fool enjoys *in his foolishness*; at best, he can have an intellectual's *ersatz*. It is unreasonable to assume, as Mill does, that humans are *of course* acquainted with "animal" pleasures as well as with other, distinctively "human" pleasures. The availability of the "higher" pleasures may change everything, so that we are no longer in a position to have, or to judge, the "lower" ones. In that case, Mill's very justification for speaking of "higher" and "lower" disappears.

Mill also assumes that broader experience (if we can have it) always puts one in a *better* position to judge. But this overlooks the possibility of decadence or corruption: perhaps the enjoyment of some pleasures may put us in a *worse* position to judge among pleasures. As Rousseau says: "Slaves lose everything in their chains, even the desire of escaping from them: they love their servitude, as the comrades of Ulysses loved their brutish condition" (Rousseau, 1762: Book One, Chapter Two).

III

In view of the difficulties attending any attempt to show that we humans are specially qualified to judge the value of animal lives, it is tempting to adopt a view that makes no such claim. Thus, E. B. McGilvary says this (1956, pp. 293–294):

> It is better to be a Socrates unsatisfied—better for whom? For Socrates or for the pig? But a pig! Who would be a pig? Is he not loathly? Assuredly he is—*to us*; but *to himself* not so assuredly. Who knows what preciousness there may not be to pigs in unadulterated piggery? Who then shall decide? To what arbiter shall we appeal?
>
> It is strange that when such a question is asked, the fact is overlooked that it is not thrown out to the universe in general. It is we men and women who are asking the question; we are asking it of ourselves; why not answer it for ourselves? We are not particularly interested in the question whether pigs like to be pigs. It matters not if they do. We are concerned with the question what *we* should like to be, what we should like to help our children become, what kind of civilization we shall lend our efforts to build up for the future.

The same sort of view appears in Philip Devine's (1978, p. 49) recent book on the ethics of homicide, where it is argued that:

> We need not be concerned with the relative value of human and animal life in the abstract, but only with their relative value in the context of decision making by human beings. Even supposing that Hume is right when he says that "the life of a man is of no greater importance in

the universe than that of an oyster,'' where the decision maker is not an oyster or God (or the universe personified), but a human being, that kind of creature which shares certain essential traits with the agent is entitled to a kind of respect to which those who do not are not. Still further, human beings—or nearly all of them—are capable of a much richer kind of life than nonhuman animals on this planet enjoy, including the very moral agency presupposed in the asking of a moral question. To be deprived of this kind of life (or to have it impaired) is a much greater harm than to be deprived of a merely animal existence. Of course, one might say that human beings are unable adequately to judge the richness of the lives of spiders or dolphins, and thus to determine whether their lives are or are not as valuable as ours. But one is forced in any case to judge matters from one's own point of view (from what other point of view might one judge them?) and there is nothing inappropriate . . . in human actions being guided by the perceptions of human beings.

It certainly sounds realistic to say that we ''are concerned with the question what *we* should like to be,'' or that ''one is forced in any case to judge matters from one's own point of view.'' But is this morally adequate? Imagine that *we* (members of a dominant race) were to offer an analogous justification for counting the lives of a subject race as of less importance than our own: 'Of course, one might say that whites are unable adequately to judge the richness of the lives of blacks, and thus to determine whether their lives are or are not as valuable as ours. But one is forced in any case to judge matters from one's own point of view, and there is nothing inappropriate in the actions of whites being guided by the perceptions of whites.' If such an argument would be nothing more than the specious rhetoric of racism, then it is difficult to see the sort of view espoused by McGilvary and by Devine as anything but self-congratulatory speciesism. If the former is morally inadequate, isn't the second as well?

IV

So far, I have rejected two views about how we are able to know that human lives are more important than nonhuman lives. The first view was that we are in a position to judge animals' lives, but they are not in a position to judge ours, because we (and not they) are, or can be, ''competently acquainted with both.'' The second view was that we know that our lives are more important because *we* are the ones judging. I want now to consider a third view, which is a bit more complex. The basic idea is that we know that our lives are more important because we can know something about our lives that animals cannot know about their lives; according to this view, the complexity of human minds is not (necessarily) valuable in itself, but it allows us to value our lives in a specially important way. There are many versions of this view.

Kant holds that nonrational beings have "only a relative value as means and are consequently called *things* (Kant, 1785, pp. 428–466). Animals, he says, "are not self-conscious and are there merely as a means to an end" (Kant, 1963, p. 239). We have no direct duties to animals, he thinks, and it follows that they have no rights, in particular no right to life. Hegel says that animals "have no right to their life, because they do not will it" (Hegel, 1952, p. 237). He, too, denies that animals are ends in themselves, and calls them *things,* remarking (p. 236) that

> the thing, as externality, has no end in itself; it is not infinite self-relation but something external to itself. A living thing too (an animal) is external to itself in this way and is so far itself a thing.

An animal lacks rationality, self-consciousness, infinite self-relation; it "lacks subjectivity" and so "is external not merely to the subject but to itself" as well. To say that an animal, as a thing, is external to itself is to say that it lacks some sort of mental complexity, some sort of reflexivity, that persons have. "An animal can intuit," says Hegel, "but the soul of an animal has for its object not its soul, itself, but something external." Henry Johnstone says this (1970, p. 138):

> The being of a person is reflexive in a way in which the being of an inert thing is not. One cannot be a person without knowing what it is like *to oneself* to be a person. It is entirely by virtue of this knowledge that persons place whatever value or disvalue they do on life . . . An animal . . . does not know what it is like *to itself*— from its own piscine, avian, feline, or canine point of view—to be a fish, bird, cat, or dog. While it enacts the behavior of its species, it does so without taking a point of view. The fish behaves like a fish—not to itself, but to us.

Currently fashionable versions of this concern for reflexivity admit that animals (as conscious beings) have desires, but emphasize that humans have something more. Harry Frankfurt (1971, pp. 6–7) and Richard Jeffrey (1974, p. 378) too, suggest that though animals have first-order desires they lack second-order (or higher-order) desires, volitions, or preferences. Animals want, but they don't want to want; they care about things, but they don't care what cares they have. Gary Watson holds that the key to understanding free agency lies in distinguishing two different sorts of motivation—desires and values—and writes: "In the case of the Brutes . . . motivation has a single source: appetite and (perhaps) passion. The Brutes (or so we normally think) have no evaluational system" (Watson, 1975, p. 220). Similarly, Charles Taylor (1976, p. 282) endorses the notion that human subjects are capable of evaluating what they are, and to the extent that they can shape themselves on this evaluation, are responsible for what they are in a way that other subjects of action and desire (the higher animals for instance) cannot be said to be.

These are all different ways of saying that animals cannot adopt a view about what they are; they cannot accept or reject their wants, and thus mold their futures, in the way humans can; nor can animals use their future to give significance, retrospectively, to their past behavior (see Harman, 1976, pp. 462–463).

Let us suppose, for the sake of argument, that there is a difference between humans and other animals of the sort that these various versions of concern about reflexivity have tried to point to. Humans, let us say, are reflexive or self-conscious in a way that animals aren't. What moral weight should such a fact have? Why should reflexive lives matter more than unreflexive lives?

V

One answer would be that reflexive capacities allow their possessor to care about life, and to mind death, in a way that animals cannot. According to a common view, animals lack the concept of death, and so cannot mind death, any more than they mind not having a ticket to the opera. Rational creatures, however, can mind, and normally do, and this is the reason why it is wrong, *prima facie*, to kill them. Is such a view correct?

You can have an interest in avoiding death if you are capable of conceiving death, and so of minding it; you can have an interest in your own continued existence if you are capable of conceiving it, and so of wanting it. But you can also have an indirect or derivative interest in life that feeds off of your other interests. If a cow likes to chew her cud, then it is, other things being equal, in her interest to be allowed to do so.[1] She is benefited by having opportunities to satisfy her desires: the more the better. But does this not give the cow an interest in continued life? When to have a desire satisfied is to be benefited, isn't one benefited more, other things being equal, the more opportunities one has to satisfy it (perhaps—where this is relevant—up to some point of satiation)? This will be so even it one lacks the concept of a future, of personal identity

---

[1] I use 'interest' in a broad sense. I have an interest in having my desires satisfied, or their satisfaction promoted, other things being equal. Also, if I desire that a particular state of affairs obtain, or a particular event or experience occur, I have an interest in the realization of that state of affairs, event, or experience, other things being equal. (But this is not a matter of what I "take an interest" in or "am interested" in.) Something is in my interest if it benefits me (or its absence harms me), if it is a good for me, if it is something I want (other things being equal). I have an interest in the things that make my life better, or prevent it from becoming worse. There may be narrower senses of 'interest', but I would argue that my broad usage is perfectly legitimate.

over time, etc. Of course, if one does have such concepts, since one will then be able to *care* about the future, that will give one an additional interest in living. But the lack of such concepts does not mean that one has *no* interest in, or claim to, life: the derivative sort of interest in life remains. Insofar as life seems likely to satisy one's desires, fulfill interests that one has, one has an interest in life.

To make this more precise, we need to distinguish between negative and positive interests. Roughly, negative interests are those that would be satisfied if one suddenly ceased to exist. Avoiding pain is a negative interest. Experiencing pleasure is a positive one. By their very nature, negative interests give one no grip on future life, since they will be fulfilled just as well if one dies as they possibly can be if one continues to live. The fact that one has an interest in avoiding pain is not enough to give one an interest in living; indeed, in some cases it may give one an interest in dying. That is the point, sometimes, of suicide and euthanasia. Since nonhumans, as well as humans, can feel pleasure, they have an interest in living, when living can provide them with opportunities to enjoy their many natural pleasures.

This view explains how animals can have an interest in living, while at the same time explaining how the interest rational beings have in living is importantly different. Humans, like animals, have an interest in life derivative from their interest in other things. But humans also can have a quite independent interest in life, since they may care about the future. If an animal's future will be one of unrelieved pain, then it lacks an interest in life, and should be killed. But we cannot immediately draw the same conclusion in the case of a human. It is possible that a human's desire to continue living may outweigh the fact that her or his life will be on balance one of pain and frustration. In the normal case, humans will have both the sort of interest in life that depends on other interests *and* the sort that depends on caring about the future.

I suggested earlier that the satisfaction of an animal's (positive) interests is a benefit to it, and that more opportunity for satisfaction is better than less, and that consequently animals can have, in virtue of their interests, a further interest in living. But is this correct?[2] There are two ways of handling any desire: one can satisfy it or get rid of it. One way of getting rid of a being's desires is to kill it: it then no longer has any unfulfilled desires. Why not handle matters in this manner? In the case of humans, we can appeal to the fact that people *care* which way their unsatisfied desires (or some of them) are handled. But this sort of appeal is not possible in the case of animals, who lack (by hypothesis) the conceptual capacities necessary for higher-order desires. If it does not matter to an animal what desires it has, or what desires it acts on, how can it matter to it whether the desire is satisfied or extinguished? All that mat-

---

[2] I owe this objection to Susan Wolf.

ters to the animal, it seems, is that it not go on having (on balance) unsatisfied desires. If so, there is no reason, as far as the animal's interests are concerned, why the animal should not simply be killed.

Notice, however, that it is not enough that persons just have higher-order desires or preferences that specify what they want to want. For the question will simply reappear at the level of the highest-order preference: Should we satisy this meta-preference or extinguish it? All that would be true is that humans would have a more complicated preference structure than animals. If there is to be a real difference between persons and animals, then, it seems that persons cannot have a highest-order volition: meta-desires must recede infinitely. (One may see in this, perhaps, part of Hegel's motivation for talking of persons in terms of "infinite self-relation.")

This would be a dark doctrine. But even if one could accept the view that humans are interminably reflexive, and so different from animals on this matter, one would still have to deal with the stubborn intuition that removing an animal's desire is not, as a general policy, just as acceptable as satisfying the desire.

Views emphasizing human reflexivity may allow that nonrational creatures have an interest in life, but they insinuate that rational creatures have *more* of an interest in life, since a rational creature can be interested in continued existence in itself. It is not clear, however, that this provides a justification for preferring human to animal life in every case of conflict. We often judge that increasing one person's opportunities to satisfy interests is more important than satisfying specific desires of another person. I don't see that we can *assume* that any desire to live on the part of a human outweighs in itself the sort of interest in living that an animal can have.

VI

A reflexive being has a kind of interest in life that an unreflexive being lacks, but it is not clear exactly why this should give the reflexive being any greater claim on life, or make its life more valuable or important. Why should mental complexity count for anything? That it *does* count is an assumption common to speciesist and anti-speciesist alike. Even Peter Singer says (1975, p. 23) that

> a rejection of speciesism does not imply that all lives are of equal worth. While self-awareness, intelligence, the capacity for meaningful relations with others, and so on are not relevant to the question of inflicting pain—since pain is pain, whatever other capacities, beyond the capacity to feel pain, the being may have—these capacities may be relevant to the question of taking life. It is not arbitrary to hold that the life of a self-aware being, capable of abstract thought, of planning for

the future, of complex acts of communication, and so on, is more valuable than the life of a being without these capacities.

Singer's rule of thumb for avoiding speciesism is that "we should give the same respect to the lives of animals as we give to the lives of those humans at a similar mental level . . . " (p. 24). But why the qualification "at the same mental level"? Why does mental complexity count? One answer would be that mentally complex beings experience greater pleasure and pain than others. Another answer would be that mentally complex beings are capable of entering into relations with one another, in a way that simple souls are not; this view requires us to see morality as radically contractual in nature. A third answer would claim that mental complexity is intrinsically valuable.

I do not have time here to discuss the faults of each of these answers.[3] Instead, let me try to indicate as forcefully as I can the difficulty raised for moral reflection by this question of the comparative value of human and nonhuman lives.

I assume that animals are conscious (see Griffin, 1976). I am willing to concede, for the sake of argument, that the consciousness of (most) animals is not self-consciousness, and that self-consciousness is a more "complex" state of mind than "mere" consciousness. But what moral weight does such complexity carry? I incline to the view that each mind can be valuable to itself. There need be nothing *intrinsically* wrong with the mentalities of those who are "mad," "retarded," or "childish." That they are not what I want for myself does nothing to show that they are not valuable to *those* beings. Shouldn't every mind have a voice, even if I cannot hear it? As Samuel Alexander says about children: "We like them because they are children, and not because they will be men" (1939, p. 115). Children are adorable, he suggests, because of, rather than despite, their mental simplicity. So, he says, is his dog. Once one starts down this path, it is difficult to know where to stop. The great microbiologist H. S. Jennings says (1906, p. 336) that he is

> . . . thoroughly convinced, after long study of the behavior of this organism, that if Amoeba were a large animal, so as to come within the everyday experience of human beings, its behavior would at once call forth the attribution to it of states of pleasure and pain, of hunger, desire and the like, on precisely the same basis as we attribute these things to the dog.

And Richard Taylor, though on less empirical grounds, suggests this (1970, p. 267):

---

[3] On the latter two especially, see my dissertation, *Species and Morality* (Princeton University, 1976), chapters four and five.

> Even the glow worms . . . whose cycles of existence over the mil-
> lions of years seem so pointless when looked at by us, will seem utterly
> different to us if we can somehow try to view their existence from
> within.

Consider, finally, the oyster, a much maligned creature. Plato argues that
"if you were without reason, memory, knowledge, and true judgment,
you would necessarily . . . be unaware whether you were, or were not,
enjoying yourself . . . You would be living the life not of a human being,
but of some sort of sea lung or one of those creatures of the ocean whose
bodies are incased in shells" (*Philebus* 21b-c). Descartes argues as follows (1970, pp. 207–208):

> The most that one can say is that though the animals do not perform
> any action which shows us that they think, still, since the organs of
> their body are not very different from ours, it may be conjectured that
> there is attached to those organs some thoughts such as we experience
> in ourselves, but of a very much less perfect kind. To which I have
> nothing to reply except that if they thought as we do, they would have
> an immortal soul like us. This is unlikely, because there is no reason to
> believe it of some animals without believing if of all, and many of them
> such as oysters and sponges are too imperfect for this to be credible.

Even Peter Singer excludes the oyster (1975, p. 188). I have no idea
whether oysters are conscious or not. My point is that, if they or any other
creatures are, if there is reason to believe that they are, there is *no reason*
to despise their consciousness as *in itself* of less value to them than our
own is to us. In the essay of Santayana's that convinced Russell to give up
his early belief in the objectivity of good and evil, Santayana (1913, pp.
147–148)[4] makes the following comment on that argument of Plato's:

> It is an *argumentum ad hominem* (and there can be no other kind of
> argument in ethics); but the man who gives the required answer does so
> not because the answer is self-evident, which it is not, but because he
> is the required sort of man. He is shocked at the idea of resembling an
> oyster. Yet changeless pleasure, without memory or reflection, with-
> out the wearisome intermixture of arbitrary images, is just what the
> mystic, the voluptuary, and perhaps the oyster find to be
> good . . . Such a radical hedonism is indeed inhuman; it undermines
> all conventional ambitions, and is not a possible foundation for politi-
> cal or artistic life. But that is all we can say against it. Our humanity
> cannot annul the incommensurable sorts of good that may be pursued
> in the world, though it cannot itself pursue them. The impossibility
> which people labour under of being satisfied with pure pleasure as a
> goal is due to their want of imagination, or rather to their being domi-
> nated by an imagination which is exclusively human.

---

[4] Santayana's influence on Russell's recantation is documented in Russell, 1956, p. 96.

If we base ethics on self-assertion, as McGilvary and Devine and (more profoundly) Santayana do, we can morally exclude oysters and pigs, which is gastronomically convenient, but we pay a price: those who are asserting themselves can, with equal justification, exclude Jews or Blacks, the retarded, me, or, even, you. One wants to draw a line here, but no rationale so far has worked. This result, if not surprising, is, for anyone who wants to believe in the coherence of ethics, deeply disturbing.

Killing Persons
and Other Beings

Dale Jamieson

What beings are we obliged not to kill? 'Persons' leaps rapidly to the lips of most. A sustained colloquy would perhaps elicit numerous exceptions: except when guilty of a heinous offense, except in times of war, and so forth. The grounds of these exceptions are worth wondering about; the grounds of the obligation still more. And we might ask whether it is only persons that we are obliged not to kill, or do we have such obligations to other beings as well? What I will argue is that consciousness, simple or reflexive, is sufficient for imposing on us a *prima facie* obligation not to kill its subject.

One might ask straightaway: What is it for a being to have reflexive consciousness? Here is a rough and ready answer: A being with reflexive consciousness has a conception of itself as a distinct being that persists through time. Such a being can have higher-order desires, volitions, preferences, and so forth. It has a point of view. To wax Germanic, such a being has "subjectivity." A being of simple consciousness has experiences, but no sense of itself as a distinct entity that is the subject of these experiences. Nor can simple beings have higher-order desires, volitions, preferences and so forth. This answer is a beginning, not an end. But it is good enough for openers.

The view that I am urging, that simple consciousness is sufficient for imposing on us a *prima facie* obligation not to kill its subject, though perhaps strange, is not novel. Edward Johnson (this volume) and Werner Pluhar (1977) advocate it as well. Although Johnson, Pluhar, and I agree that simple consciousness is sufficient for imposing an obligation not to kill its subject, our grounds differ drastically.[1]

---

[1] Here and elsewhere, the obligations that I discuss are *prima facie* ones, and it is persons who are under them. Stylistic considerations lead me to speak simply of obligations when referring to *prima facie* obligations that people are under. Furthermore, the "other beings" with which I am concerned are simple beings of simple consciousness.

Johnson finds the source of our obligation not to kill simple beings in murky considerations concerning the value of their consciousness to them.

> . . . there is no reason to despise their consciousness [the simple consciousness of simple beings] as *in itself* of less value to them than our own is to us. (p. 132)

Surely we can agree that we ought not "despise consciousness." Still, it is unclear what Johnson means to deny in this passage. Am I to compare the value of my consciousness to me with the value of a cockroach's consciousness to it? How could I do that? I could perhaps take up an evolutionary point of view. I might speculate that whatever consciousness cockroaches have is more valuable to them than pterodactyl consciousness was to the pterodactyl. After all, the pterodactyl is extinct; cockroaches thrive. Does this show that the consciousness of a cockroach is more valuable to a cockroach than pterodactyl consciousness was to the pterodactyl? Probably not. Survival involves more than the value of a being's consciousness. Whatever the case with cockroaches and pterodactyls, I doubt whether the evolutionary approach is what Johnson has in mind. But aside from this approach, what could it mean to ask about the value to a simple being of its simple consciousness? I can make some estimation of the value of my IBM Correcting Selectric II to me, but that is because I am a being with reflexive consciousness. About the value to cockroaches and pterodactyls of their respective consciousnesses, I am forever ignorant. And if they are truly simple beings of simple consciousness, so they are.

Perhaps Johnson just means to say that such questions are absurd, or that such comparisons are impossible. That would be quite sensible. But what would follow? The judgment that many make, and Johnson and I deny, is that reflexive consciousness is necessary for imposing an obligation not to kill its subject.[2] Whatever one thinks of the truth of this claim, surely its defense does not require answering Johnson's unanswerable questions.

Pluhar provides little in the way of direct argument for the claim that simple consciousness is sufficient for imposing an obligation not to kill its subject. He does say that this claim has "a good deal to recommend it" and gives us some examples. Here is the first (Pluhar, 1977, p. 160):

> For example, most people who agree with the above-mentioned triple position [the moderate position on abortion that holds that there is a point between conception and birth such that abortion is permissi-

_____

[2]Operating at this rarefied level of generality poses certain difficulties that are best noted and ignored. One could not hope to defend, without qualification, the doctrine that it is necessary for a being to have reflexive consciousness if we are obliged not to kill it. You are obliged not to kill my azaleas because they are my azaleas, even though the azaleas do not have any consciousness, simple or otherwise.

ble before that point, but not after] do so at least in part because they empathize with the entity whose destruction is being considered; i.e., they imagine themselves in its position and hence as having its experiences.

This claim is as difficult to evaluate as most claims about what "most people" believe. Even if most moderates on abortion hold the position that they hold for the reasons we are told, still, the inference from what most people "empathize with" to what is the case is problematical. Perhaps most people "empathize with" baby seals, but not with cows. Nothing much follows about the morality of killing either. Nor does Pluhar's shift to talk of what people "imagine" help. Someone might object to the clubbing of baby seals because she or he can imagine what it is like to be a baby seal in pain.[3] But can she object to the painless killing of baby seals because she or he can "imagine" what it is like to be a baby seal that is painlessly killed?

An ambiguity presents itself. In one sense, to be a baby seal that is painlessly killed is to be a dead baby seal. To imagine what it is like to be a dead baby seal is to imagine what it is like not to exist. Some have argued that we cannot imagine what it is like not to exist (Freud, 1963). Whether or not we can seems irrelevant to the morality of killing.

In another sense, to be a baby seal that is painlessly killed is to be a baby seal that is being painlessly killed. What would it be like to be such a baby seal? Were I such a baby seal, I would have no conception of myself as a distinct being that begins, persists for awhile, and then ends. I would therefore have no conception of myself as dying. Were I a baby seal that is being painlessly killed, I would merely experience some painless sensations. Either I cannot imagine what it is like to be a baby seal that is being painlessly killed, or if I can imagine what it is like and if I have rightly described what it is like, then it seems not unpleasant. Either way, such speculations provide no support for the claim that simple consciousness is sufficient for imposing an obligation not to kill its subject.

Here is Pluhar's second example (1977, pp. 160–161):

> Or consider how much more important we consider even a person so senile as to be hardly self-aware, but who does of course still have experiences, than we do the mere body once the person has died and all sentience has ceased.

One could argue analogously that we consider a living, breathing dog to be "much more important" than the mere body of a dog once it has died. Does this show that we think it wrong to kill dogs? Hardly. Every year in this country alone we kill tens of millions of dogs, many of them in extremely painful ways.[4] Whatever moral concern most of us have for dogs

[3]I am supposing that baby seals are simple beings for illustrative purposes.
[4]See various publications available from the American Humane Society, 5351 S. Roslyn, Englewood, Colorado 80100.

flows from our relationships with their owners. Similarly, it may well be that concern for humans reduced to simple consciousness is largely motivated by our concern for those persons whose lives are bound up with these simple beings. Again, Pluhar's second "example" provides scant support for the claim that simple consciousness is sufficient for imposing an obligation not to kill its subject.

Pluhar's method of reasoning from common conceptions to moral principles has little to recommend it. That people have the beliefs that they have, though of great sociological interest, is of only marginal philosophical interest. The principles that animate the beliefs, the relations between these principles and others, the evidence that bears on the truth of the principles, these are the proper objects of philosophical concern. Furthermore, arguments from common conceptions to moral principles stack the deck against those who are "moral revolutionaries." The movement on behalf of animals, like other contemporary "liberation" movements, is an attempt to change common moral conceptions. Defenders of such movements would be and should be unmoved by such arguments. Perhaps nothing could be more obvious to most people than that there is nothing wrong with experimenting on animals or raising them for food. But very little of philosophical interest follows from this.

Let us carefully examine the view that Johnson, Pluhar, and I deplore: the view that we are obliged not to kill beings of reflexive consciousness, but that we are under no such obligation with respect to simple beings of simple consciousness.

Johnson nicely traces this view to Hegel (1952, p. 237):

> [Animals] have no right to their life because they do not will it.

Hegel's sentiments are echoed in recent writings by Michael Tooley and Peter Singer. Here is Tooley (1972, p. 49):

> To sum up, my argument has been that having a right to life presupposes that one is capable of desiring to continue existing as a subject of experiences and other mental states. This in turn presupposes both that one has the concept of such a continuing entity and that one believes that one is oneself such an entity. So an entity that lacks such a consciousness of itself as a continuing subject of mental states does not have a right to life.

Here is Singer (1979, p. 153; cf. Singer, 1980):

> Given that an animal belongs to a species incapable of self-consciousness, it follows that it is not wrong to rear and kill it for food, provided that it lives a pleasant life and, after being killed, will be replaced by another animal which will lead a similarly pleasant life and would not have existed if the first animal had not been killed.

Though there are obvious differences between the views of Hegel, Tooley, and Singer, there are common themes. On the basis of these

common themes, I will construct a view that I will refer to as "HTS"—
short for 'Hegel/Tooley/Singer'.

HTS is the view that we are obliged not to kill beings of reflexive
consciousness because such beings can conceive of themselves as distinct
entities that persist through time, and therefore can have preferences regarding their continued existence. HTS also holds that we are under no
such obligation to simple beings of simple consciousness because they
cannot conceive of themselves as distinct entities that persist through
time, and therefore they cannot have preferences regarding their continued existence.

Is HTS speciesist?[5] Singer denies that it is (1979, p. 153):

> I am sure that some will claim that in taking this view of the killing
> of some non-human animals I am myself guilty of "speciesism"—that
> is, discrimination against beings because they are not members of our
> own species. My position is not speciesist, because it does not permit
> the killing of non-human beings on the ground that they are not mem-
> bers of our species, but on the ground that they lack the capacity to
> desire to go on living. The position applies equally to members of our
> own species who lack the relevant capacity.

Though perhaps HTS is not speciesist, it is puzzling that Singer resorts to
talk of species in formulating the conditions under which it is permissible
to rear and kill a being for food. If HTS is not speciesist, the permissibility of rearing and killing a being must turn on the nature of its consciousness and not on what species it happens to be a member of. If under certain conditions it is permissible to rear and kill simple nonhumans for
food, then under certain conditions it is also permissible to rear and kill
simple humans for food. This conclusion is unavoidable if Singer is to
espouse HTS and remain nonspeciesist.[6]

One should also bear in mind that as formulated, HTS is only a view
about the morality of killing. Singer, for one, holds that simple consciousness is sufficient for imposing an obligation not to cause pain to its
subject, and that this prohibition is sufficient for requiring us not to eat
meat that is produced by present methods (1979, pp. 150–151, 145;
1975, Ch. 1).

Having thus characterized HTS, it is useful to introduce a more perspicuous vocabulary. Following the example of Locke, I will distinguish
the extension of 'person' from that of 'human being'.[7] I will use 'person'

---

[5]The term 'speciesism' was introduced by Richard Ryder. The term was popularized by
Peter Singer (1975).

[6]Michael Lockwood suggests that Singer tends to understate "the true radicalness of
his views." Perhaps this is what leads Singer to invoke the notion of species in
formulating the conditions under which simple beings can be reared and killed for food.
Lockwood, 1979, pp. 167–168.

[7]Locke uses 'person' and 'man' to mark (roughly) the same distinction that I mark with
'person' and 'human being.' See Book 2, Chapter 27 of Locke, 1690.

to refer to all and only those beings of reflexive consciousness. I will use 'human being' to refer to all and only those members of the species *homo sapiens*. Probably many persons are not human beings. Probably many human beings are not persons.

Employing this more perspicuous vocabulary, we can say that HTS holds that we are obliged not to kill persons, but that we are under no such obligation with respect to other beings.

Johnson provides a critique of HTS. First he argues that a simple being of simple consciousness can have an interest in continued life. According to Johnson:

> Something is in my interest if it benefits me (or if its absence harms me), if it is a good for me, if it is something I want (other things being equal).[8]

As a rough elucidation of the semantics of 'interest,' Johnson's remarks are unobjectionable. What is objectionable is what he does with this elucidation.

Johnson argues that Bessie the cow, to take an example, has an interest in continued life in virtue of her interest in chewing her cud. The idea seems to be that having an interest in something gives one a "derivative" interest in those things that are necessary for the satisfaction of one's "primary" interest. But even if it is granted that simple beings of simple consciousness have a "derivative" interest in continued life, it does not follow that a being's having an interest in continued life is sufficient to impose on us the obligation not to kill it. Yet this premise that Johnson fails to argue for is precisely the premise that a defender of HTS would deny. After all, the heart of HTS is the doctrine that a being's reflexive consciousness is the source of our obligation not to kill it. Talk of mere interests is of little concern to one who holds to HTS. It is talk of higher-order preferences that is crucial.

Not only does Johnson fail to argue for the premise on which his refutation of HTS turns, the premise in question is vulnerable to an obvious *reductio*. Bessie the cow has certain interests in things that are good for her; adequate food, room to roam, and so forth. Ludwig the geranium too, has certain interests in things that are good for him; adequate water and sunlight, an occasional squirt of plant food and so forth. Just as Bessie's primary interests generate her interest in continued life, so Ludwig's primary interests generate his interest in continued life, or so it would seem. Since having an interest in continued life is, *ex hypothesi*, sufficient for imposing on us an obligation not to kill, it looks as though we are obliged not to kill Bessie and we are obliged not to kill Ludwig. This conclusion would make most philosophers wilt. It can be avoided by

---

[8]Here Johnson is employing a sense of 'interest' first noted by Tom Regan in Regan 1976b. Regan invokes this sense of 'interest' only to show that on McCloskey's conception of a right, it is logically possible that animals have them.

rejecting the premise that a being's having an interest in continued life is sufficient for imposing on us an obligation not to kill it.

The general principle, that a being's interest in something in this weak sense is sufficient to impose an obligation on others not to take it, is clearly indefensible. Exxon has an interest in maintaining its ownership of a substantial portion of America's energy. I have an interest in continuing to drive the BMW that I stole last night. Yet the fact that Exxon and I have these interests does not imply that anyone is morally obliged to respect them.

Perhaps the principle should be amended in the following way: A being's interest in something is sufficient to impose an obligation on others not to take it, on the condition that the being has a right to the thing in question. This version is plausible, but here plausibility is purchased at the price of circularity. How can we determine what things a being has a right to without determining what things we are obliged not to take from him? Until this question is answered, the revised principle remains unilluminating.

One critique of HTS has been examined and found wanting, but no argument in favor of HTS has yet been considered. Peter Singer gives an argument in favor of HTS. Here it is:

> Classical hedonistic utilitarianism does not support this distinction between personal and impersonal life, but a different variety of utilitarianism, preference utilitarianism, does (Singer, 1979, p. 151).
>
> This version of utilitarianism is sometimes known as 'economic utilitarianism' because it is the form of utilitarianism used by economists who work in the area known as 'welfare economics'; but a more accurate name would be 'preference utilitarianism' (Singer, 1980, p. 238).[9]
>
> If I imagine myself in turn as a self-conscious and a merely conscious being, it is only in the former case that I could have a desire to live which will not be fulfilled if I am killed. Hence it is only in the former case that my death is not balanced by the creation of a being with similar prospects of pleasurable experiences . . . Self-conscious beings therefore are not mere receptacles for containing a certain quantity of pleasure and are not replaceable (1979, pp. 151–152).

Classical hedonistic utilitarianism permits the replacement of any being, so long as the balance of pleasure over pain in the universe remains the same or is made greater. That is clear enough. What is unclear is why

---

[9]The appeal to a 'form of utilitarianism used by economists who work in the area known as "welfare economics" ' is quite misleading. Welfare economists disagree about the characterization of preference, about the possibility of interpersonal utility comparisons, and even about the measurability of utility. On the former point, contrast J. R. Hick's account of preference (1969) with that of von Neumann and Morgenstern (1946). On the latter two points contrast the ordinalism of K. Arrow (1963) with the cardinalism of J. Harsanyi (1976). A good account of all these issues can be found in Majumdar 1958.

Singer believes that preference utilitarianism permits replaceability for simple beings but forbids it for persons.

Perhaps he has the following picture in mind. Rudy, a person, is the only being in the universe. At time t_1 Rudy's utility score is $+5$. At time t_2 Rudy is replaced by another person, Rosie. At time t_2 Rosie's utility score is $+5$. Since Rudy is a person whose preference for continued life has been frustated, his death is an evil of say -5. The replacement of Rudy by Rosie has resulted in a net reduction in utility from $+5$ to 0. Thus from a utilitarian point of view, it would be wrong to replace Rudy with Rosie. But now consider a universe in which there is only a simple being, Bessie. At time t_1 Bessie's utility score is $+2$. At time t_2 Bessie is replaced by another simple being, Barney. At time t_2 Barney's utility score is $+2$. Since Bessie is a simple being with no preference concerning her continued existence, her death is not an evil, but only the extinguishing of her utility score. The replacement of Bessie by Barney would not result in a change of total utility. Thus from a utilitarian point of view, the replacement of Bessie by Barney would be a matter of moral indifference. A generalization of this argument would seem to show that it is wrong to replace persons, but not wrong to replace simple beings.

Perhaps this is the picture that Singer has in mind. It is, however, just a picture and not an argument. It does not show that it is permissable to replace simple beings, but not persons. It does not even show that preference utilitarianism permits the replacement of simple beings, but not persons.

In order to generate the conclusion that persons are not replaceable, we need to suppose that the value of satisfying preferences is roughly equal for all persons. Look again at Rudy and Rosie. At time t_1 Rudy's utility score is $+5$. At time t_2 Rudy is replaced by Rosie. Suppose, as we did before, that the death of Rudy is an evil of -5. But now suppose that at time t_2 Rosie is racking up a utility score of $+11$. Perhaps Rosie stands to Rudy in roughly the same relation that a person stands to Mill's pig. Perhaps Rosie's satisfactions are more sublime, more complex, and so forth. In this case replacement of Rudy with Rosie has increased net utility from $+5$ to $+6$. Thus, from a utilitarian point of view, it would be permissible and perhaps obligatory to replace Rudy with Rosie (see Lockwood, 1979, pp. 162–163).

In order to generate the conclusion that the replacement of simple beings is permissible, we must suppose that the killing of a simple being is merely the extinguishing of a utility score rather than the chalking up of a negative one. Contrary to this, one might argue that the killing of a simple being is of negative utility because it is the destruction of a life, or because it is the destruction of a consciousness, or because it is the destruction of a unique entity, or for some other reason.

Look again at Bessie and Barney in the light of these considerations. At t_1 Bessie is scoring $+2$. At t_2 Barney is scoring $+2$. If the killing of Bessie is counted an evil, and scored at say -1, the replacement of Bessie with Barney is impermissible because it diminishes utility from $+2$ to $+1$. On this view, killing persons is *ceteris paribus* worse than killing simple beings; yet it is *prima facie* wrong to kill either persons or simple beings.

Other variations can be constructed without much difficulty. Suffice it to say that preference utilitarianism is compatible with virtually any view concerning the morality of killing. The hard work is in motivating utility assignments for satisfying and frustrating preferences, and for killing.[10] Until that hard work is done, Singer's preference utilitarianism remains a mere hint of a theory, rather than something more substantial.

We have looked at one failed critique of HTS as well as a faulty argument on its behalf. Now I want to argue, contrary to HTS, that we are under a *prima facie* obligation not to kill even simple beings.

When a person dies, more than mere life has been lost; that which makes life valuable has been lost as well. What makes life valuable? The beginnings of a plausible answer are provided by L. W. Sumner (1976, p. 159)[11]:

> The value of life is therefore to be sought in two things: pleasure (or at the very least, freedom from suffering) and the fulfillment of goals which the agent himself believes to be important.

There is much that is puzzling in this passage. Sumner seems to oppose pleasure to suffering, yet the natural opposition is between pleasure and pain. Simple beings can experience pleasure and pain; only persons can suffer. Suffering requires a conception of one's self as a distinct being that persists through time. To suffer is to be deeply harmed, and to know that one is being deeply harmed. But even if we revise Sumner on this score, difficulties remain concerning the relation between pleasure and freedom from pain. It is clear why a classical utilitarian would consider pleasure one of the goods that makes life valuable. On the classical scheme, pleasure either is utility, or is closely associated with it. It is not so clear, however, why a classical utilitarian would count freedom from pain as one of the goods that makes life valuable. We might well expect that an experi-

---

[10]J. Broome (1978) has argued that no reasonable view concerning the causing of death follows from any version of cost/benefit analysis because of insoluble problems in trying to value a life. I am indebted to James Buchanan for this reference. See also my unpublished paper, "Valuing Lives".

[11]I am using 'the value of life' as a rough semantic equivalent of 'that which makes life valuable.' I am aware of the fact that there are contexts in which these expressions are not even rough semantic equivalents.

ence that involves neither pain nor pleasure would be assigned a value of
0 by a classical utilitarian. Sumner, as we have revised him, evidently
believes that freedom from pain should be assigned a positive utility
score.[12] Why?

I believe that Sumner has gotten hold of an important insight, though
it is not expressed clearly in the quoted passage. Pleasure and the
fulfllment of one's goals are important goods that make life valuable.
Consciousness is another important good that makes life valuable.

This is not to say with Epicurus and Moore that what is good necessarily involves consciousness.[13] Rather, on the view that I am urging,
consciousness itself is a good.

The classical utilitarians often spoke as if the value of an experience
is just the amount of pleasure (or happiness) or pain (or unhappiness) encompassed by it. Pleasant experiences are valuable; painful ones are not.
Down this path lies one of the famous quandries of classical utilitarianism: Why not euthanasia for the slightly unhappy?

But a confusion lurks here. Suppose that Masami is lying on the
beach, soaking up the sun. Masami's consciousness can be distinguished
from the object of his consciousness, and both can be distinguished from
the pleasure (or happiness) or pain (or unhappiness) encompassed by the
experiences.[14] Once these distinctions have been made, the value of
Masami's consciousness can be distinguished from the value of the object
of his consciousness, and both of these can be distinguished from the
pleasantness or unpleasantness of the experience.

There is a grammatical form that serves as a useful guide in distinguishing consciousness from its objects. In sentences of the form, ''α is
conscious of P,'' often the expression that replaces 'α' can be thought of
as referring to a subject of consciousness, and, often the expression that
replaces 'P' can be thought of as referring to an object of consciousness.

Making these distinctions does not commit one to a ''causal'' theory
of pleasure, nor to a ''sensation'' theory of pleasure, nor to the currently
fashionable ''adverbial'' theory of pleasure.[15] What making these distinctions does do is to allow us to see something that has often been overlooked in the utilitarian tradition.[16]

[12]T. Nagel hints at a similar view in the third paragraph of ''Death,'' reprinted in
Nagel, 1979.

[13]A caveat: Moore grudgingly admits that some things are of ''negligible value'' independent of any consciousness of them. See Moore, 1903, Chapter 6. For the views of
Epicurus, see the selections reprinted in Melden, 1950. Discussions with Jan Narveson
prompted this paragraph.

[14]Henceforth I use 'pleasure' or 'pleasantness' to refer to whatever the hedonic calculus is supposed to measure.

[15]For a survey of some theories of pleasure, see D. Perry, *The Concept of Pleasure*
(The Hague: Mouton and Co., 1967).

[16]Sidgwick's position is close to mine, however, He distinguishes consciousness from
its ''objective conditions'', but argues that only ''desireable consciousness'' is an ultimate
good. See Sidgwick, 1907, Book 1, Chapter 4, and all of Book 2.

My claim is that consciousness is itself a good, whatever its object, and whatever the pleasantness of a particular experience. This explains why most of us most of the time would rather spend a day slightly depressed than a day drugged into unconsciousness. Of course this is not to deny that there are funks so black and holes so deep that unconsciousness is a sweet release. The good of consciousness can be overcome by the evil of its object, or by the unpleasantness of a particular experience.

That consciousness is itself a good and that we implicitly recognize it as such, explains (at least part of) what is right about preference utilitarianism. Most of us most of the time prefer continued life. As to the reason, preference utilitarianism remains silent. Yet the reason is obvious. We prefer continued life because we prefer consciousness to its extinction.

That consciousness is itself a good and that we implicitly recognize it as such, also explains why most of us want consciousness to persist even after we are dead. Jonathan Bennett echoes the sentiments (1978, p. 66) of most of us when he claims that he is

> . . . passionately in favour of mankind's having a long future, and not just because of the utilities of creatures who were, are or will be actual.

Bennett denies that this "attitude," as he calls it, is principled, moral, or esthetic. Yet he claims that he would work, suffer, and fight for the continuation of the species. Rather than supposing that Bennett's attitude has some strange and foreign logical status as he seems to suppose, it is easily explainable on the assumption that consciousness is itself a good, and that Bennett implicitly recognizes this. On this assumption, Bennett's willingness to work, suffer, and fight for the continuation of the species is founded on moral principle.

What I am saying is this: Because consciousness is itself a good, we are under a *prima facie* obligation not to kill its subjects.

Of course this does not imply that all obligations not to kill are equally stringent. In fact in my view the obligation not to kill persons is more stringent than the obligation not to kill simple beings.

But one might ask: Why are the lives of persons more valuable than the lives of simple beings? And I might answer: Because pesons and simple beings can both experience pleasure, but only persons can anticipate experiencing pleasure, and feel pleasure at the anticipation of experiencing pleasure, and so forth; and because persons and simple beings can both experience pain, but only persons can fear the experience of pain, and be ashamed of fearing the experience of pain, and so forth. In short, the mental complexity of persons counts for something because persons can have the joys and sorrows of simple consciousness and those of reflexive consciousness as well.

Where does this leave the problem of practical decision-making? How do we weigh the lives of persons against those of simple beings? I don't know. I do know, however, that any such decisions, rightly made, are excruciatingly difficult.

Classical utilitarianism provided great insight into ethical decision-making. The price it paid for the clarity of its insight was its allegiance to primitive and simplistic theories in psychology and philosophy of mind. As these theories are replaced by more complicated and more adequate ones, simple utilitarian decision procedures become increasingly scarce. Such is the price of truth.

Finally, I want to return to the beginning and say something about consciousness, simple and reflexive. I have assumed that some beings have simple consciousness and others have reflexive consciousness. Indeed, this is the common conception. But how can we tell whether a particular being has simple or reflexive consciousness? Is the adoption of an efficient means for the pursuit of a desired end a mark of reflexive consciousness? If so, praying mantises have reflexive consciousness. Is the use of language such a mark? Zoological and psychological studies suggest otherwise (see Griffin, 1976). Perhaps observing the distinction between an object language and a meta-language is such a mark. But if we suppose it is a necessary condition, what are we to say of Russell?

Although it is clear that there are degrees of consciousness, it is not so clear that there is a distinction between beings of simple consciousness and those of reflexive consciousness. Perhaps there is a deep truth here that Kant should have expressed. Perhaps simple consciousness, however dim, presupposes reflexive consciousnes, however slight. Perhaps. But further investigation awaits another occasion.[17]

[17]Many friends and colleagues have assisted in the preparation of this paper. I would like especially to thank Peter Singer for sending me advance copies of forthcoming publications, and for attempting to clarify has views in conversation. Most of all I would like to thank Tom Regan with whom I have discussed these issues for years, and who made careful comments on an earlier draft of this paper. Thanks also to David Rosenthal, who pointed out some mistakes in the penultimate draft.

Interspecific Justice and Animal Slaughter

Donald VanDeVeer

One should not be misled, then, by the somewhat unusual conditions which characterize the original position. The idea here is simply to make vivid to ourselves the restrictions that it seems reasonable to impose on arguments for principles of justice, and therefore on these principles themselves. Thus it seems reasonable and generally acceptable that no one should be advantaged or disadvantaged by natural fortune or social circumstances in the choice of principles. It also seems widely agreed that it should be impossible to tailor principles to the circumstances of one's own case.

Rawls, 1971, p. 18

Surely every sentient being is capable of leading a life that is happier or less miserable than some other possible life, and therefore has a claim to be taken into account.

Singer, 1975, p. 251

Recent accounts of how we may or ought to treat nonhuman animals tend to be troublesome in three respects.[1] First, they tend to disturb our dogmatic slumbers on the question of how animals ought to be treated. As many have said, to whom I have recommended the reading of Singer's *Animal Liberation,* "I'd better not; I like steak too much." It is surprising that the parallels are not recognized, e.g., an antebellum, American plantation owner saying "I'd better not read Harriet Beecher Stowe's *Uncle Tom's Cabin*; I like the profits derived from slaves too much." So much for the first way in which recent accounts of animal treatment are troublesome. Of greater philsophical interest, such accounts tend to defend the "equality of all animals," an "equal consideration of interests," or an equal assignment of rights to certain animals if they are to be assigned to human beings (Singer, 1975, p. 23). When push comes to shove, in interspecific conflicts of interest, such accounts often sanction limited prefer-

[1]For example: Singer 1975 and Regan 1975.

147

ential treatment for humans. At some points Singer seems to suggest that the lives of animals are *as important* as those of humans. At one juncture, he speaks pejoratively of "the view that human lives are *more important* than the lives of animals—a view the flesh-eater would surely accept" (my italics) (p. 242). He describes the claim that "humans come first" as the assumption hardest to "overcome" in arousing public concern about animals and as being an "indication of speciesism (p. 232). These remarks suggest that he is opposed to the view that the interests of humans take precedence over comparable interests of animals. Singer's strong remarks here appear to be in some tension with his admission elsewhere that we "may legitimately hold that there are some features of certain beings which make their lives more valuable than those of other beings . . . . " (p. 20). My later remarks attempt to show that this last claim is correct and compatible with the view that "human lives are more important than the lives of animals"—though generally and contingently, and not necessarily. Thus, I attempt to develop a view that Singer recognizes, but does not himself explore; as he states "—a rejection of speciesism does not imply that all lives are of equal worth" (p. 21). Why such a non-egalitarian view is reasonable and how it is to be reconciled with the mentioned egalitarian prescriptions is not made as clear as one would like. Although this is a philosophically unsatisfactory state of affairs, it is not surprising in initial attempts to reconsider an enormously neglected issue and one that, in its own right, presents recalcitrant difficulties (like other cases concerning permissible treatment of those falling outside the domain of existing paradigmatic members of Homo sapiens, e.g., abortion and obligations to future generations). A third troublesome set of features involves the not entirely satisfactory appeals to intuitions about rights and presumptive duties, what counts as "unnecessary killing" and "needless suffering," and the considerable gap between a utilitarian ideal of miminizing suffering and the rather specific exhortation that, e.g., vegetarianism is morally obligatory (at least for the agriculturally affluent) (Singer, 1975, pp. 22, 244). In short there is a need for a more abstract, systematic account that will provide an impartial (non-anthropocentric) basis for adjudicating interspecific conflicts of interest. Such an account is, predictably, hard to come by, and the reader will not find a fully developed one here. Still, I shall consider an approach that will either take a few steps in that direction or, if it fails, may constitute an instructive failure. It purports to provide *part of* what may be called an interspecific theory of justice, and sets forward a justification for *both* a condemnation of many of our current practices involving animals as well as approval of a (limited) subordination of their interests to those of Homo sapiens (in familiar cases).

In Part I, I suggest a basis for discounting the value of the lives of many animals. In Part II, I propose the use of a modified heuristic device, a variant on John Rawl's (1971) employment of the "original position" in order to find a neutral basis for ascertaining what principles might be chosen to regulate interspecific social interactions. In Part III, I make use of two principles (whose reasonableness emerges in Part II) to assess the widespread practice of breeding animals for slaughter. There I also consider some objections that I believe would be posed by Peter Singer.

I. THE VALUE OF SENTIENT LIFE

It is hard to make sense out of talk about the "infinite value of life" or even the "infinite value of human life." Is there any reasonable way of assessing the value of human life? With regard to human life this is both a theoretical and a practical problem. It is a practical one in that answers to certain practical questions hinge on it. For example, how far ought people to go in taking risks or expending resources to save another human life? In addition, judges and juries are often required to set compensation for the estate of a decedent in cases of wrongful killing. In one recent civil liability suit (1978) parents were awarded 821,000 dollars in cash and monthly payments (to be annually increased) as a settlement for the negligence of swimming instructors, the result of which was the semi-comatose condition of the parents' five-year-old daughter. Should the girl live to be seventy-one the total payment will be 147 million dollars.[2] The basis for determining the sum to be awarded was not reported. In other cases one approach that is taken is to calculate the loss of the stream of income that a family would have enjoyed had it not been for the death of the breadwinner. To make the family as well off, financially at least, as they were before requires transferring to them a sum equivalent to the foregone income. The value of the decedent's life then, prior to his or her death, may be thought to be equivalent to that sum. Such an approach, whatever its merits as a practical rule of thumb, is troublesome. It suggests that the life of an executive making two hundred thousand dollars year is worth twenty times that of a secretary making ten thousand dollars a year, other things remaining constant. Further, such an approach tends to devalue the lives of those whose work has no market determined price, e.g., a housewife (or a "househusband"). More bizarre yet, it suggests little or no value for the lives of the extremely retarded or the retired who earn no income. In any case such an approach seems to focus, in large part, on the value of a life *for others*—as determined by wage and market considerations. More promising is the proposal that we try to ascertain how valuable a person's life is *to that person*. That is, what sort of value does one place on preserving and extending one's own life? On this view, we may ask a person how much life insurance he or she is willing to purchase. However, since death will exclude *the insured party's* receiving

---

[2]The case was reported in the *Raleigh News and Observer* (May 18, 1978), p. 29.

any compensating benefits, this question cannot be a reliable guide to how an individual measures the value of his or her own life (as opposed to the value of its loss for others). The question of how much one is willing to insure one's eyes for might be a better clue to how much value one attaches to having one's eyes. Even so, willingness to pay for insurance would be a function of seemingly contingent factors, e.g., the relative wealth of the individual in question. A related approach is to inquire about what sort of added compensation an individual might require before taking a job involving certain risks to one's life—assuming there was a wide variety of choices, full apprisal of the risks, and so on. If we found that someone was willing to clean windows on the tenth floor of a building, as opposed to the first floor, for double the wages of cleaning on the first floor, we might be able to calculate the monetary value that person places on his or her life. Still, this approach involves assumptions of a dubious nature, and it would again provide no useful way of ascertaining what value young children or the severely retarded place on their own lives. Studies of suicides, nevertheless, suggest that many people do not regard their own lives as *infinitely* valuable and come to a point where they prefer their own death to continuation of life (Hamermash and Soss, 1974). For fairly straightforward reasons, such a choice may be rational. Suppose an individual, *A,* contemplates the choice of suicide or continuing to live. *A* may reason as follows. If I continue to live, my further life will either be one of net utility to me or my further life will be one of net disutility. The former is preferable to the latter. If I commit suicide, my life ends and there is no continuation of it that will involve utility or disutility to me—for I will not be. Still, in choosing death I will *forego* what might have been—either a continuation of life, which would have involved a net utility to me or a net disutility. Terminating my life when it would, if continued, have involved a net utility to me is worse than terminating my life and avoiding a continuation involving a net disutility. In general I have two alternatives and each has (to simplify) two basic outcomes

	Alternatives		*Outcomes*
A.	Continue to live:	1.	A further life of net utility
		2.	A further life of net disutility
B.	Commit suicide:	3.	Foregoing a further life of net utility
		4.	Foregoing a further life of net disutility

Clearly, the reasoning continues, (1) is preferable to (4), (4) is preferable to (3), and (3) is preferable to (2). If no probabilities could be assigned to the four outcomes, a maximaxer (one who believed rationality requires, in situations of uncertainty, choosing that alternative whose *best* possible outcome would be better than the best possible outcome of any other available alternative) would decide to continue to live (A); a maximin follower (one who believed that rationality requires, in situations of uncertainty, choosing that alternative whose *worst* possible outcome is better than the worst possible outcome of any other available alternative) would choose suicide (B). Of course, there is some controversy as to whether the maximax or the maximin principle is a proper criterion of rational decision-making. This problem may be circumvented by our contemplator of suicide, A, who may reasonably believe that if one continues to live, outcome (2) *is* far *more probable* than (1). If that is so, and given that (3) or (4) is preferable to (2), it is not unreasonable to choose B and commit suicide. In some cases it may be an unfortunate truth that a continued life is not worth living, that no life continuation is preferable to a continuation involving a serious net disutility. It, thus, may be *reasonable* for A to commit suicide even if A may fail to carry out the requisite calculations, which, if they were carried out, would dictate suicide. In real life, of course, it may be a rare occurrence for anyone to be able reasonably to assign probabilities to outcomes in the manner supposed. It is certainly intelligible, however, that someone faced with life imprisonment, with no prospect of parole and little chance of escape, might reasonably suppose suicide is the proper choice. Much the same might be said of someone seriously impaired by disease or injury and who had little prospect of regaining capacities essential for a minimally decent life.

There are compelling reasons, then, when we focus on the value of continued life to a given individual—as expressed by actual choices and hypothetical rational choices under certain conditions—to conclude that the value of continued organic functioning is not ''infinite,'' but is rather a function of its prospective quality or prospective satisfactions. There are things worse than death, e.g., a life of continued dissatisfactions not counterbalanced or outweighed by satisfactions. This conclusion is not unfamiliar, but is one incompatible with the claim, often used as a weapon in debates over abortion and mercy-killing, that life is ''sacrosanct'' or of ''infinite value.'' It seems a not unimportant truth that the value of continued life is a function of prospective satisfactions (or their absence) and that this cnclusion is reached by considering the way in which a rational individual might choose between alternatives involving life or no life for himself. After all, judgments of *others* regarding the value of A's life might be arbitrary and based solely on the value of A's continued existence *for them*.

It should be noted that the discussion thus far has focused on considerations relevant to arriving at some understanding of the idea of rationally assessing the value of a *continuation* of a life of a certain type of being with some sort of post-natal life history behind him (or her).[3] I wish to now shift our point of view to another hypothetical perspective,

---

[3]To avoid tedium, I sometimes use gender tied pronouns but need not. Context should make this clear as well as the fact that later 'him' or 'her' may refer to nonhuman beings as well.

namely, one where a rational being is contemplating, prenatally or prior-to-conception as it were, a choice between no life at all (not even the beginning of such) or, on the other hand, the living of certain possible lives.

II. THE PRE-ORIGINAL POSITION

With the aim of trying to formulate principles of just treatment toward *animals* or, more accurately, just interaction between human and non-human sentient beings, I propose a thought experiment that involves a non-trivial modification of the one proposed by John Rawls in *A Theory of Justice* (1971, Ch. 1). To avoid excessive repetition I shall assume some familiarity with Rawls' approach to ascertaining principles of justice, in particular, his heuristic appeal to the situation of the original position. Still, a summary account will be useful. Rawls asks us to suppose that a number of rational, self-interested persons are to propose and decide on principles that would determine and regulate the design and function of the basic institutions of a society (e.g., its constitution and basic political institutions) in which they will be participants.

The "pre-societal" situation in which these principles are to be determined is dubbed "the original position" and we may refer to *its* participants as "participants OP." Individuals in a society in accord with the chosen principles of just interaction we may refer to as "participants JS." Participants OP, by stipulation, know general facts of human history, psychology, economics, and so on, but they are under what Rawls characterizes as a "veil of ignorance" for they do not know what their sex, social or economic position, or race will be qua participants JS. Nor do they know into which generation they will be born or the effects of the natural lottery on them, e.g., whether they will possess the capacities of Einstein, Woody Allen, Joan of Arc, or a Downs Syndrome or microcephalic human being. They do not know what particular conception of a good life they will have as participants JS, only that they will have one. Ignorance about such individuating features is to serve as a guarantee, on Rawls' view, that they, as participants OP, will not be able to choose principles whose effects would be to create special advantages for themselves qua participants JS. The principles they choose will then, it is presumed, be *impartial* in assigning rights and duties or allocating primary goods among participants JS. Given the self-interest of participants OP, neither will the principles they choose exhibit *indifference* to the well-being of participants JS.

For our purposes here, it is worth noting that participants OP *do* know, in Rawls' scenario, that as participants JS they (1) will live, and (2) will be human beings (and, presumably, will be sentient at least for some period of time). It is possible to imagine a situation like Rawl's original position in which the veil of ignorance is even greater. Suppose that the participants OP do not know as much about their own "genetic wiring," namely, they do not know that they will be human beings, but only that qua participants JS they will be *sentient creatures*. So, participants OP are contemplating a society in which they might occupy the "position" of being a normal or subnormal human, cow, pig, and so on. In passing, someone might suggest "why not go further and allow the possibility of being an orchid or poison-ivy—since we can sensibly speak of their well-being as well?" I want to exclude this possibility since, while a situation might be "in the interest of" a plant, only sentient creatures are capable of satisfaction or dissatisfaction.[4] It is not obvious that self-interested participants OP would have any reason to care about the well-being of entities incapable of satisfaction or dissatisfaction (for their own sake). I shall suppose, then, that participants OP will be concerned about the interests of *sentient* creatures JS since that is a position they may occupy. Indeed, if Rawls's strategy in designing the original position is to have a veil of ignorance sufficient to preclude its participants from choosing principles arbitrarily favoring themselves, and hence preferential to some relevant group to which they might belong, why should the veil *include* knowledge of species membership (in Homo sapiens)—when it is unreasonable to claim that many nonhumans are incapable of satisfactions and dissatisfactions?[5] So, a more impartial thought experiment may require the more extensive veil of ignorance I am proposing and be a part of what it is tempting to call the "pre-original position." The original position abstractly recognizes the varying effects of the natural lottery on those participants JS who will have, at least, a human genotype; our pre-original position goes a step further (or "backward") and recognizes the varying results of the natural lottery on participants JS who will have, at least, a genotype for a sentient being.

What sorts of considerations would receive attention among the rational, self-interested participants in our pre-original position? One would

[4]Is it odd or false to claim that something is in a plant's interest, i.e., conducive to its well-being? I do not see why. A remark of Carl Sagan seemed natural enough: " . . . it is in the interest of the orchids not to be consumately attractive . . . " To find out why, see Sagan 1977, p. 71.

[5]The contractors in the Original Position, as Rawls describes them, have sophisticated intellectual capacities. Since this is so it is not possible for some humans to "enter into" such a position, e.g., the severely retarded. Similarly, neither could nonhuman animals. Further, participants in the Original Position cannot assume they shall have, as participants JS, the capacities of contractors in the Original Position. After all, the natural lottery may allot them the position of being severely retarded. So participants in the Original Position must choose principles governing interaction among participants JS, some of whom may be unable to reason about or have a "sense of justice." If this is so, should they not consider the possibility of occupying the role of creatures with capacities for satisfaction (even if lacking a capacity to enjoy all primary goods, e.g., self-respect) and even fewer capacities than retarded humans, e.g., many animals. The above is the sketch of a reply to a possible objection of the sort espoused by David A. J. Richards (1971, p. 182).

surely be the recognition of the brute good or bad fortunes imposed by the natural lottery. Here I have in mind primarily the effects of the genetic lottery as opposed to post-natal natural fortunes or misfortunes. Participants in the pre-original position (''participants PP) might hope to be beings JS with the genotype of beings with highest capacities for satisfaction, i.e., (normal) Homo sapiens. But they have no choice about this and, following Rawls, do not know the likelihood of becoming Homo sapiens or that of belonging to some other species. Supposing that the set of the least advantaged members of a just society consists almost solely of the genetically least capacitated sentient beings, it is doubtful that they would choose a principle to regulate the distribution of basic goods that would bear much resemblance to one of Rawls' principles, namely, The Difference Principle (which prohibits inequalities unless they redound to or maximize the benefits of the least advantaged). This last claim could stand more elaboration than I wish to give it here. However, it would be clear that the range of satisfactions obtainable by certain sentient creatures (e.g., rabbits, chickens, or pigs) is radically limited by the genetic lottery. Hence, there would be severe limits on possible attempts at having social arrangements that would increase the level of satisfaction for certain creatures above a certain point (a rather low one from the standpoint of a typical member of Homo sapiens; perhaps, an even lower one from an angelic viewpoint).[6] Somewhat similar remarks (which suggest *other* worries) might be made about the plight of Tay-Sachs or microcephalic members of Homo sapiens). Further, the costs, to beings with greater capacities, of improving the lot of the genetically least advantaged, beyond certain levels, would be enormous (imagine ''medicare'' for squirrels). Given the possibility of becoming a member of any species JS, including Homo sapiens, and the enormous losses of satisfactions that would be required for beings capable of comparatively great satisfactions, it is more likely that participants PP would not choose anything like The Difference Principle. Overcoming the inequalities of satisfactions among members of different species would be too costly, and might further generate a radical inequality of sacrifices to be made, in that rational, productive beings would have obligations to underwrite the well-being of many sentient creatures incapable of voluntarily acting in some sort of reciprocal fashion. Setting aside this rather fantastic option, it seems, then, that *within certain limits* participants PP would choose to accept their genetic lot. There is no effective way of raising the level of a satisfied pig to that of a somewhat dissatisfied human being (and, if I understand him correctly, J. S. Mill was right about which is the preferable position).

[6]Some genetic defects are remedial to an extent and there is now the prospect of alleviating many monogenic diseases. Still a healthy pig will not have the capacities of a human or a chimpanzee. Some genetic alterations may mitigate inequalities among a species (e.g., mitigating human dwarfism by gene implantation) without mitigating interspecific inequalities.

More positively, it would seem that participants PP would recognize, as suggested earlier, that the life of a human or any sentient being is not necessarily worth living, that the value of some lives is so low as to make death preferable. Whether this is so depends partly on the irremediable effects of the natural lottery (not *all* its effects are irremediable) and partly on matters that can be socially controlled by deliberate choice (depending on the availability of resources as well). Focusing on these latter considerations, participants PP might find it reasonable to accept a principle regulating social arrangements that required, so far as possible, that no sentient creature should have forced upon it (by deliberate act of a rational being) treatment that makes its life not worth living, i.e., a life such that no life at all is preferable, on balance, for that creature—unless such treatment is a defense against unprovoked threat (a factor introducing complications I wish to set aside here). Let us call this the Life Preferability Requirement. Such a principle seems modest and would be, presumably, only one of a number of principles that might be accepted by participants PP. It may be understood as generating a set of negative duties, duties to refrain from treating sentient creatures in certain ways. Further, such a principle might be part of an ordered set (as are Rawls' principles) and may not be highest ranking, or may be subject to a weighting of other relevant considerations. I wish to leave many of these very difficult questions open. Nevertheless, it is plausible that participants PP would accept such a principle. Their rational self-interest, along with recognition of irremediable genetically based inequalities, would seem to require guaranteeing the opportunity, free of direct destructive (or disutility promoting) interventions of rational beings, to live a life preferable to no life at all. The more specific implications of such a principle remain to be examined. Such a principle, however, does not itself *require* positive intervention to rescue or aid sentient creatures, e.g., from predators or disease. Participants PP, being self-interested, would want this bare minimum, even though the vicissitudes of the genetic lottery and post-natal natural fortunes would limit the possible satisfactions of any participant JS and, perhaps unavoidably, render the life of such a participant not preferable to no life at all. There should be a social arrangement designed to restrain any rational being (one capable of reasoned voluntary restraint) from making the life of another sentient being not preferable to no life at all. Policies in accord with only this principle would not necessarily be just, but if participants PP were to accept such a principle it is plausible to believe that it would be a necessary condition of just interaction among sentient beings. For the reasons mentioned it seems that the rational, self-interested participants PP would accept such a principle.

The Life Preferability Requirement, as stated, would place constraints on what may be done to *existing* sentient creatures. The reasons participants PP would have for accepting it would require the further acceptance, I propose, of a related principle that I shall call the Creation

Requirement. The Creation Requirement is that no rational being should deliberately *cause to exist* a sentient creature when it is certain or highly probable that such a creature would have a life not preferable to no life at all. Just as an existing being would (if it could rationally decide) prefer no further life to a further life of serious net disutility, so would (I hope this does not impose an excessive tax on the reader's imagination) a rationally empowered gamete prefer not to come into being as a sentient creature if it is certain or highly probable that its life prospects were such that no life at all was preferable.[7]

This principle might justify, or be part of a justification for, avoiding conception of a human zygote when the prospect is that it would, on developing naturally, become a most seriously defective human being, e.g., an anencephalic, a microcephalic, or Tay-Sachs infant. Further, it provides a compelling reason to abort fetuses with such defects—on the assumption that the prospective lives of such beings would be such that no life at all is preferable.[8] In spite of controversies over abortion, such an implication is not at all implausible (and even those generally opposed to abortion frequently allow, often paradoxically, that "exceptions" are permissible in cases such as these). The Creation Requirement has implications also about questions regarding the breeding of animals for the purposes of experimentation and provision of food for other sentient creatures. And it is here—finally—that our prior theoretical wanderings begin to link up more directly with current disputes about proper and just dealings with sentient creatures outside of Homo sapiens.

III. HOMO SAPIENS AS KILLERS AND BREEDERS

Many animals substantially lack capacities requisite for important satisfactions, some of course more than others. Hence, in many cases an animal's dying does not involve a *foregoing* of substantial satisfactions that a continued life sometimes involves. Thus from the standpoint of the *opportunity cost* imposed on an animal by its being killed, the cost is greater, presumably, for a chimpanzee than for a chicken (at, say comparable stages in their prospective life spans). This fact might well be recog-

---

[7] I am not sure that my thought experiment is more taxing than that of Rawls. If it is, we must contemplate the White Queen's advice to Alice's complaint about believing impossible things, "I daresay you haven't had much practice . . . sometimes I've believed as many as six impossible things before breakfast" (Carroll, 1872, Ch. 5).

[8] I shall not pursue them all, but the Creation Requirement has implications for not only (1) responsible human parenthood, (2) the breeding of animals, but also (3) the claim that God is beneficent—indeed, if an omniscient God foresees that some will be eternally damned and, thus, no life at all is preferable, the Job-like complaint arises "why was I allowed to be born?"

nized by participants PP and it suggests that there is no compelling reason to adopt principles that would prohibit the killing of animals with meager capacities for satisfactions, especially after they have lived a substantial part of their normal life span (or, possibly, even earlier). There is then a basis for thinking that participants PP would not choose principles prohibiting actions (such as killing certain creatures) when the "losses" to the animal would, for reasons mentioned, be comparatively small—especially when nontrivial goods would accrue to other creatures on doing so and the participants are ignorant of the probability of assuming the role of any particular type of sentient being, qua participants JS.

People, of course, do not only kill other animals; they bring them into existence with the *express intention* of killing them at a later point and making use of their bodies before or after the killings. Breeding animals for such purposes is, then, different in a key respect from killing animals who live in the wild. As many are inclined to say "if it had not been for human beings, many animals would not exist at all." The usual suggestion implicit in such a remark is that, hence, since such animals would not have had the blessings of *any* satisfactions whatsoever had we not caused their existence, there are no moral constraints (or virtually none) on what we may do to them. Humans may say "only we give life, so only we can take it away" or "since we created these lives, we can take them away whenever we choose." In some respects, breeders play the role of God to those bred.

Are there, then, no constraints on what breeders of animals may do to the animals they have bred? An affirmative answer is problematic. If we bred, a la Jonathan Swift, human babies for purposes of organ transplantation or increased food supply, such an argument would be regarded suspiciously. Similarly, if we said of a starving, abandoned human infant whom we had saved that "if it had not been for us, you would not exist; hence you have no legitimate complaint about our later employment of your parts in a recipe for Kentucky Fried Homo sapiens," many would have reservations. These examples, however, focus on our judgments about breeding or using *human beings* for certain purposes, and part of our repulsion may result from the nature of the being exploited, in part icular, the magnitude of the loss (the foregone satisfactions) caused to *humans* when they would be so treated. A tacit assumption here may be that the humans in question are relatively normal. That is, they possess capacities like those of paradigmatic humans. So one might (and, I think, ought to) feel repulsion about such treatment of any sentient creature with capacities similar to more or less normal humans. I have suggested that the magnitude of the loss (imposed by killing) to many animals would be far less. So, it is not obvious that it is wrong to kill animals (many) for such purposes. Still, if it is not wrong, it seems doubtful that it is not wrong *because* many animals would not exist at all if it had not been for the breeder's causing them to exist. Being causally responsible for the existence or continued existence of another creature does not seem to be a fact in virtue of which one has a moral license to either (a) kill it wh enever it suits one's purposes or (b) impose whatever deprivations on it which one chooses.[9]

I have argued, albeit sketchily, for the reasonableness of the Life-Preferability Requirement and the Creation Requirement, and such principles do entail (with other assumptions) constraints on breeding. Further, such principles rightly recognize that some types of life are not preferable to no life at all. Breeders exercise control over the type of life that the animals they breed will live as well as its duration. It is wrongheaded to loosely generalize here about "animals raised for slaughter" since the kinds of lives they enjoy or suffer varies enormously with the way they are bred. As Peter Singer observes, pigs, chickens, and cattle had, on old-fashioned, more traditional styles of rearing, room to move about in the open air, access to roughage, sunshine, more natural foods, and less crowded conditions than do such animals when raised under, increasingly common, intensive rearing or "factory farming" practices. Among animals intensively reared, the situation of calves raised for veal is striking.[10] It is now possible to raise calves until they weigh three hundred and twenty-five pounds and retain their pale, tender flesh (for which there is considerable consumer demand, and, hence, a high price). To do so a calf is kept in a stall about two by four and a half feet tethered by a chain around its neck to prevent it from turning around (to prevent the development of tough muscle). The stall is without straw since if the calf ate it, its flesh would turn darker. It is fed an all liquid diet. The calf is deprived of virtually all iron in its diet, for what is desired is pale and tender flesh. Hence, no grass is permitted; nor are metal stalls which the calf might lick. Barns are kept warm to prevent the calf from becoming cold and, hence, burning more calories. To reduce restlessness among calves deprived of any stimulation barns are commonly kept darkened. The calf, during its thirteen to fifteen week life, does get to go outside—at least once—on its way to slaughter. The actual situation seems more bleak than that briefly conveyed here. Those interested should consult Singer's account.

For the reasons mentioned it would seem that the life of a veal calf under intensive rearing conditions is not preferable to no life whatsoever. If, as argued, participants PP would accept the Creation Requirement,

[9]I was initially provoked into thinking about this issue by a stimulating discussion in Nozick, 1974, pp. 35–41.

[10]My comments on the treatment of calves raised for veal is drawn from Singer, 1975, pp. 121–128. It may be mentioned here that I make little attempt in this essay to convey the extreme deprivations suffered by animals raised for food and other purposes. Anyone not familiar with such matters would do well to read *Animal Liberation* rather than this essay, which is substantially motivated by concern over questions about practices well-described by Singer.

that no rational being should deliberately cause to exist a sentient creature when it is certain or highly probable that such a creature would have a life not preferable to no life at all, then a reasonable principle of interspecific justice prohibits breeding veal calves in the way commonly done currently. In view of further reasonable empirical assumptions, it would seem that the creation of and consequent radical deprivations imposed on other animals raised for both culinary, experimental, and other purposes would be unjustified. Is all breeding for slaughter unjustified? It is difficult to defend an affirmative answer. The Life-Preferability and Creation Requirements do not require such an answer. If such principles are accepted, they are, admittedly, comparatively weak and *at best* only part of a more complete conception of interspecific justice. Perhaps a more thorough elaboration of interspecific justice would prohibit all breeding for slaughter. However, this prospect seems doubtful for a number of reasons. As noted earlier the disvalue of the loss of life for many animals seems minimal from the standpoint of the magnitude of satisfactions foregone. Similarly, *some* of the confinement imposed on animals that are bred is probably not conducive to experienced dissatisfaction of the animal during its life (though some certainly *is*). Severe crowding, of course, surely results in an impediment to the satisfactions of which certain animals are capable. The conditions under which many animals are commonly raised for slaughter, however, involve not only factors that lower an animal's quality of life, but also some that improve it (by comparison with an alternative life in the wild) or generate a floor of wellbeing below which it is not allowed to fall. For example, domestically raised animals are typically protected from (1) natural predators (except their breeders, at any rate), (2) the painful deaths that they might undergo as a result of such predation, (3) serious illness, (4) death from starvation, (5) certain impairments resulting from lack of shelter, and so on. That certain domesticated animals are given certain benefits and not harmed in certain ways by their breeders *only* because of a breeder's interest in bringing to market a profitable ''product'' is no doubt true and may be lamentable; however, the important issue here is how animals ought to be treated and why; if some do the right thing for the wrong reasons, or with less than ideal motives, that is another matter. That an animal is given some quid for its quo, its deprivations and premature loss of life do not by themselves justify certain radical deprivations imposed on certain animals—any more than similar considerations would justify human slavery. Nevertheless, the issue of whether it is justified to breed animals for food is kept in a more objective perspective by keeping such facts in mind. For *some* of the burdens of domestication animals receive compensating benefits. When these securities are provided for animals, they go *some* distance toward making its life preferable to no life at all and, thus, satisfying the Life-Preferability and the Creation Requirements. That they do not go far enough by themselves is a point I have defended with regard, for example, to the case of the way in which veal calves are treated.

Given the small disvalue of the deaths of *certain* animals and on condition that the Creation Requirement is satisfied, it is not obviously wrong, in all cases, to breed certain animals for slaughter.[11] The satisfaction of these minimal constraints would guarantee that the interests of certain animals are taken into account and given weight in determining permissible treatment of them. If so, they would not be used as a *mere* means to human ends. Their interest in not dying is given some weight though not as much as that of a like interest of a (typical) Homo sapiens, but reasons for assigning a lesser weight to a comparable animal interest have been set forward—reasons not couched in terms of species membership per se.[12]

Since the view defended here conflicts with Peter Singer's account it is worth considering some objections posed by Singer to the view that it is permissible to raise animals for slaughter, even if they were not deprived as severely as they commonly are. Singer would object, I conjecture, by appealing to his principle that we have a duty to give *equal consideration* to the *interests* of sentient beings. Although his account tends to focus on an interest in not suffering, he must, I believe, assume that sentient beings have an interest in not being killed as well. It is important to be clear about what it *means* to say that we must give "equal consideration to the interests of sentient beings." Singer's principle has implications for situations involving conflicts of interests between animals and humans, and further, cases where "like interests" are at stake. First, consider cases where unlike interests are at stake. If I could save my dog's life by spending twenty dollars on medicine, my dog's interest in continuing to live is a basic one and my interest in retaining my twenty dollars is a peripheral one. The interests do not seem to be "like interests" and Singer would not, I believe, object to weighting them unequally. Conversely, Singer would view Jones' interest in eating bacon as peripheral compared to the pig's interest in continued existence or in not suffering. On his view, it would be wrong to sacrifice the pig's greater interest for a lesser interest of Jones. Such an act would fail to extend an equal consideration of interests—so it appears.

---

[11]Singer, in attempting to undermine stereotypes about wolves as vicious beasts, notes that they "almost never kill anything except to eat it" (1975, p. 235). One might compare here human *breeding* of animals for food—as opposed to, for example, the killing that occurs in recreational hunting, bullfighting, and many experiments on animals, experiments whose value is dubious.

[12]Compare Singer's remark about his book: "The core of this book is the claim that to discriminate against beings solely on account of their species is a form of prejudice, immoral and indefensible in the same way that discrimination on the basis of race is immoral, and indefensible" (1975, p. 255).

Where *A* and *B* are sentient beings and each has an interest in not dying, those interests seem to be "like interests." I would suggest that they clearly *are* like interests in a purely *descriptive* sense. The same relevant descriptive phrase may characterize the respective interests of *A* and *B*, namely, an "interest in not dying." Are they, however, *evaluatively* like interests, i.e., ought descriptively like interests be given *equal moral weight*? Is this what is required by "equal consideration" of descriptively like interests? One can give equal "consideration" of such interests in a weak sense by simply impartially *recognizing* that the interests of *A* and *B* are descriptively like interests. Singer, however, must mean something stronger; it would seem that he must mean, by "equally considering" like interests that they be given equal moral weight. An implication of such a view would seem to be this: that when other things are equal, and I and my dog are in a life raft about to be swamped, and one of us must go, there is no basis for choosing which of us should die. Perhaps I must flip a coin and if I lose I must jump overboard. What else would be fair given that our interests are descriptively like and have the same moral weight? The "equality principle," as I understand it, seems most counterintuitive here. Although Singer is not fond of anyone's moral intuitions, he at one point allows that there may be a basis for allowing that the value of the lives of different sentient beings may be unequal (Singer, 1975, p. 20). Thus, he would probably not suggest that the resolution of the life raft problem is to be found in flipping a coin. However, it is not clear that one can consistently (1) allow an assigning of unequal moral weight to the preservation of the lives in question, and (2) insist on the principle of equal consideration of like interests—as it seems reasonable to understand the latter.

Singer might try to insist that the equality principle applies only to consideration of interests other than the interest in not dying. It is not clear why such a restriction would not be arbitrary. Further, if it were allowed, it is not clear why it might not be reasonable to assign different moral weight to other pairings of like interests. There are, then, at least these difficulties for the equality principle, and it seems to me that they are not easily remediable. Much of Singer's critique of current treatment of animals might succeed if he appealed to a *weaker* principle, e.g., that we have a duty to not sacrifice *basic* interests (e.g., in not being prematurely killed, or in not suffering) of sentient beings for the sake of promoting the (comparatively) *peripheral* interests of others. For example, we ought not to kill wildebeests for flyswatters, minks for fur coats, or whales for margarine. Such a principle would not bar the door to the intuitively permissible resolution of the life raft problem. It would also provide a basis for disallowing much current treatment of animals—as well as prohibiting Swiftian breeding of human babies for the sake of satisfying the tastes of cannibalistic gourmets (belonging to Homo sapiens).

Participants PP, accepting the inevitabilities of the genetic lottery, would, I think, simply accept their genetic lot and its limits on possible satisfactions. Further, their rational self-interest would dictate acceptance of the Requirements discussed, and others no doubt. Given the possibility of becoming a member of any species, qua participants JS, they would accept principles like those mentioned, principles which identify a basis for certain duties toward animals (and others). These principles also provide reasons for concluding that much current treatment of animals *is* unjust, but *also* that there are non-arbitrary reasons for permitting certain interspecific inequalities, e.g., for permitting the raising of certain animals for slaughter under certain conditions and not extending comparable treatment to virtually any human beings. This last point, however, should provide, generally speaking, little support for public complacency about much of our current dealings with other sentient creatures.

SECTION IV

Humans and Other Animals—Linkages and Likenesses

Introduction

We are like the other animals and unlike them, tied to them and separate, in many ways. Like the others, Stephen R. L. Clark points out, we *are* animals, our nature is an animal nature, and our desires are, at least in large part, animal desires. Peter S. Wenz links us and the others as members of ecosystems, with, he argues, an obligation not to destroy such systems. Patrick F. Scanlon agrees with Clark and Wenz that we are animals, but insists that we are naturally predators, hunting animals. Duane Rumbaugh and Sue Savage-Rumbaugh report their research linking us to nonhumans in the one way Descartes thought decisive—language.

Stephen Clark contrasts utilitarian and 'Stoic' moralities on human nature and the realtion between the human and the nonhuman. Utilitarianism, of course, sets the maximization of utility (usually pleasure) as the highest good, whereas Stoicism and the Kantian and Thomist positions of which it is an ancestor stress virtue and reason. If either of these views were widely and consistently followed, human demands upon nonhuman animals would be noticeably decreased. But both views seriously distort our perception of animals, and of ourselves as animals, contributing to the stereotypes that make it so difficult to *see* animals (as Benson urged in Section I).

Stoicism rejects our animal nature and sets us apart from the 'beasts,' but in fact we *can* sympathize with our fellow animals, and naturally do so unless prevented by dogma or hardened by cruelty. (The natural sympathy of humans for nonhumans was, of course, one of Annette Baier's Humean points.) We naturally see patterns of behavior as evincing mental states, in nonhumans as in humans. Stoicism, followed to the end, becomes an ethics of 'rational' detachment that despises our animal nature. But we, as naturally rational animals, cannot reject Nature in the name of Reason.

On the other hand, utilitarianism urges upon us an equal weighting of the pleasures and displeasures of all members of all species. But this too ignores our nature, for we (and other animals) naturally are more concerned with our conspecifics. We can, Clark argues, continue to distinguish between our species and others without being oppressors and without being predators. Distinguishing (naturally) between one's own group and other groups need not involve exploiting the latter for the good of the former.

Peter Wenz's paper changes the focus of discussion somewhat, since he gives a central place to the moral importance of ecosystems rather than individuals. Most justifications for ascribing moral import to the environment are in terms of human interests, with nonhuman objects having only derivative moral value. Wenz tries to provide a 'nonhomocentric' justification for assigning value to ecosystems. Surely (the argument goes) other things being equal, a course of action that does no harm to an ecosystem is preferable to one that destroys such a system, even if the system contains no intelligent members. It is not a matter, Wenz argues, of derivative values for humans (even though we all seem impelled so to argue), but of an intuited value of the ecosystem in itself.

Arguments against hunting that rely on a moral right to life ascribed to animals have the embarrassing corollary that we should intervene to protect prey from predators. Wenz bases his argument on the claim that modern hunting is ecologically unsound, stripping ecosystems of large predators and of 'nongame' animals. Hunting by industrial humans is prima facie wrong because of its destructive effects on good ecosystems.

But we are not, Patrick Scanlon emphasizes, simply observers of ecosystems; we are participants in the global ecosystem. We consume, and thus we compete with other consumers. We consume much more than what we eat, and our wastes are of a volume and diversity unlike those of any other species. The division of labor in civilization has now brought us to the stage at which most of us no longer gather or produce our own food. Thus, though we are still predators, the majority of us are not so overtly. But even so, a substantial amount of world-wide protein consumption is from wild animals.

Regulated hunting provides meat (a renewable resource) without the further conversion of wild lands to farmland, and is a direct form of participation in the ecological system. The vegetarian, too, competes with other animals, and if he or she competes successfully it will be to the detriment of some other species. We compete not only by eating but by driving, flying, building, consuming electricity and so on. If we honestly acknowledge our status as competitors we will see hunting as one among many natural activities of our species. It is not a matter, Scanlon insists, of choosing whether or not to affect ecosystems, but rather of admitting and controlling our inescapable effects on those systems of which we are members.

The other primates are our closest kin in the animal world and thus the performance gap between them (especially the great apes) and ourselves has been taken as conclusive evidence of humanity's 'platonic' special status. Apes do not speak or write, and Descartes and many others have taken the absence of language to be proof of irremediable 'lower' status. But perhaps the conclusion is hasty.

Duane Rumbaugh and Sue Savage-Rumbaugh here report on some of their research, concerned not primarily with what primates naturally do, but with what they can do. Chimpanzees do not naturally use language, and they seem neither as well equipped nor as predisposed as humans to do so. But they *can*. (Recently some have claimed that this behavior by apes does not constitute genuine language use, and serious problems arise in distinguishing linguistic from nonlinguistic behavior in many contexts. The experimental design of Rumbaugh, Savage-Rumbaugh, and their colleagues, briefly described here, eliminates the most suspect factors. If this is not language, neither is the greater part of human oral communication.)

If chimpanzees *can* use language, why do they not do so naturally? It may simply be that the genetic/environmental roulette wheel did not come out that way. If humans should say of the great apes "There, but for the breaks of our great[n]-parents go I," something important about the boundaries of our moral community may follow.

The sharing behavior of the chimpanzees Sherman and Austin, when compared to the typical feeding behavior of chimpanzees, and the increased 'civility' of the seriously impaired human children as they gain some sort of linguistic facility, suggest that an animal that becomes linguistic simultaneously becomes a different sort of creature. (Come to think of it, is this not confirmed by millenia of observation of young humans?) This may strengthen the "There but for the breaks . . . " position and also may raise a perplexity. If linguistic apes live a better sort of life than their nonlinguistic conspecifics, are we obliged to provide language training for as many apes as possible? (And if so, at what acceptable level of damage to existing ecosystems?)

Humans, Animals, and 'Animal Behavior'

Stephen R. L. Clark

Attitudes to nonhuman animals and to 'animal behavior' in human animals are perplexingly entangled, and both utilitarian and Stoic theories of morality are vitiated by failure to understand animal behavior and motivation. The discovery that nonhuman, like human, animals are subject to ethical constraints can provide the basis for a morality that allows due weight to both Sentiment and Reason. Such a morality discriminates on subjective as well as objective grounds, but can give us no license to regard creatures not of our own species as beyond the moral pale.

THE UTILITARIAN AND THE STOIC

Utilitarian theory proposes that pleasure is good and pain evil, and that actions are right insofar as they lead to pleasure, wrong insofar as they lead to pain. In sum, we should act so as to achieve the greatest happiness of the greatest number.' How this is to be calculated, and how far the principle's specious simplicity is real, are matters beyond my present brief. What concerns me is that the classical utilitarians included nonhuman animals in their calculations, and were attacked for so doing (Whewell, 1852, p. 223; see Mill, 1852).

> The morality which depends upon the increase of pleasure alone would make it our duty to increase the pleasure of pigs or of geese rather than those of men, if we were sure that the pleasure we could give them were greater than the pleasures of men.

Such an approach, Whewell believed, disregarded the distinctively human attributes that were the proper objects of moral endeavor. Virtue, not pleasure, should be the end of action.

I have some sympathy with this criticism. There are notorious problems about taking pleasure, just as such, to be the only good. If that were so, for example, any pleasure produced by sadistic cruelty would compensate for the victim's suffering, and might even be so great as to outweigh that suffering. The case is not improved by hurried appeals to longterm imponderables. Not all pleasures are worth pursuing, and a worthwhile activity is not simply one that produces the best available ratio of pleasure to pain, either in the individual or in the universe.

But though I sympathize with Whewell's attack on naive utilitarianism, and think the question of our duty, or supposed duty, to increase the happiness of nonhuman animals is a real one (Clark, 1979), I do not share his disdain for nonhuman animals and for 'animal' pleasures. Nor do I share Kant's fears (1963, p. 164):

> Sexuality exposes us to the danger of equality with the beasts.

And vice versa. To think of 'animal pleasures' as of moral significance is to 'lower ourselves' to the animal level. To recognize nonhumans as seriously deserving of our consideration is to acknowledge their goals as ones we could share. To be concerned for animal welfare is to be a libertine. Really moral persons are not to sympathize with the trivial, ignoble, or disgusting aims of our nonhuman kin. Strangely, these aims are often perceived as sexual. Almost the only animals permanently in rut, and almost certainly the only ones ever to make a life's work of it, are humans, but sexual desire in particular is held to turn people into beasts. A rather similar sequence of thought results in the belief that males who like the company of women are effeminate. True manhood is displayed in disregarding the softer, soggier feelings of women, as nonhumans, and the male's own feelings. To give in to a woman is, for a male, to give in to his own lower nature. Spinoza attributed concern for animal suffering precisely to womanish sentimentality (Spinoza, 1677, 4.37.1). Interestingly, again, it is women who are traditionally held to be sexual beings, and males supposed to be tempted by them against their own real natures. Traditional oppression of women is linked with fear of their sexuality and distrust of 'womanish' sentiment. The male embodies passionless reason, if only he is not corrupted by bad company.

Utilitarian propaganda has played a part in the slow erosion of such value systems. We are readier to admit the value of innocent pleasure, readier to spare each other pain. The discovery of anesthetics and analgesics, by making it possible to evade (some) pains, has decreased our need to imagine that pain is good for us. Whatever we have lost in the last century, we cannot now regret our increased susceptibility to sympathetic and other distress, our increased readiness to enjoy sexual and other delights. Although I think utilitarian theory has helped to excuse many evils, of realpolitik and laboratory science, I gladly acknowledge that utilitarians have helped to produce legislation on behalf of animals, children, women, and the poor that would have been laughable—was laughed at—two centuries ago. But perhaps something should be said on the other side.

What is paradoxical about the association of animal welfare and libertinism is that it contradicts the ancient association of animal welfare and asceticism. The charge made against zoophiles by nineteenth century moralists was that they were libertines (Austin, 1885), or at best sentimentalists. The charge made against them by Augustine was that they were Manichaeans, dedicated to the belief that the world and the flesh were evil (Augustine, DMM 35f, CD I.19). It is still quite commonly assumed that vegetarians in particular are always moved by sheer distaste for incorporating the unclean. The assumption is, of course, sometimes correct. I have seen vegetarianism advocated, by a schoolmaster, precisely on the grounds that meat enflames the passions of the animal boy. This was also one of Gandhi's reasons. Meat especially, perhaps: there was a medieval tradition that fish (and barnacle-geese) alone were to be eaten on Fridays and during Lent, since they were not sexually generated.

Ascetics have often adopted vegetarian ways as part of their progressive detachment from the luxuries and corruptions of this naughty world. Moral behavior lies in controlling and frustrating the desires of the flesh, 'the beast within' (Plato, *Republic,* 589a-b). We should follow Reason, embodied in the moral law, and not concern ourselves with the desires and feelings we share with the nonhuman. It is the mark of a slavish man to follow out his animal desires (Aristotle, *Eudemian Ethics,* 1215634): good people have their beasts well-trained. Both Cynics and Academics were often vegetarian, and in refraining from the luxuries and medicaments of their day were sometimes enabled to *see* the animals they no longer needed to oppress. Such asceticism is not to be despised: I have no doubt that our health would be improved, our duties to the human world made easier, and our eyes opened to much of value if we made more effort to practice the moral discipline of our predecessors. It would also have some good effects for animal welfare. Cosmetics, furs, expensive meats would give way to simpler tastes, and fewer of our nonhuman kin would be oppressed to produce them. We might also manage to endure more readily the pains and problems of this mortal life, and cease to seek 'cures' and 'palliatives' with such casual brutality. The only recipe for a long and healthy life yet known to us is simply moral discipline (and even that is not a guarantee).

But though the general adoption of a more disciplined approach to living, the abandonment of the absurd pursuit of pleasure and avoidance of every slight distress, would have some good effects, the 'beast-controlling morality' has at least two drawbacks. First, that it involves a disrespect for what are taken to be 'animal' concerns. If we do not think *our* pains and pleasures are to be taken seriously, how can we trouble ourselves about theirs? Second, that we may find ourselves engaged in symbolic behavior. Animals symbolize things to us, in particular human capacities. So indeed do our fellow humans. How many victims of rape or marital violence or child abuse were raped or beaten simply because of their actual, individual natures? It is at least likely that in raping or beating the agent was conquering or hitting out at something for which the victim was only a symbol: personal weakness, insecurity, or memory of failure. The list is not exhaustive. A moral code that despises 'animal behavior' may also lead to symbolic action against the animals that seem to us to embody that behavior. Even self-hatred is a dangerous game, given the human capacity for projecting undesirable qualities into an external object (Epictetus, D,I.3.7ff):

> It is because of our kinship with the flesh that those of us who incline toward it become like wolves, faithless and treacherous and hurtful, and others like lions, wild and savage and untamed; but most of us become like foxes, that is to say, rascals of the animal kingdom. For what else is a slanderous and malicious man but a fox, or something even more rascally and degraded? Take heed, therefore, and beware that you become not one of these rascally creatures.

Epicetus was concerned above all to seek the good and worthwhile in what is not shared with the nonhuman. All other creatures are made for our use, for they lack understanding (D, II.8.6ff). Strictly, of course, they cannot then be treacherous, savage, or rascally: they are only following out their God-given natures. But since we should not give way to the sort of impulses that we think operate in them, it is easy to regard them as positively evil, and therefore as suitable objects of correction.

Popular moralists continue the tradition. To behave in a particularly hurtful, rapacious, or lustful way is to behave 'like an animal.' Carnivores are 'vicious killers' for no better reason than that they feel no particular compunction about killing their prey. Epictetus went rather further. It is not only the sensual or aggressive passions that work against true human dignity, but also sentiment and the body (Epictetus, E. 3; 26):

> If you kiss your own child or wife, say to yourself that you are kissing a human being; for when it dies you will not be disturbed.
> Some other person's child or wife has died; no one but would say "Such is the fate of man." Yet when a man's own child dies, immediately the cry is "Alas, woe is me!" But we ought to remember how we feel when we hear of the same misfortune befalling others.

There is of course some merit in this advice, but the general message is disturbing. We ought not take special ties of affection seriously, but strive to view events and creatures as we would view any objectively similar event or creature. One way of doing this might be to practice universal love, but human creatures being what they are, a universal indifference is more likely.

We are born to contemplate God and His works, and should not allow passion or private affection or bodily needs to corrupt our vision. These things do not matter.

> On earth there is nothing great but man; in man there is nothing great but mind!

—a quotation from the journal of the British Research Defence Society, *Conquest*.[1] A few years ago I wrote that this was sub-Hegelian gibberish (Clark, 1977, p. 7), which was careless of me. In fact it appears as the epigraph of W. Hamilton's edition of the *Works of Thomas Reid* and is attributed by him (I, p. 217) to a 'forgotten philosopher.' It appears to be of Stoic origin, though I have not yet located the exact source.

> The Stoics . . . saw and said that in the world, after God, there is nothing so important as man, and in man nothing so important as mind . . . They starved and blighted human nature by finding no place or function for passion and worshipping as their ethical ideal apathetic wisdom. They shut their eyes to patent facts of experience in pretending to regard outward events as insignificant and pain as no evil. They silenced the voice of humanity in their hearts by indulging in merciless contempt for the weak and the foolish (Bruce, 1899, p. 387).

This judgment is not entirely fair to the Stoics, but it has an alarmingly exact application to the character and practice of some moderns. It is worth noting here that we do not, as Stoic or rationalist theory requires, call someone 'inhuman' because that person is stupid, but because he or she lacks certain fundamental emotional responses.

Like utilitarianism, Stoicism has its good points, as have the Thomist and Kantian philosophies that are its descendants. It has helped to inculcate courage and temperance in many generations sorely in need, as are we, of these virtues. Even the seemingly pathological alienation from the body that such philosophies sometimes produce has its uses (Epictetus, D, I.1.23ff):

> "Tell your secrets!" I say not a word, for this is under my control. "But I will chain you." What is that you say, man? Chain *me*? My leg you will chain, but my will not God Himself can chain. "I will throw you into prison." My paltry body, rather. "I will behead you!" Well, when did I ever tell you that mine was the only neck that could not be severed? These are the lessons that philosophers ought to rehearse.

If philosophers, and others, did now rehearse them, our demands upon the environment, upon nonhuman animals and upon our human kindred would grow less, and our susceptibility to threat and blackmail likewise. Even if a Stoic community continued to regard animals as exploitable,

---

[1]*Conquest,* January 1970; quoted in Ryder 1975, p. 146f.

and contemptible, material, it would still in fact exploit them less than we do. In a Stoic community, vivisection would not be defended as leading to release from pains, for pains as such are indifferent. But it would be defended as leading to knowledge, just as such, and quite apart from any sensual advantage. Human reason is operative in the pursuit and contemplation of Truth, and this is the ultimate value.

Utilitarians pursue Happiness, and Stoics Truth. Both offer objective codes of behavior, and urge us to act to anyone as we would to everyone, leaving all private affection and subjective difference aside. Utilitarian theory concedes that pleasure and pain are not morally indifferent nor animals of quite a different kind from ourselves, but licenses the exploitation of animals for our pleasure and weakens our hold on virtue. Stoic theory reminds us of the need for virtue and bids us endure with equanimity the post to which God has appointed us, but conceives all nonrational creatures to be there entirely for our use and systematically disregards the nonrational impulses we share with them. Oddly, both tend to assume that only pleasure and pain are of any significance to our nonhuman kin. By criticizing this last assumption we may work towards a more satisfactory moral system.

UNDERSTANDING ANIMALS

What are animals actually like? Kenneth Dover remarks in passing (1975, p. 75) that

> . . . to judge from the pronouncements made throughout human history on the subject of animals, it would seem that ours is the first culture actually to observe animals in their natural state and perhaps the first to care whether what it says about them is true or false.

Understandably, he exaggerates a little: there have been exact observers in the past, and there are considerable confusions in our present understanding. It is really extraordinarily difficult to find out what animals are actually like, to move beyond the stereotypes and projections and sentimental misapprehensions to what the creature itself feels and desires. This is not a problem special to ethology. It is always difficult to disentangle our perception of a creature's quiddity from our own emotional involvement: if we desire someone, the erotic charge makes it very difficult not to see her or him as, precisely, inviting. If we like cuddling a furry animal, we tend to assume that the animal is cuddly. Lambs are sweet and innocent; cows are bovine (obviously); pigs are greedy. The problem is made worse with animals because our social conditioning is deliberately (I do not say 'cynically') unrealistic. It is very interesting to watch parents painfully coping with their children's immediate sympathy for animals when that sympathy comes too close to home. Our children's books are full of pretend-animals with whom our young can identify; we encourage them to care for 'pets,' to take an interest in living creatures. Farms and zoos and circuses are presented as occasions for community with animals—but what happens when the child realizes what he or she is eating, or what is being done to the animals so that he or she may enjoy their 'company'? Kantians are forced to dissimulate: for them, animals are a superior sort of toy, to be cared for only as practice for caring for people; but they cannot reveal this to the child without subverting the program. The child must respond to, and care for, the animals it pets, but must not in the end take its own concern seriously. The result is a sort of schizophrenia well exemplified in C. S. Lewis' *Narnia* saga: on the one hand, talking animals to be respected as companions and helpers; on the other 'dumb' animals who can be killed and hunted at will.[2] To be fair to Lewis, he was against causing even dumb animals to suffer, and perhaps never realized how much his supper had suffered in farm, lorry, and slaughterhouse.

It is entirely understandable that serious investigators have reacted against this sentimentalism (see Diamond, 1978) by seeking to eliminate all identifications, emotional attachments, and fantasies from their observations of animals. It is a methodological rule amongst ethologists that descriptions of behavior should be purely 'objective', without appeal to any mental element imagined into being. A better term would perhaps be 'impersonal': scientific or scholarly descriptions should obviously be 'objective' (i.e., realistic) and 'objective' (i.e., referring to qualities that exist independently of our attitude). It should also be 'objective' (i.e., unbiased), but cannot hope to be 'objective' (i.e., lacking in any emotive force) when the facts described are such as to move any normal person. There is no necessary conflict between objectivity and emotive force. It is not obvious that such description should exclude all reference to the motives and attitudes of the creatures described. But the policy is understandable, and it is also understandable that laboratory scientists, and even some of the laity, have concluded to the Cartesian dogma, that animals are merely mechanisms. If we do not need the hypothesis of subjectivity or consciousness to explain the phenomena we will manage without it.

It is understandable, but it is no better an argument for all that (Bradley, 1897, p. 18)[3]:

> It is doubtless scientific to disregard certain aspects when we work; but to urge that such aspects are not real, and that what we use without regard to them is an independent real thing—this is barbarous metaphysics.

[2]See especially Lewis, 1953.
[3]See also Clark, 1975, pp. 191ff.

Barbarous metaphysics, and barbarous psychology. Like Descartes, the impersonalist rejects the immediate impulse of sympathy for what is felt to be animal distress, in the name of a rational theory. The impersonalist has less excuse than Descartes, for it is an axiom of modern research that we are animals, evolved along with other animals, and our responses are adapted to the world we and our ancestors have inherited. Descartes could believe that human consciousness is detached from all things bodily, though he found no satisfactory way of expounding the obvious connections between Mind and Body. Modern impersonalists, who have mistaken methodology for ontology, are heirs of two quite different attitudes to humanity: the Rationalist and the Naturalist. If we are essentially Reason, then we may justly disregard the motions of our, or other, animal frames, but cannot hope to explain anything of our psychology by observations or experiments on animals. If we are essentially Animal (of a particular kind), we cannot so easily disregard our immediate responses or believe that our close kindred are unfeeling mechanisms.

> An epistemological lobotomy, which prevents an intelligent man from using the normal cognitive functions nature gave him, does indeed constitute an act of dehumanization.[4]

I emphasize that our ability to recognize patterns of behavior, to sympathize with other creatures, is indeed a cognitive function, part of the way we know the world. It is unfortunate that professional biologists and animal psychologists seem, on this point, to be several decades behind the philosophical times. Few philosophers would now describe 'mental states' as things uneasily inferred from purely impersonal descriptions, as though we could only *see* a creature twisting about, sweating, emitting high-pitched noises (and so on) and must thence *infer* that there is a mental cause of these motions, namely pain. Such an inference would be very weak indeed, for I have no reason to believe that such motions are associated with such mental states unless I can recognize the latter directly. I cannot infer the existence of fire from the sight of smoke unless fire and smoke are at least sometimes observed together. On such terms I can have no good reason to believe that any other creature is ever in what I call 'pain.' Nor are humans in better case because they (some of them) can say 'it hurts.' They could not have been taught the words, nor could I understand them, if they, if their parents, if we could not recognize a pain state nonverbally expressed.

The attribution of mental qualities to a creature is not, in its normal form, an unreliable inference on the basis of securely established physical information. On the contrary, such attributions are the framework of all human endeavor and scientific research. If I did not believe that a fellow researcher *is* a fellow researcher, a being with cognitive capacities and ethical responsiveness, how could I take that scientist's reports as evi-

---

[4]Attributed to K. Lorenz, but without exact location.

dence (Ritchie, 1964)? Of course, we are sometimes in error; of course, there might be creatures who always pretend (but not many); of course, there might be things that looked like creatures but are not (there are, called 'toys'). But these are particular problems that give us no good reason to doubt the general accuracy of our assessments and recognitions. In saying that a creature likes or dislikes, fears or loves, is envious or spiteful or sympathetic, we are not, in general, indulging an illicit animism, nor in 'mysticism' (as E. O. Wilson supposes: Wilson, 1975, p. 176). We are recognizing patterns of behavior, and may do so legitimately among adult humans, infants, and our nonhuman kin (Midgley, 1978, pp. 344ff). It is of course the more difficult to do so the further removed, in genealogical terms, the creatures are.

Decent ethologists in fact know all this well enough: their practice is far better than their naively Cartesian theories would suggest. Thanks to the efforts of exact observers who have sought not to jump to conclusions too early, or too late, we have some detailed information about our kindred that does not support the view that animals are lawless, sexually rapacious, or murderously violent. Nor, of course, are they gentle, law-abiding, and supremely altruistic. But the post-Aristotelian, and Stoic, view that there are 'impulses to the noble' in animals (as in children) (Aristotle, *Magna Moralia* 1206b17) has received support. They are moved to act in an affectionate manner, to care for the small and defenseless, to avoid hurting familiars, to respect authority, to do their duty when they are themselves authorities, to honor sexual taboos, to fight for their young, and to feed their old. All this was indeed familiar to the ancients, some of whom concluded that Nature was a principle superior to the Stoics' vaunted Reason (Plutarch *Gryllus, or Beasts are Rational,* 991). Natural law is *quod natura omnia animalia docuit* (Justinian, *Institutiones,* I.8); it is therefore natural law at least for mammals (and others) that parents care for their children, that we be loyal to our clan-mates, respect status, acknowledge territorial claims, and the like. What is fascinating about the 'territorial instinct' is not that animals are possessive, but that other animals acknowledge their claims. Even butterflies, it appears, give the occupant of a patch of sunlight the moral edge over an intruder (Smith, 1978, p. 144).

I do not imply that animals do not also cheat, kill, conspire, and desert. Jane van Lawick-Goodall's observations of the Gombe chimpanzees make it clear both that chimpanzees are social and responsive creatures, and that they can fail as blatantly as ourselves to be moved by distress to anything other than irritated neglect. It is worth remembering that chimpanzees are like humans also in this, that they are disgusted by obvious physical illness, and that they are ready to kill creatures, even of their own species, for their own purposes (Van Lawick-Goodall, 1971).[5] We

---

[5] Later reports suggest that the Gombe chimpanzees also indulge in infanticide and tribal warfare.

need not react against traditional views of animals by adopting romantic fantasies about 'the innocent lusts of the unfallen creatures!'

Stoic moral theory, whatever its failings, acknowledged that nonhuman creatures could be moved by considerations that also move good people. Human morality arose by reflection upon the ethical values I have mentioned, with a view to achieving the sort of universal, objective outlook that frowns upon personal attachment and nonrational sentiment. It is here, I believe, that they made their error. A morality that systematically denigrates and denies its own roots in ethical responsiveness is doomed. The clearest case of this is Monod's 'ethic of objectivity' (i.e., of impersonality). The primary command of that morality is 'thou shalt not "participate" in the world's workings.' Everything is to be seen without emotional affect, and the conclusion must be that of Epictetus, that we should not even identify with our own bodies. But in so detaching ourselves, we lose all basis upon which to argue scientifically, to pick out any particular aspect of the phenomena, to seek the truth. The end is emptiness (Monod, 1972).[6]

A sound morality must be based upon, and not denigrate, the personal ties and attachments into which we are born. We may properly extend our respect for human creatures, as Stoics would have us do, from our immediate families to creatures of the same objective kind, but we cannot quite eliminate subjective discriminations without destroying the natural roots of our morality. We dare not wholly despise Nature in the name of 'Reason,' for Reason (a compendium of natural faculties) is itself a product of Nature and draws its premises from the stock of natural sentiment (Lewis, 1943).

NATURAL LAW

If these natural sentiments of affection, loyalty, and service are the necessary roots of morality, what are we to say about our relations with creatures not of our species? If a fully objective code of behavior is not to be hoped for, are not zoophiles on the wrong track when they point to the objective similarity between creatures of another species and some of our fellow humans (infants or imbeciles or the aphasic)? Speciesists need not claim that all human beings are objectively different (save, of course, in species) from all nonhuman beings, but only that they are subjectively different: they matter more to *us*, not always to the universe. Stoics and Utilitarians, of course, should not rely on this distinction: only objective value matters, value rationally acknowledged. My own naturalism cannot simply override actual feelings.

Again, if it is natural sentiment that lies at the root of morality, if natural law is *quod natura omnia animalia docuit,* how can I denounce

---

[6]See Clark, 1977, pp. 145ff.

the predatory aspects of nature? Is it not natural for creatures to prey on one another? How can we so self-righteously stand aside from the fray? A rationalist can reply with Plato (Philebus, 67b):

> Those who appeal to such evidence are not better than those who put their trust in the flight of birds. They imagine that the desires we observe in animals are better evidence than the reflections inspired by a thoughtful philosophy.

Or, as one of Aristophanes' heroes put it, to someone who claimed that it was only natural for him to beat his father, since cocks did it (*Clouds* 1420f):

> Why don't you feed on dung, then, and sleep on a perch?

The fact that chimpanzees grab young baboons to eat them no more justifies us in doing so than the fact that they neglect and repulse sick members of their own tribe justifies us in doing so. Their sentiments may not allow any weight to certain distresses; their behavior may sometimes even run counter to their own natural sentiments—we are not the only animals sometimes to be torn between desire and proper feeling, or proper desire and misplaced feeling. What is natural is not simply what happens always, or for the most part. If it were, disease and error would be natural (Epictetus, D,I.11.7). The form of life natural to a creature helps to define what happiness is for that creature's kind, what capacities are there to be filled, what occasions are needed for it wholly to be itself. This Aristotelian concept helps us to see what is wrong in imprisoning and frustrating, say, a veal calf or in declawing a cat: not that they suffer pain sensations, but that their lives are systematically prevented from being full lives according to their kind. Disease and injury, though natural in the sense of being likely enough to happen, are precisely against the victim's nature. Similarly, not every sentiment is natural; some are pathological manifestations of glandular imbalance or social corruption. Sentiment is the root of morality, as sense experience is of science; neither sentiment nor sense experience can be wholly condemned without destroying the systems grown from them, but equally they are not unquestionable.

This is as much as to say that the natural world does display corruption when judged by informed sentiment. How exactly we are to tread the narrow track between Manichaean rejection of the natural world (and consequent destruction of the ethical responsiveness that arises from that nature) and a pantheistic adoration of whatever happens (and consequent destruction of all moral discrimination) is a matter I have discussed elsewhere (Clark, 1976). My suspicion is that only traditional theism can cope. But leaving that aside, it is surely proper to admire the beauty and order manifest in the world without incurring any obligation to copy predators or to refrain from change. What has been need not always be, even if we cannot now see any clear way wholly to eliminate past evils.

But the challenge still stands. Even if it is possible for people to respect, befriend, and care for creatures not of their own species, might this not be pathological? Is it not properly natural to be speciesist? A lion who really made no difference between his treatment of lions, lionesses, gazelles, and men with guns would not be a lion (nor long survive as a monster). Creatures are of the kind they are because they do, and their ancestors did, discriminate between species, between sexes, between ages in just the ways they do. How can it not be the sentiment of a deeply disturbed human to seek an end to such discrimination?

Is it not natural, some would add, for human beings, evolved as social predators, to hunt? Are not our distinctive characteristics fitted precisely to such a predatory life? If they are, of course, we are doomed: for only a post-cataclysmic world could give our descendants any chance of returning to a hunter/gatherer economy. If we cannot be happy except as small bands of naked wanderers, ganging together to run down antelope, we are not going to be happy, no matter what ersatz substitutes we devise. Of course, we can learn from such societies, but we must also hope that, as it seems, we are a sufficiently adaptable species to find other ways of happiness.

Could these other ways be such as to satisfy a radical zoophile? The question is similar to one posed to radical feminists: how could we end discrimination between male and female without wholly destroying the social and biological bases of society? Science-fantasists may speculate of a time when people select, at will and temporarily, child-bearing or begetting modes; when the pronouns 'he' and 'she' are as obsolete as the pronouns betokening social status with which some languages are laced. A society that really did not discriminate between chimpanzees and human morons would be even more bizarre.

But of course feminists need not believe that the wholly nonsexist society is desirable even if it is possible. The evils of sexism are that females are denied the opportunity to live authentic lives, that girls are conditioned to take a subordinate role, that males are isolated from their feelings and encouraged to behave in grossly competitive ways, that women are raped, beaten, and insulted. A nonsexist society is one in which people are not oppressed, exploited, and manipulated to fit sexual stereotypes. The evils of speciesism, similarly, are that creatures are robbed, assaulted, and killed, not simply that they are distinguished from members of our own species. It is right so to distinguish them, both objectively (for their species-specific qualities differ) and subjectively (for some species are more readily understood and liked by us). What is wrong is to use them with cruelty and disrespect.

It is natural to find sexual partners chiefly in the opposite sex of one's own species. In that sense, it is natural (and proper) to be both speciesist and sexist. But it is not natural to be concerned only for creatures who are members of one's own species. As Richard Dawkins has said, 'the ethics of speciesism . . . has no proper basis in evolutionary biology' (Dawkins, 1976, p. 11). The reason is simple: a species is a reproductively isolated population, such that its members rarely or never breed outside that population to produce fertile offspring. What is significant for neo-Darwinian evolution is not the species, but the gene lines tangled together in the gene pools. Characteristics are embedded in the genotype if they result, in a given community, in phenotypes that multiply, relative to competitors, the responsible genotype. This is how altruism, most probably, becomes a stable instinctual pattern: that those who benefit from altruistic acts generally carry the same genes as the altruist. It does not follow, as some careless writers have implied, that animal altruism is really selfishness, or even really a desire to multiply one's genes. The evolutionary explanation does not explain altruism away. Nor does it follow that only those who carry the altruist's genes are ever benefited: the altruist has no way of discovering who does and who does not. The altruist responds rather to the familiar cues of sight, smell, and evocative posture. Other creatures can 'exploit' this fact, as do cuckoos. Put it differently: altruists can be concerned for creatures not of their own species, if only they are familiar or sufficiently like what is familiar.

It is from these literally familiar roots that our concern for others may blossom. Stoics urged that

> We should regard all men as our fellow countrymen and fellow citizens, and there should be one life and one order like that of a single flock on a common pasture feeding together under common law.[7]

Advocates of Earth's Household employ a parallel rhetoric. Just as a household may contain creatures of many species, each of whom evoke feelings of concern and affection (unless suppressed), so does Earth's Household contain many creatures with whom we can often find mutually helpful symbioses, against whom we must sometimes struggle, whom we need not hate nor despise.

In the household of earth and heaven, Aristotle taught (*Metaphysics*, 1075a19ff), the free men have more responsibilities than the children or slaves. The Stoics were doubtless right at least in this, that God has called us out to know the world in ways not open even to our nearest kin (as they, it may be, in ways not open to us), and that we do in fact hold their lives and happiness in our hands. As later philosophers have taught, we are in the position of stewards, bound by duties experienced as feeling and expounded by reason (Darling, 1970, p. 122):

> . . . to serve the lesser creation, to keep our world clean and pass on to posterity a record of which we shall not feel shame.

Fraser Darling's words provide my peroration, but I will add one thing: our office as stewards, often enough, is not to reform and tame the world,

---

[7]Ascribed to Zeno by Plutarch, *De Alexandri Fortuna aut Virtute* 329ab.

but to allow our cousins their own ways forward. Anarchism may not be the best political system for humans, but I suspect that it may be for the world at large.

I have wielded many mighty opposites in this paper: Pleasure and Virtue, Subjectivity and Objectivity, Nature and Reason. If they remain strong to move us, it is perhaps because all are necessary to the good life.

> Man is a lumpe where all beasts kneaded bee,
> Wisdom makes him an Arke where all agree.
> (Donne, 1610)

Where exactly the wise man will put the Mean (Aristotle, *Nicomachean Ethics*, 1106b.35f),[8] I must leave others to decide.

[8]See Clark, 1975, pp. 84ff.

Ecology, Morality, and Hunting

Peter S. Wenz

A defense of human obligations toward the environment based neither on theism nor on considerations of long-term human interests cannot be found in contemporary western philosophy. Aldo Leopold, for example, calls for "a land ethic [that] changes the role of *homo sapiens* from conqueror of the land-community to plain member and citizen of it" (Leopold, 1970, p. 240). He immediately points out, however, that "the conqueror role is eventually self-defeating." This leaves the impression that the reason for adopting the land ethic is for human beings to avoid their eventual self-defeat and demise. Further evidence that this is what Leopold had in mind comes from his conception of an ethic "as a mode of guidance for meeting ecological situations so new or intricate, or involving such deferred reactions, that the path of social expedience is not discernible to the average individual" (p. 240). The point seems to be that a land ethic is needed to protect human beings, along with other members of the biotic community, from the deleterious long-term effects of actions in disregard of environmental integrity. Leopold's essay contains no defense of the view that one has a *prima facie* obligation to protect the environment for its own sake, apart from any possible advantage to human beings.

Christopher Stone's *Should Trees Have Standing?* is similar, though he does confront the problem directly. He writes (Stone, 1974, p. 43):

> The time is already upon us when we may have to consider subordinating some human claims to those of the environment *per se*. Consider, for example, the disputes over protecting wilderness areas from development that would make them accessible to greater numbers of people. I myself feel disingenuous rationalizing the environmental protectionist's position in terms of a utilitarian calculus, even one that takes future generations into account.

But the two arguments Stone offers in defense of protecting the environment per se are clearly homocentric. He begins the first by noting that "the strongest case can be made from the perspective of human advantage for conferring rights on the environment'' (p. 45). For the only way we can avoid ecological difficulties that would be disastrous to humans is by changing our styles of life, for example, by ''forfeiting material comforts'' and ''giving up the right to have as many offspring as we might wish'' (p. 45). ''Such far-reaching social changes are going to involve us in a serious reconsideration of our consciousness towards the environment'' (p. 46). One such change is granting rights to environmental objects, another is concern for the good of the environment per se. Human well-being clearly carries the weight of this argument for protecting the environment per se.

Stone's other argument is this (1974, p. 48, 49):

> If we only stop for a moment and look at the underlying human qualities that our present attitudes toward property and nature draw upon and reinforce, we have to be struck by how stultifying of our own personal growth and satisfaction they can become when they take rein of us.
>
> To be able to get away from the view that Nature is a collection of useful senseless objects is . . . deeply involved in the development of our abilities to love—or, if that is putting it too strongly, to be able to reach a heightened awareness of our own and others' capacities in their mutual interplay.

Here again, a human interest is paramount, the interest in securing richer, more satisfying experiences.

The absence of an argument based neither in theism nor humanism for protecting the environment seriously hampers environmental advocacy. Theistic arguments are largely ineffective in a secular world. Homocentric arguments, including those that are only indirectly and subtly homocentric, such as Leopold's and Stone's, suffer in the following way. Suppose the environmentalist is opposing the construction of a new road by appeal to Leopold's land ethic or Stone's concern for the environment per se. Because the principles on which the environmentalist is relying are themselves supported primarily by considerations of human advantage, those wishing to construct the road may reasonably suggest that these principles be bypassed in favor of calculating directly all the human advantages and disadvantages of the road's construction. This is like the response that act-utilitarians make when confronted with rules that are supposed to promote maximum utility. If maximizing utility is really the goal, then in any particular case it makes sense to suspend those rules which generally maximize utility in favor of a direct calculation of the utilities and disutilities of the acts available in the situation at hand. Just as the rule becomes superfluous for the act-utilitarian, so do the principles of the land ethic and concern for the environment per se to those who defend these principles by appeal, however indirect, to human advantage. If human advantage is the bottom line anyway, why not calculate it directly in each situation? The land ethic, if retained at all, becomes just another tool through which human beings take maximal advantage of the nonhuman environment.

I try to release environmental advocacy from this liability. In the first section I characterize good or healthy ecosystems. In the second I argue that we have a *prima facie* obligation to protect such ecosystems irrespective of all possible advantage to human beings. The result of this argument is used in the third and fourth sections to show that hunting is morally wrong when practiced by agricultural, industrial, and post-industrial peoples.

GOOD ECOSYSTEMS

An ecosystem is what Aldo Leopold referred to as a "biotic pyramid." He describes it this way (1970, p. 252):

> Plants absorb energy from the sun. This energy flows through a circuit called the biota, which may be represented by a pyramid consisting of layers. The bottom layer is the soil. A plant layer rests on the soil, an insect layer on the plants, a bird and rodent layer on the insects, and so on up through various animal groups to the apex layer, which consists of the large carnivores.
>
> Proceeding upward, each successive layer decreases in numerical abundance. Thus, for every carnivore there are hundreds of his prey, thousands of their prey, millions of insects, uncountable plants.
>
> The lines of dependency for food and other services are called food chains. Thus soil-oak-deer-Indian is a chain that has now largely converted to soil-corn-cow-farmer. Each species, including ourselves, is a link in many chains. The deer eats a hundred plants other than oak, and the cow a hundred plants other than corn. Both, then, are links in a hundred chains. The pyramid is a tangle of chains so complex as to seem disorderly, yet the stability of the system proves it to be a highly organized structure.

It is so highly organized that Leopold and others write of it, at times, as if it were a single organism which could be in various stages of health or disease (p. 274):

> Paleontology offers abundant evidence that wilderness maintained itself for immensely long periods; that its component species were rarely lost, neither did they get out of hand; that weather and water built soil as fast or faster than it was carried away. Wilderness, then, assumes unexpected importance as a laboratory for the study of land-health.

By contrast,

> When soil loses fertility, or washes away faster than it forms, and when water systems exhibit abnormal floods and shortages, the land is sick (p. 272).

The disappearance of plant and animal species without visible cause, despite efforts to protect them, and the irruption of others as pests despite efforts to control them, must, in the absence of simpler explanations, be regarded as symptoms of sickness in the land organism (pp. 272–273).

In general, a healthy ecosystem consists of a great diversity of flora and fauna, as ''the trend of evolution is to elaborate and diversify the biota'' (p. 253). This flora and fauna is in a relatively stable balance, evolving slowly rather than changing rapidly, because its diversity enables it to respond to change in a flexible manner that retains the system's integrity. In all of these respects a healthy ecosystem is very much like a healthy plant or animal.

A description of one small part of one ecosystem will conclude this account of the nature of ecosystems. It is Leopold's description of a river's sand bar in August (1970, p. 55):

The work begins with a broad ribbon of silt brushed thinly on the sand of a reddening shore. As this dries slowly in the sun, goldfinches bathe in its pools, and deer, herons, killdeers, raccoons, and turtles cover it with a lacework of tracks. There is no telling, at this stage, whether anything further will happen.

But when I see the silt ribbon turning green with Eleocharis, I watch closely thereafter, for this is the sign that the river is in a painting mood. Almost overnight the Eleocharis becomes a thick turf, so lush and so dense that the meadow mice from the adjoining upland cannot resist the temptation. They move *en masse* to the green pasture, and apparently spend the nights rubbing their ribs in its velvety depths. A maze of neatly tended mouse-trails bespeaks their enthusiasm. The deer walk up and down in it, apparently just for the pleasure of feeling it underfoot. Even a stay-at-home mole has tunneled his way across the dry bar to the Eleocharis ribbon, where he can heave and hump the sod to his heart's content.

At this stage the seedlings of plants too numerous to count and too young to recognize spring to life from the damp warm sand under the green ribbon.

Three weeks later (pp. 55–56):

The Eleocharis sod, greener than ever is now spangled with blue mimulus, pink dragon-head, and the milk-white blooms of Sagittaria. Here and there a cardinal flower thrusts a red spear skyward. At the head of the bar, purple ironweeds and pale pink joepyes stand tall against the wall of willows. And if you have come quietly and humbly, as you should to any spot that can be beautiful only once, you may surprise a fox-red deer, standing knee-high in the garden of his delight (pp. 55–56).

HUMAN OBLIGATIONS TO
ECOSYSTEMS[1]

Let us now consider whether or not we, you and I, have *prima facie* obligations towards ecosystems, in particular, the obligation to avoid destroying them, apart from any human advantage that might be gained by their continued existence. My argument consists in the elaboration of two examples, followed by appeals to the reader's intuition. The second, Case II, is designed to function as a counter-example to the claim that human beings have no obligations to preserve ecosystems except when doing so serves human interests or prevents the unnecessary suffering of other sentient beings.

Some clarifications are needed at the start. By "*prima facie* obligation" I mean an obligation that would exist in the absence of other, countervailing moral considerations. So I will construct cases in which such other considerations are designedly absent. A common consideration of this sort is the effect our actions have on intelligent beings, whether they be humans, extraterrestrials, or (should they be considered intelligent enough) apes and aquatic mammals. Accordingly, I will construct my cases so that the destruction of the environment affects none of these. Finally, the obligation in question is not to preserve ecosystems from any and every threat to their health and existence. Rather, the obligation for which I am contending is to protect ecosystems from oneself. The differences here may be important. A duty to protect the environment from any and every threat would have to rest on some principle concerning the duty to bring aid. Such principles concern positive duties, which are generally considered less stringent than negative duties. The duty to protect the environment from oneself, on the other hand, rests on a principle concerning the duty to do no harm, which is a negative duty. Those not convinced that we have a duty to bring aid may nevertheless find a *prima facie* duty not to harm the environment easy to accept.

Case I

Consider the following situation. Suppose that you are a pilot flying a bomber that is low on fuel. You must release your bombs over the ocean to reduce the weight of the plane. If the bombs land in the water they will not explode, but will, instead, de-activate harmlessly. If, on the other hand, any lands on the islands that dot this part of the ocean, it will explode. The islands contain no mineral or other resources of use to human beings, and are sufficiently isolated from one another and other parts of the world that an explosion on one will not affect the others, or any other part of the world. The bomb's explosion will not add to air pollution because it is exceedingly "clean." However, each island contains an ecosystem, a biotic pyramid of the sort described by Aldo Leopold, within which there are rivers, sandbars, Eleocharis, meadow mice, cardinal flowers, blue mimulus, deer, and so forth, but no intelligent life. (Those who consider mice, deer, and other such animals so intelligent as to fall under some ban against killing intelligent life are free to suppose that in their wisdom, all such creatures have emigrated.) The bomb's explosion will ruin the ecosystem of the island on which it explodes, though it will not cause any animals to suffer. We may suppose that the islands are small enough and the bombs powerful enough that all animals, as well as plants, will be killed instantly, and therefore painlessly. The island will instantly be transformed from a wilderness garden to a bleakness like that on the surface of the moon.

Suppose that with some care and attention, but with no risk to yourself, anyone else or the plane, you could release your bombs so as to avoid hitting any of the islands. With equal care and attention you could be sure to hit at least one of the islands. Finally, without any care or attention to the matter, you might hit one of the islands and you might not. Assuming that you are in no need of target practice, and are aware of the situation as described, would you consider it a matter of moral indifference which of the three possible courses of action you took? Wouldn't you feel that you ought to take some care and pay some attention to insure that you avoid hitting any of the islands? Those who can honestly say that in the situation at hand they feel no more obligation to avoid hitting the islands than to hit them, who think that destroying the balanced pyramidal structure of a healthy ecosystem is morally indifferent, who care nothing for the islands' floral displays and interactions between flora, fauna, soil, water, and sun need read no further. Such people do not share the intuition on which the argument in this paper rests.

I assume that few, if any readers of the last paragraph accepted my invitation to stop reading. I would have phrased things differently if I thought they would. Many readers may nevertheless be skeptical of my intuitive demonstration that we feel a *prima facie* obligation to avoid destroying ecosystems. Even though no pain to sentient creatures is involved, nor the destruction of intelligent life nor pollution or other impairment of areas inhabited by human beings or other intelligent creatures, some readers may nevertheless explain their reluctance to destroy such an ecosystem by reference, ultimately, to human purposes. They can thereby avoid the inference I am promoting. They might point out that the islands' ecosystems may be useful to scientists who might someday want to study them. No matter that there are a great many such islands. The ecosystem of each is at least slightly different from the others, and therefore might provide some information of benefit to human beings that could not be gleaned elsewhere. Alternatively, though scientists are studying some, it might be to the benefit of humanity to establish Holiday Inns and Hilton Hotels on the others. Scientists have to relax too, and if the accomodations are suitable they will be more likely to enjoy the companionship of their families.

I believe that such explanations of our intuitive revulsion at the idea of needlessly destroying a healthy ecosystem are unhelpful evasions. They represent the squirming of one who intellectually believes ethics to concern only humans and other intelligent creatures, perhaps with a rider that one ought not to cause sentient creatures unnecessary suffering, with the reality of his or her own moral intuitions. The next case will make this clearer.

Case II

Suppose that human beings and all other intelligent creatures inhabiting the earth are becoming extinct. Imagine that this is the effect of some cosmic ray that causes extinction by preventing procreation. There is no possibility of survival through emigration to another planet, solar system or galaxy because the ray's presence is so widespread that no humans would survive the lengthy journey necessary to escape from its influence. There are many other species of extraterrestrial, intelligent creatures in the universe whom the cosmic ray does not affect. Nor does it affect any of the non-intelligent members of the earth's biotic community. So the earth's varied multitude of ecosystems could continue after the extinction of human beings. But their continuation would be of no use to any of the many species of intelligent extraterrestrials because the earth is for many reasons inhospitable to their forms of life, and contains no mineral or other resources of which they could make use.

Suppose that you are the last surviving human being. All other intelligent animals, if there were any, have already become extinct. Before they died, other humans had set hydrogen explosives all around the earth such that, were they to explode, all remaining plant and animal life on the earth would be instantly vaporized. No sentient creature would suffer, but the earth's varied multitude of ecosystems would be completely destroyed. The hydrogen explosives are all attached to a single timing mechanism, set to explode next year. Not wishing to die prematurely, you have located this timing device. You can set it ahead fifty or one hundred years, insuring that the explosion will not foreshorten your life, or you can, with only slightly greater effort, deactivate it so that it will never explode at all. Who would think it a matter of moral indifference which you did? It seems obvious that you ought to deactivate the explosives rather than postpone the time of the explosions.

How can one account for this "ought"? One suggestion is that our obligations are to intelligent life, and that the chances are improved and the time lessened for the evolution of intelligent life on earth by leaving the earth's remaining ecosystems intact. But this explanation is not convincing. First, it rests on assumptions about evolutionary developments under different earthly conditions that seem very plausible, but are by no means certain. More important, as the case was drawn, there are many species of intelligent extraterrestrials who are in no danger of either extinction or diminished numbers, and you know of their existence. It is therefore not at all certain that the obligations to intelligent life contained in our current ethical theories and moral intuitions would suggest, much less require, that we so act as to increase the probability of and decrease the time for the development of another species of intelligent life on earth. We do not now think it morally incumbent upon us to develop a form of intelligent life suited to live in those parts of the globe that, like Antarctica, are underpopulated by human beings. This is so because we do not adhere to a principle that we ought to so act as to insure the presence of intelligent life in as many earthly locations as possible. It is therefore doubtful that we adhere to the more extended principle that we ought to promote the development of as many different species of intelligent life as possible in as many different locations in the universe as possible. Such a problematic moral principle surely cannot account for our clear intuition that one obviously and certainly ought not to reset the explosives rather than deactivate them. It is more plausible to suppose that our current morality includes a *prima facie* obligation to refrain from destroying good ecosystems irrespective of both the interests of intelligent beings and the obligation not to cause sentient beings unnecessary suffering.

It is not necessary to say that ecosystems have rights. It is a commonplace in contemporary moral philosophy that not all obligations result from corresponding rights, for example, the obligation to be charitable. Instead, the obligation might follow from our concept of virtuous people as ones who do not destroy any existing things needlessly. Or perhaps we feel that one has a *prima facie* obligation not to destroy anything of esthetic value, and ecosystems are of esthetic value. Alternatively, the underlying obligation could be to avoid destroying anything that is good of its kind—so long as the kind in question does not make it something bad in itself—and many of the earth's ecosystems are good.

Our intuition might, on the other hand, be related more specifically to those characteristics that make good ecosystems good. Generally speaking, one ecosystem is better than another if it incorporates a greater diversity of life forms into a more integrated unity that is relatively stable, but not static. Its homeostasis allows for gradual evolution. The leading concepts, then, are diversity, unity, and a slightly less than complete homeostatic stability. These are, as a matter of empirical fact, positively related to one another in ecosystems. They may strike a sympathetic chord in human beings because they correspond symbolically to our personal, psychological need for a combination in our lives of both security and novelty. The stability and unity of a good ecosystem represents security. That the stability is cyclically homeostatic, rather than static, involves life forms rather than merely inorganic matter, and includes great diversity, corresponds to our desires for novelty and change. Of course, this is only speculation. It must be admitted that some human beings seem to so value security and stability as to prefer a purely static unity. Parmenides and the eastern religious thinkers who promote nothingness as a goal might consider the surface of the moon superior to that of the earth, and advocate allowing the earth's ecosystems to be vaporized under the conditions described in Case II.

My intuitions, however, and I assume those of most readers, favor ecosystems over static lifelessness and, perhaps for the same reason, good ecosystems over poorer ones. In any case, the above speculations concerning the psychological and logical derivations of these intuitions serve at most to help clarify their nature. Even the correct account of their origin would not necessarily constitute a justification. Rather than try to justify them, I will take them as a starting point for further discussion. So I take the cases elaborated above to establish that our current morality includes a *prima facie* obligation to avoid destroying good ecosystems, absent considerations of both animal torture and the well-being of intelligent creatures.

HUNTING

In this section I will argue that in view of our *prima facie* moral obligation to avoid destroying good ecosystems, agricultural, industrial, and post-industrial peoples have a *prima facie* moral obligation to refrain from the practice of hunting.

Contemporary thinkers often consider environmental degradation to be a byproduct of the industrial revolution and its aftermath. But although it is true that industrialization exacerbated environmental difficulties, it did not create them. Historically, the transition from hunter–gatherer or horticultural societies to agricultural societies caused good ecosystems to be degraded. Paul Shepard writes (1973, p. 20):

> In official history 3000 BC marks the beginning of civilization, corresponding to the rise of equatorial valley irrigation monocultures, the city–rural complex of specialized, single-crop farms and ruling bureaucracies of the great river valleys. In the archaeological residue of the Mesopotamian states there is evidence of ox-drawn carts, trade, writing, slaves, wars, and theocratic kingships. During this same period there was debilitation of the total natural complex, pillaged ecosystems that never recovered. The signs of this were the local extinction of large wild mammals, deserts replacing forests, the degradation of grasslands and the disappearance of soil, the instability of streams and drying up of springs, and the depletion of land fertility— all of which affected water supply, climate, and economy.
>
> Eroded slopes surrounding ancient cities, the burial of these cities under successive layers of silt, periodic floods, and the explosion of the

human population, all testify that in both the near east and the far east agricultural civilizations upset the balances characteristic of good ecosystems. For the first time human beings had a disruptive environmental impact unlike that of any other species. Subsequent expansion of agricultural systems and the growth of industrial civilizations, accompanied by unprecedented growth in the human population, simply made matters worse and worse.

The deleterious environmental impacts of these developments suggest that human beings are part of and yet opposed to the biotic pyramid in the way that cancer cells are a part of and yet opposed to the body in which they are growing. Cancer cells arise naturally (as opposed to supernaturally) from cells that play a normal part in the body's functioning. When normal, these cells interact with cells of their own and other types in ways that promote bodily health. Similarly, agricultural, industrial, and post-industrial human beings arise from hunter–gatherer and horticultural humans who play a normal part in the functioning of the ecosystems of which they are a part. These humans interact with one another and other environmental constituents in ways that promote the ecosystem's health.

Cancer cells, unlike the normal ones from which they arise, multiply so quickly and absorb so many of the body's resources that they upset the balances constitutive of bodily health. Left unchecked, they eventually destroy the bodily system from which they arose. Similarly, agricultural and subsequent human beings, unlike hunter–gatherers, multiply so quickly and absorb so many of the environment's resources that they upset the balances constitutive of healthy ecosystems. Left unchecked, they destroy the ecosystems from which they arose.

In the last section it was argued that people have a *prima facie* moral obligation to avoid destroying good ecosystems. Hunter–gatherer and horticultural people fulfill this obligation, since their ways of life are not ecologically damaging. So there is nothing in the present paper to suggest that these people ought to cease hunting.

Agricultural and industrial people, on the other hand, do tend to destroy good ecosystems. Their *prima facie* moral obligation to avoid causing such destruction therefore gives them a *prima facie* obligation to alter their behavior so as to minimize their destructive impact. Because their practice of hunting is one of the ways they damage the environment, agricultural and industrial people have a *prima facie* moral obligation to cease this practice.

Advocates of hunting would challange this reasoning by questioning its claim that hunting is ecologically damaging. They give two reasons for thinking, quite to the contrary, that ecological considerations favor hunting. First, they argue, our ecological difficulties are (Scanlon, this volume):

. . . due in part to the fact that most agricultural and industrial people lack an ecological perspective; they tend to think of themselves as overlords and observers of the ecosystems in which they live rather than as participants co-equal with other species. What is needed, therefore, is for more agricultural and industrial people to participate knowingly and explicitly in the primary relationship that we have with other species. That relationship is competition. By consuming space, water, food and energy, we deprive individuals of other species of their lives. Hunting is a singularly appropriate way for agricultural and industrial people to internalize a realization that human beings fill a competitive ecological niche like all other species. For in hunting, the competitive aspect is most explicit and the experience of other species most personal. So ecological considerations suggest that hunting ought to be encouraged, rather than discouraged among agricultural and industrial people.

A second ecological reason for encouraging hunting is that hunters often help preserve the balance of nature. The hunter's ethic, in concert with government regulations, insures that hunted populations are neither decimated nor allowed to so increase as to jeopardize other species.

I will now counter these arguments in the reverse order of their presentation. The second argument grossly misrepresents the ecological situation. Modern hunters have a significantly different impact on the ecosystems in which they hunt than do hunter–gatherers. They are more numerous per acre of land suitable for hunting because in modern society so much land has been allocated for other purposes, farming and urbanization. Also, the human population per habitable acre is so much greater in modern than in hunter–gatherer societies as to much more than compensate for the fact that a smaller percentage of moderns than of hunter–gatherers participate in the practice of hunting. Finally, the modern methods of hunting facilitate considerably the successful consummation of the hunt. The result is that the modern hunter disrupts rather than helps maintain natural balances in the ecosystem. To keep hunters from decimating hunted populations, the natural predators of those populations are decimated, such as the wolf, formerly a predator of deer in Wisconsin. Leopold describes this process (1970, p. 268):

> It works thus: wolves and lions are cleaned out of a wilderness area in the interest of big-game management. The big-game herds (usually deer or elk) then increase to the point of overbrowsing the range. Hunters must be encouraged to harvest the surplus, but modern hunters refuse to operate far from a car; hence a road must be built to provide access to the surplus game. Again and again, wilderness areas have been split by this process. (p. 268)

> Damage to plant life usually follows artificialized management of animals—for example, damage to forests by deer. Overabundant deer, when deprived of their natural enemies, have made it impossible for

deer food plants to survive or reproduce. Beech, maple, and yew in
Europe, ground hemlock and white cedar in the eastern states, moun-
tain mahogany and cliff-rose in the West, are deer foods threatened by
artificialized deer. The composition of the flora, from wildflowers to
forest trees, is gradually impoverished, and the deer in turn are
dwarfed by malnutrition. There are no stags in the woods today like
those whose antlers decorated the walls of feudal castles.

On the English heaths, reproduction of trees is inhibited by rabbits
over-protected in the process of cropping partridges and pheasants. On
scores of tropical islands both flora and fauna have been destroyed by
goats introduced for meat and sport. It would be hard to calculate the
mutual injuries by and between mammals deprived of their natural
predators, and ranges stripped of their natural food plants (pp.
287–288).

Thus, the argument that hunting is ecologically beneficial is entirely unsound.

The other pro hunting argument related directly to ecology was that
hunting serves to give agricultural and industrial people a better appreciation for their places in the biotic community. But if hunting is actually
destructive of the biotic pyramid, whereas hunters consider it beneficial,
then the appreciation hunters gain for their ecological niche is ill-
informed, misdirected, and therefore a danger to the ecosystem. This is
not to deny the importance for the ecosystem of agricultural and industrial
people visiting and traveling in wilderness areas. Such communion with
nature is necessary for many people to appreciate good ecosystems and
therewith the obligation to refrain from destroying them. But hiking and
backpacking can serve this purpose as well as hunting. And if a trophy is
needed, then photographs can be taken.

Thus, neither argument purporting to show that hunting is beneficial
to the environment is sound. The overall ecological effect of hunting is, in
fact, very negative. So our *prima facie* obligation to avoid destroying
good ecosystems yields a *prima facie* obligation to refrain from hunting.

THE STRENGTH OF THE OBLIGATION

Now it is necessary to investigate the countervailing moral considerations
that might override one's *prima facie* obligation to refrain from hunting.
Several candidates will be discussed, only some of which will be found
acceptable. We begin with the unacceptable suggestions.

1. Camping and wilderness photography, it might be maintained,
have a detrimental effect upon the environment. So if the benefits of these
pursuits override one's obligation to avoid destroying good ecosystems,
the benefits of hunting do so as well. Why discriminate in favor of
backpacking photographers and against hunters when considering the
ways in which people can commune with nature?

The reason is simple, hunters have a much greater detrimental effect on the environment than do backpacking photographers. Leopold put it this way. "The camera industry is one of the few innocuous parasites on wild nature. We have, then, a basic difference in relation to mass-use as between two categories of physical objects pursued as trophies" (Leopold, 1970, p. 228). If, as I have argued, agricultural and industrial people have an obligation to reduce the destructive impact of their ways of life on the world's ecosystems, then they have an obligation to prefer backpack wilderness photography to hunting.

2. Another objection to the substitution of photography for hunting is based on considerations of human psychic health. We humans have, presumably, been hunter–gatherers throughout ninety-nine percent of our evolutionary history. Hunting must come naturally to us. We must have natural tendencies to hunt, engrained during our evolutionary history, just as we have natural tendencies to respond to certain secondary sex characteristics in members of the opposite sex. In the one case, as in the other, natural selection would have favored members of our species with these sentiments over those without them. The frustration of natural tendencies is detrimental to people's mental, and sometimes physical health. So just as people should be encouraged to be sexually active, barring overriding, countervailing considerations, so should they be encouraged to hunt. This is the consideration stressed recently by Paul Shepard and by José Ortega y Gasset before him (Ortega y Gasset, 1972). Shepard quotes with approval Ortega's claim that hunting is "a deep and permanent yearning in the human condition," adding that as a result "there is a chronic fury in all people to whom it is denied" (Shepard, 1973, p. 150). "But photography as a substitute for hunting is a mockery" (p. 151).

In an age of increasing complexity and artificiality, it is easy to sympathize with the view that people will only be at peace and feel comfortable with themselves and others if the conditions of their existence more closely approximate those present during the major part of our evolutionary history. Surely a society without extended family ties suffers greatly. Nor can humans expect to find fulfillment working on assembly lines in modern factories. But to admit this is not to admit that because people used to hunt animals, they must do so now in order to promote their mental health. A mean must be found between the extremes of slavery to our distant past and acquiescence in whatever conditions of life are dictated by the need for efficiency in a growth economy.

Difficult as it is to locate this mean in many specific cases, I will venture at least this generalization. Given the unparalleled adaptability and flexibility of human beings among animals, attention should not be focused on detailed similarities between specific activities or relationships in the modern and pre-historical human environments. Rather, the basic human needs met by a specific activity or relationship in the pre-historic period should be sought. Then we should consider what substitute activities or relationships might fill these same basic needs. Finally, attention must be paid to the experience of how life feels to people and how they function socially with given substitutes. Only then will it be known whether or not we have cheated or impoverished our sensibilities by accepting these substitutes.

On this measure, adoptive parenthood, for example, fares well. The ties of affection formed, the socialization of the child and the fulfillment of the adults' desires for parenting are indistinguishable in adoptive parent–child relationships from those in the historically more natural relationship in which parent and child are biologically related. Yet even here, recent evidence suggests that it is important for the adopted child to have full knowledge of his or her biological parents. With this in mind, I would guess that Paul Shepard is in error when he claims that wilderness photography is an inadequate substitute for hunting. The photographer must still study animals' habits and adjust to their habitats as would a traditional hunter-gatherer. The major differences concern the consequences of "shooting," and the nature of the trophy obtained. Given the massive differences in myriad areas between modern and hunter–gatherer life, it is implausible that the differences between hunting and wilderness photography are the ones that the human psyche cannot tolerate. If it is maintained that it is precisely because of the massive differences in so many other areas that hunting is necessary, the reply is that we should, rather than hunt, modify our life-styles in some of these other areas. Unlike hunting, many modifications that would make modern life more "natural", such as walking short distances instead of driving a car, do not run afoul, other things being equal, of any *prima facie* moral obligations.

3. A third contervailing consideration that might be thought to override one's *prima facie* obligation to refrain from hunting concerns the economic effects of such a change in people's recreational habits. Gun and ammunition manufacturers, distributors and retailers would experience the human misery of economic dislocation. So people should continue to hunt in order to spare their fellows such misery.

The weakness of this argument is apparent upon the reflection that all changes in technology and consumer tastes have similar effects. Yet, we do not think ourselves morally obligated, for example, to continue to buy large, gas guzzling cars, automatic popcorn poppers, and hula hoops because a change in purchasing habits disrupts the lives of the manufacturers, distributors, and retailers of these products. Nor do we consider the inventor of the telegraph to have immorally deprived workers on the pony express of their employment. In short, we do not normally think we have an obligation to arrest change simply to avoid inconveniencing those benefiting from the status quo. And we reject such an obligation for the sound reason that its acceptance would have a stultifying effect upon society and thereby do a great deal of harm. In fact, the argument that one ought to continue hunting, and thereby destroying the environment, in order to avoid disrupting the gun and ammunition industries is a perfect case in point. Since we do not ordinarily find this line of reasoning acceptable, there is no reason to accept it here.

4. Because some people enjoy hunting more than (they suppose) they will enjoy any alternative recreation, it might be argued that their loss of enjoyment constitutes a moral consideration that overrides their *prima facie* moral obligation to cease hunting.

But this reasoning is very weak. A non-utilitarian would point out that one *prima facie* moral obligation can be overriden only by another. But we do not have any *prima facie* moral obligation to maximize our own enjoyment. So the loss of enjoyment here in question does not constitute a countervailing, much less an overriding moral consideration.

Utilitarians, of course, would maintain to the contrary that the loss of enjoyment is a countervailing moral consideration. However, they would not consider very small losses to override a *prima facie* moral obligations. If every small loss overrode one's *prima facie* moral obligations, people would always be obligated to do what is in their individual interests (no matter how trivial those interests might be), and utilitarianism would become equivalent to egoism. So a utilitarian who includes healthy ecosystems among those things considered good cannot simply accept the claim that the loss of enjoyment on the part of hunters overrides one's duty to avoid destroying ecosystems. This claim would need to be accompanied by a demonstration that this loss is quite considerable (and therefore of greater importance than the obligation in question). The only serious attempts to show this, Ortega's and Shepard's, have already been examined and found wanting. So from any non-egoistic ethical perspective, the loss of enjoyment on the part of would-be hunters does not override the moral obligation to cease hunting.

There are, of course, considerations that are overriding. In the short run, to meet an emergency, it may be permissible for people to hunt in order to feed those who would otherwise starve. It may similarly be permissible in the short run, to meet an emergency, to hunt in order to eliminate animals that threaten crops that people must raise in order to eat. But, in the latter case, hunting must be conducted only until natural predators are returned to the ecosystem. The predators will then take care of the problem. This will work for most insect control, as well. In cases of human starvation, hunting should be used only until a combination of population control and local agriculture eliminate the problem. In no case should hunting be a sport or recreation. If it is pursued as a sport, there will be incentive to extend its practice beyond that necessary to meet temporary emergencies, including a tendency to perpetuate those problems which create the emergencies in the first place. Such hunting is immoral.

Humans as Hunting Animals

Patrick F. Scanlon

Henry Cooper was a British heavyweight boxer who, during his career, was the British and European champion. He achieved some fame by knocking the then Cassius Clay off his feet at an early stage of Clay's career. On a television program Cooper was confronted by Dr. Edith Summerskill who was an advocate of a total ban on boxing. She asked him had he looked in a mirror lately and seen the state of his nose. He retorted by asking her the same question and quipped that boxing was his excuse. Cooper told the story so presumably he felt he made his point. Dr. Summerskill probably felt she had made her point also. The sequence is not atypical of discussions between animal users and their critics. Both sides seem to make their points without ever convincing the other.

One might ask why the lack of success in convincing others of the seemingly compelling arguments on both sides of the issues on animal use. In my opinion much of the reason relates to fundamental misunderstanding of the position of man as a biological entity in the global ecosystem or biosphere. (Ecosystems may be considered the functional unit derived from the physical, chemical, and biological relationships of biological communities and their physical surroundings.) Unquestionably man is a part of the global ecosystem. He depends on the physical and biological resources of the earth for his survival. He has demonstrated the ability to manipulate the global ecosystem towards his own ends. Yet he is still a biological entity totally dependent on the global ecosystem for survival. To survive he must be a participant in the ecosystem and in fact he is.

HUMANS IN THE ECOSYSTEM

Many who fail to understand the human place in the global ecosystem fail to recognize its role as a participant; rather they see humans as observers or somehow set apart as nonparticipants in the ecological scheme. But humans clearly are participants and as such behave, or ought to, as other biological entities. They consume, excrete, grow, reproduce, and die. To consume, they *compete* with other biological entities. Human consumption has impact on plant growth and succession. Humans compete with animals for plant growth. The majority of the members of the species choose to exercise the option of being predators, though they may not necessarily recognize themselves as such. It could be argued that humans would not readily survive without adopting the role of predator.

The human role is somewhat more complex than that of other species. Human consumption involves more than merely what people eat, and likewise, human waste products are more complex and exert major impacts on other species. In order to provide food, humankind has adopted agricultural and animal husbandry practices that necessitate its protection of the cultured species, and as a result, humans compete with unwanted would-be consumers and predators of their crops and animals. The remarkable continued population growth of humans exerts tremendous competitive influences on other species.

The above few points are made to emphasize humankind's role as a *participant* in the global ecosystem rather than as a detached observer. The arguments may be based on human needs for food and habitat (which are just as those of any other species) and human willingness and ability to compete for them. The basic difference between humans and other animal species lies in the level of intelligence and development as a rational being. Further differences result from the human species' development of the ability to husband crops and animals to supply its food. Domestication of crops and animals has been employed to increase the efficiency of food gathering and predation, and humans have domesticated only a relatively few species of plants and animals for food and/or fiber production. Although the range of animals so domesticated is impressive (from silkworms to elephants), the vast majority of animal species have presumably been found unsuitable for domestication or have successfully resisted domestication and remained wild. It could be argued that many domesticated animals are born "wild" and are subsequently tamed during continuing domestication. For example, all horses have to be trained for use as draft or riding animals. The ability to domesticate animals has allowed the further development of human society and led to a division of labor in food production. This has meant that humans are no longer necessarily required to be gatherers of food as individuals and so are no longer overt predators in most cases. In many countries and for a substantial portion of humankind, food gathering is reduced to that done within the four walls of grocery stores. The participation of the great majority of the human population in food, and particularly protein, production has greatly decreased in developed countries. Few people are alive today who recall the production of domestic animals, milk, and so on within city limits. In such circumstances it is easy for people to lapse into seeing themselves as nonparticipants in the ecosystem and to fail to recognize their roles as predators.

WILD ANIMALS AS FOOD

All meat production does not necessarily come from domesticated animals. Production of protein foodstuffs as a result of fishing and hunting is substantial on a world-wide basis. The vast majority of fish produced in the world are recovered from natural ecosystems (i.e., not subjected to husbandry), yet they amount to 10–20 percent of the animal protein consumed by man (Ehrlich et al., 1973). Much meat is produced from a wide range of mammal, bird, and reptile species, though the statistics on production are difficult to evaluate. Many African large mammal species produce substantial quantities of meat that is harvested essentially by hunting techniques (Talbot, 1963, 1966; Talbot and Talbot, 1963; Skinner et al., 1971). The potential exists to produce more meat efficiently from wild ruminants, especially under African conditions. The species involved are not particularly suited to domestication. Talbot (1966) considers meat from game animals the only reliable food crop capable of being produced in the savanna regions of Africa because of those regions' unpredictable climatic conditions. Species of Cervidae that were introduced into New Zealand are important sources of meat that are accessible only by hunting methods. In the USSR, wild ungulates are harvested as meat animals (Talbot, 1966).

The hunting of large ungulates in North America produces substantial quantities of meat. An example is the estimated yield of meat from white-tailed deer in Wisconsin, which amounted to 5300 tons in 1977 and was roughly equivalent to 10 percent of the swine production of that state (Bishop, 1978). Not all species can withstand commercial hunting, however regulated—waterfowl and upland game bird species of North America are examples here. Yet they, and all game animals of North America, yield substantial quantities of usable protein despite the fact that most hunting in North America is considered by some to be sport hunting.

HUMANS AS HUNTERS AND THEIR ATTITUDES TO ANIMALS

The continued development of the role of predator by humans probably evolved from the role as gatherers; presumably this meant that humans took what they could get without regard for the age, reproductive status, population status, or species of the animals involved. But humans are not particularly well-equipped predators, though our ingenuity has enabled us to develop the equipment necessary to be successful. The development of weapons, nets, traps, and snares have transformed us from opportunistic gatherers into full-fledged hunters.

With development as a rational being, questions of human relationships with animals developed. Judaism, and later Christianity, recognized humankind as unique, created in the image of God, and with total domination over nature (Spotte, 1976). Though it was acknowledged that there was a common bond of life between humans and animals, the animals were considered valuable property or a resource if wild. The development of such an attitude would be considered consistent with Genesis 1, 26–31. In large measure, the Judeo-Christian ethic is prevalent in North America today in regard to animals. It shares the stage, to some extent, with the panmictic ethic that does not accept the killing of animals for human benefit. In discussing these opposing ethics, Bider (1979) noted that they developed under contrasting living conditions. The Judeo-Christian ethic developed when people needed to survive; the panmictic ethic emerged only when they clearly had very great capability of destroying other organisms. He argues that both are extreme in nature, that both could be extrapolated to ridiculous extremes, and that some middle ground is necessary. From the standpoint of hunting and the use of wild populations, one such middle ground is the conservation ethic articulated by Aldo Leopold.

The conservation ethic is reasonable from the standpoint that it recognizes the production characteristics of the species and advocates the *wise use* of populations that can sustain limited harvest. Harvest by hunting is conducted on harvestable surpluses of the species prior to times when portions of the populations could be expected to be lost to the combined rigors of winter seasons or prior to times when populations start to exceed carrying capacity of ranges. The management associated with such an approach to animal harvesting can be designed to accomodate sustained yield of the species without allowing populations to proceed to points of major conflict, such as excessive crop damage, with the best interests of humans.

The management of species could be deemed imperative under modern conditions because of the major changes that have been wrought in the environment by human activities. Species respond differently to the changes humans induce—some respond with greatly increased numbers when areas are converted to agricultural cropland, while others fail to thrive under such changed circumstances. In any event it is unlikely that disturbed habitats and environments will revert to prior pristine conditions given the size of the present human population. Consequently, some management of species must be exercised in the absence of natural balancing mechanisms. If not provided, humans will continue to exert their adverse influences on the animal biota in a variety of directions, not all of which are desirable.

HUNTING AS A NATURAL PHENOMENON

From the standpoint of humankind's natural role as a predator, hunting is a realistic means for humans to involve themselves in the ecological process. Given our knowledge of the biology of individual species and our vastly improved means of harvesting them, restraint is obviously necessary. Proper hunting regulations provide a considerable means of control among those less than willing to restrain themselves, while many feel honor bound to observe their personal principles within the framework of existing laws.

A major problem in human relationships with animals in general, and wildlife in particular, is that many of us have, as suggested earlier, lost realistic contact with the natural world. Humans function as biological entities, yet often are far removed from the processes that ultimately determine their survival and from those processes that are an essential part of life in general, such as the provision of shelter and food. Usually the production of shelter is delegated to others. Many individuals never experience involvement in food production, be it gathering of crops or harvesting of animals. Yet they are practicing omnivores who derive the benefits of applied ecological competition and applied predation. Agriculture cannot be practiced for human benefit without conflicting with the interests of some other animal species, particularly wild species. Seed-eating birds and crop-damaging rodents, for example, are unwelcome as competitors to humans in crop production systems and efforts are taken to minimize competition from them. The provision of meat or animal products requires someone to undertake the slaughter of animals, remove eggs from chickens, or wean calves from dairy cows. Those that partake of the product are, by implication, delegating others to do their animal husbandry (which is essentially a form of applied predation). Modern human societies delegate many of the vital processes that a rural, or urban for that matter, family of a century ago would have routinely undertaken, such as the production of food, the supervision of the birth of offspring, and the burial of the dead.

The point of the above is that, by living as we do today in North America, we have to some extent fallen out of touch with the realities of life, at least from an ecological standpoint. We have put ourselves in a position where we see ourselves as observers of ecological processes when we are participants, albeit sometimes far removed from the ultimate realities. Those who involve themselves in food production at the primary producer level participate directly in ecological events. They are acting as participants, usually changing some "balance" in nature, but producing necessary food. Food producers are currently relearning some forgotten biological principles, fundamental ideas abandoned with the widespread use of chemicals for plant protection (Giese et al., 1975). The recognition

of the value of an integrated pest control approach, one emphasizing biological control, is necessary if we are to realign production systems with
natural processes. Those who harvest wild animals *participate* in and exert humankind's role in ecology, at least to some extent. The fact that
their role is modified by considerations of modern wildlife management
for the species being taken does not lessen the fact that they are exercising
a realistic participatory role in ecology.

The hunting of wild animals is a realistic use of the species being
hunted. Modern game regulations are designed to avoid adversely
influencing the species involved. The hunted species may yield considerable quantities of meat, which is in itself a valuable resource, one that
also is renewable if the species is properly managed. Use of this meat
prevents the further conversion of wild lands to lands supporting the production of domestic animal meat, and keeps meat demand satisfied.
Hunting provides unique opportunities for humans to participate in the
natural ecological scheme in such a way that one does not delude oneself
concerning the degree and nature of the participation.

A condemnation of hunting while advocating vegetarianism has
been made by Wenz (this volume). In attempting to augment both components of his arguments, Wenz failed to understand the human role as a
participant in the system. Participating in the ecological scheme implies
surviving within it, and to survive humans must successfully compete for
its resources. Adoption of the role of "conquerer" is not synonymous
with the role of "participant," though participants may sometimes perform conquering acts. Thus, to participate in the global ecology, humans
must eat. If we kill other animals to eat, that is one form of participation.
It involves use of individuals of other species, but not necessarily the killing of the entire species, to do so. The wise use of a species allows for use
only of a harvestable surplus. Moreover, the level of crop production necessary to support humans as vegetarians necessarily involves competition
with other animal species. Usually, the competition is less well-defined in
terms of competition with individual members of species; rather whole
species tend to be adversely influenced, usually being completely dis-
placed by the crop production necessary to achieve the aims of vegetarianism. Put very simply, the position of universal vegetarianism can only
be argued from positions of ignorance regarding global ecology or of naive interpretation of ecological realities.

Concerning the delusory aspects of human participation or
nonparticipation in ecological processes, many are unconvinced concerning their roles in ecological processes. They see themselves as
"nonconsumers." Such a concept is self-delusory. Many examples of the
influences of "nonconsumers" on wild animal populations have been
identified (Weeden, 1977; Wilkes, 1977). Humans cannot live without
exerting influences on other species. From an ecological standpoint, to
live is to consume, to consume is to compete, to compete successfully is

to out-compete, i.e., to work to the detriment of something else, be it plant or animal. Whether one competes by spewing automobile exhausts along highways, or by using electricity the generation of which disrupts anadromous fish runs or produces the sulfur fumes that cause acid rain, or by using food grown on land made available by the displacement of some native species, or by wisely using the animal products of the land, one competes and other organisms feel the competition. Understanding and honest acknowledgement of such competition renders humans capable of staying in touch with their real needs as biological entities on this planet. The more closely and honestly we participate, the more likely it is we will understand our true role in the global ecology and apply those restraints necessary to avert major disasters in the planetary ecosystem.

Apes and Language Research

Duane M. Rumbaugh
and
Sue Savage-Rumbaugh

People are always amused when they note others who bear resemblances to their pets. Humankind has had a long standing preoccupation with animals, has found great delight in them, and has concluded that it might learn some things about itself by observing their behavior.

When it comes to language, however, we tend to view ourselves apart from animals because it is thought that through language we can achieve many rather remarkable things that they cannot. We know that language helps us to communicate about abstract ideas that are not present in time and space, and that with language we are able to summarize the basic principles of our behavior and world. With language it is also possible to communicate with others in ways that allow us to solve problems cooperatively and in novel ways. If dogs had a valid comprehension of what a ladder really was, had a word for ladder, and could ask for and use ladders, they would be equipped to get at the neighbor's cats in more effective ways than by just chasing them on the ground and barking at them when treed.

Herein lies one of the basic dimensions of quantitative differences between humans and animals; humans are far more plastic in their definition of the options available for coping with their problems than are animals. On the other hand, the comparative psychologist or the animal behaviorist who studies animals within the context of laboratory and domestic settings is frequently impressed with what animals come to do when they are given free time and spared from the pressures of survival in the field. Even captive birds have been known to do some rather clever things, such as taking morsels of food and dropping them through their cage's floor directly onto the dog that is lying underneath.

One of the mistakes in studying animal behavior in the past has been to assume that their behavior is totally restricted to specific responses being conditioned to particular stimuli. There is no doubt that learning at this simple level can account for many behaviors acquired by animals. However, in our view the behaviors of many mammals are not limited to simple stimulus–response contingencies. We are intrigued with *what it is that animals can do, as opposed to what it is they typically do.*

The first author started with a squirrel monkey colony for breeding purposes in the early 1960s and one Saturday afternoon received a call that there had been a birth. Unfortunately, the baby was born dead. Interestingly, there was no obvious way of telling which of the three females in the cage, along with a single male, was the mother of that baby that lay dead at the bottom of the cage. The cage was six feet tall, and all of them stayed on perches at the top of the cage. After some 30 minutes, a simple experiment was performed. The lifeless baby was probed with a stick, whereupon a very dramatic thing happened. One of the four adult animals dropped from the top of the cage down to the dead baby. It was one of the females; she thrust her abdomen onto it in a way that squirrel monkey mothers do when they are retrieving their babies. When they do this, the baby normally attaches itself to the mother's ventrum, whereupon the mother runs off with the baby clinging to it. In the context of reference, this did not happen. Apparently in response to the baby's failing to grasp, the mother sat back on her haunches, then with both hands picked up the baby and walked off bipedally for a distance of about two feet! Why was this of significance? It was because the squirrel monkey is a very profound quadruped, not a biped. She is also a very passive mother. Typically she does not pick up her babies with her hands; she does not cradle or groom them. The baby essentially uses her as a bio-carrier, and the baby does just about everything for itself. But, in this instance, a strong compensatory mechanism came into play.

Working on the hypothesis that this compensatory behavior was probably a function of incompetence in the infant, the birth of a viable squirrel monkey was eagerly awaited. At the appropriate time we took the baby from the mother, handicapped it by carefully taping its arms to its body so that it could not cling, allowed the mother to come to it, and again we saw the whole panorama. The mother (not the same as above) again put her abdomen on the baby, but the baby could not cling. She then sat back on her haunches, picked up the baby, and walked away carrying it as she walked bipedally (Rumbaugh, 1965). So, once again, another very passive quadruped squirrel monkey mother was converted into a very active, care-giving biped.

It is possible that this was one of the major dimensions of pressure for the evolution of the erect posture in man; the human baby is relatively more dependent upon its mother than is any other primate form. Interestingly, the great ape babies are really quite dependent (almost as much as is the human child). Some of the longest instances of apes in the field walking erect entail mothers carrying sick babies. If one has a dependent baby, dependent because of profound immaturity at birth, then how is that baby going to survive? It survives to the degree that the mother can compensate for that. How can she best compensate for it? By carrying it with her, and that might be facilitated if she walks erect. The babies that would thus survive would more likely than not carry some of the genes that allowed the mother to assume this walking posture. Thus, there might have been a co-evolution of dependence of the infant, because of its increased immaturity at birth, and an erect and bipedal posture by the carrier-mother.

Now that is not to say this is the only factor that brought about the erect posture and bipedalism. We ever so frequently make the mistake of attributing major developments to single events. It can be, but only rarely is that likely the case. Rather, there is an orchestration of selective pressures that urge forth the evolution of attributes, and that is what makes it, in the final analysis, impossible to find out why some fundamental capabilities such as language are what they are.

In 1971 we received our first grant support for what is now known as the Language Formation Study Project (Rumbaugh, 1977). When that proposal for the grant was submitted, the intent was to develop a totally automated computer-based, language training situation, to eliminate the human factor, and thereby get down to the hard science of language-like behaviors in the ape. That effort was misconceived. Lana, our original chimpanzee, learned very little unless a human being was working with her and attempting to teach her symbols in an interactive-social context. Under those conditions, she learned some things quite rapidly. Thus, although the introduction of social factors had initially been resisted in order to achieve as ''pure'' an experimental setting as possible, it rapidly became clear that with the removal of the social factors, there appeared nothing even like language in Lana's behaviors.

Chimpanzees have great similarities to humans in terms of their affective behavior. Baby chimps like to be held, they like to be comforted, they like to be hugged, even kissed, and they will hug and kiss back, and they become very distressed when people with whom they have close bonds go away from them. If you want to punish an ape, just take away from it its social security attachments and you have a very disturbed creature.

In addition to maintaining close relationships between our teachers and ape subjects, we involve the apes in the various activities of the daily routine. The day begins with the chimps asking for their brushes, soap, and water and they help scrub down the cages in which they spent the night. And they will focus upon the fecal material and dirt as targets for scrubbing, and then they will help hose down the room. So it is within the scenario of cleaning the room then that they learn some things related to cleaning utensils and activities. They also help with the preparation of the dishes for washing, and in various tasks, some of them rather unpleasant, such as unplugging of drains in colony rooms and cleaning out of bark from the track, which is usually necessary so that the outside door may be closed at night.

The project is situated at the Yerkes Regional Primate Research Center of Emory University. The research takes place in a 1500 square-foot Butler building to the rear of the main center. It has two high-rise exercise yards, not only to take advantage of limited ground space but also to avail to the apes the opportunity to climb and to get up to where they can get a good view of things.

The typical training situations appear in Figs. 1A, 1B, 1C. The heart of our approach has been the employment of a computer-based keyboard. This enables us to avoid many of the problems inherent in a manual sign-based communicative system. The only way signs can be recorded is either to tape or film everything or to have an observer there dictating each sign as it occurs. Such observers might not see all the signs, or might introduce a bias, intentionally or otherwise, in terms of what it is that they report. Correct and interesting behaviors tend to catch the eye, more so than nonsensical, inappropriate behaviors. It is also expensive to go over video records in order to extract the information and then get it into some kind of a computer record. Chimpanzees are not necessarily very facile signers, and frequently the chimps' signs are not clearly interpretable by people who know sign language. Thus, one constantly runs the risk of projecting one's self onto the rather amorphous behavior of the chimpanzee.

Figure 1A

Figure 1B

The keyboard that currently is in use in our project is the result of nine years of engineering effort. The symbols on the board can be activated or deactivated at will by the experimenter. As keys are activated, they become backlighted. This brightly lighted board is very colorful and inherently appealing to both the chimpanzees and the intellectually disadvantaged alinguistic children with whom we also work at the Georgia Retardation Center. They are attracted to the color, to the change of the lights as keys are activated, and to the facsimilies of the geometric patterns on the surface of the keys as they appear on the projectors above the keyboard as the keys are used.

Now, by having duplicates of word-keys on the boards and differentially activating them, it is possible to move the keys about, even between trials, so that the subject must attend to the lexigram and not the location of the key. This is important because primates are often predisposed to use location as a cue. Mentally retarded children also tend to use the location of keys as a cue for selecting them; however, they apparently are not as strongly compelled to do so as are the apes.

As the apes become increasingly sophisticated, location tends to fall out of their own preference, and they begin to concentrate more on the new lexigrams as they are introduced to be learned. The computer-based system also allows the experimenter to record, in real time, who used the keyboard and whether what was said was correct or incorrect. The capability of the keyboard is exercised through use of a modified hand calculator that has been encased in a stainless steel box to make it more rugged

Figure 1C

for laboratory use. This device also enables the investigator to rapidly activate or deactivate the keyboard, and to enter codes that define the context of the utterance.

The terminal in the computer area allows the experimenters to retrieve data in a systematized form at the end of any training session. In

terms of real time activity, as soon as an utterance is made by a subject, be it chimp or human, there is a printout of it when the period key is depressed. The period key is not for punctuation purposes; it serves as a signal to the computer that a transmission has been completed and to activate the high-speed printer to print out what has been uttered along with the time (in hours, minutes, and seconds) of that utterance. It is possible to have a number of keyboards functioning concurrently, attended to by the computer on a priority-interrupt basis, and all handled so quickly that there is no noticeable delay. At the end of the session, the experimenter may call out a record of what has occurred in each test room, a rank ordering of utterances in terms of *frequency* of use, the relative frequency with which any subject used a given word, or just those utterances that are adjudged correct when "uttered" as opposed to those that were adjudged incorrect.

Chimpanzees are not the furry, little animals that just want to learn everything and be good subjects that they are frequently thought to be by the general public. Chimpanzees are rather temperamental, and can be extremely dangerous. The chimpanzees, Sherman and Austin, are of the species *Pan troglodytes,* the so-called common chimpanzee, and they tend to be rather emotional, highly reactive, and easily upset. Sherman is now 7, Austin is 6. This is the first time that people have worked so closely with males of this age and, to date, in relative safety. Our intent is to work with the chimpanzees throughout adulthood if possible. Whether this will be possible is a question to be answered day-by-day.

WORDS AND CHIMP-TO-CHIMP COMMUNICATION THROUGH THE USE OF LEARNED SYMBOLS

Teaching chimpanzees that people, objects, places, and events can be represented by symbols has proven far more difficult than originally anticipated. For example, the chimpanzees did not initially grasp the concept that objects could be replaced, in a communicative sense, by lexigram symbols even though they had learned *to associate* particular symbols with specific contexts. To be more specific, if a door between two rooms were locked, and the chimpanzee needed a key in order to unlock the padlock, it learned very rapidly to say GIVE KEY, in order to get the experimenter to hand it a key. The chimpanzee would then unlock the padlock and go through the door to its companion or to the food on the other side. But if that same chimpanzee were then asked to simply name a key, without using it, it appeared to be totally ignorant. "Key," in short, was *what the chimps said* when they faced a particular problem situation, only part of which was this padlock. Key as a word was not specifically focused upon a little brass object, and it was only after a thorough, extended training that intermeshed a *variety* of contexts, in order to focus upon the particular object itself, that the chimpanzees became able to separate the "name" of an object from its function.

The acquisition of even a single word is therefore, a gradual process that involves the accumulation and interdigitation of a variety of symbol-linked skills. The chimpanzee does not spontaneously bring to the task of language-learning the ability to use a symbol to talk about things when the original exemplar or reference of those things is not present in some time and space. But chimpanzees are, indeed, capable of doing that.

The interanimal communicative use of symbols was first studied between Sherman and Austin in a situation where foods and drinks were put in sealed containers in a paradigm that ensured that only one of the chimpanzees knew what was in the container on a given trial (Savage-Rumbaugh, Rumbaugh, and Boysen, 1978). That chimpanzee, through use of the keyboard, would name the contents of the container, e.g., THIS ORANGE-DRINK. This gave his companion the information needed to request the hidden food or drink, by saying, for example, "GIVE ORANGE-DRINK." If the container, when opened, was found to contain orange-drink, then the chimpanzees shared it.

A number of control tests were administered to make sure that the second chimpanzee was *not* just matching-to-sample by selecting the same key on the board that had been used by the first chimpanzee. These controls included the use of two keyboards. Each keyboard was located in a separate room, but a window between the rooms enabled the second chimp to see the specific keys used by the first chimp. However, the symbols on the two keyboards were in different locations, thereby, precluding the second chimpanzee from simply lighting the key touched by the first chimp.

In order to determine that the second chimp was not just matching symbols (irrespective of position), an attendant outside of the room who *did not know* what was in the container gave a predetermined set of three pictures to a person in the room with the chimp. One of the pictures was of the hidden food. These pictures were laid out in a row by that person, who also did not know the contents of the container. The second chimp (who had not seen the food placed in the container) then had to choose one. In the example at hand, only if the chimp pointed to a picture of orange drink was the container opened, because that was what he asked for.

Thus, there was room for many errors. The first chimp might have misperceived or misremembered what he saw put into the container. Or, he might have misstated at the keyboard. The second chimp might have misread or misrequested, or might have pointed to the wrong picture, and thus led to the animals' failure to be correct. Those problems notwithstanding, the chimps were better than 90% correct in this task, including their first inter-chimpanzee communications of this nature and throughout all the control tests. When the keyboards were turned off, their ability to communicate the container's contents dropped to chance. They could not, when left to their own natural communications reliably tell one another what had been placed in the container.

After these two chimpanzees learned to identify for one another what food or drink they had seen put into a sealed container, they transferred these skills to a task in which they simply asked one another for foods that were not hidden. In this situation, one chimpanzee was placed on one side of a window containing a small porthole. A tray of foods was also located near that window. The chimpanzee on the other side had no food, but it did have a keyboard. It could see what was in the first chimp's tray of food, and it would use its keyboard to request any food which it saw. The chimp that had the tray of food was encouraged to pick up the requested food and hand it to the chimp that asked. The chimpanzee that had the keyboard would ask for any of the various foods on the tray, and the chimpanzee that had the food tray picked up a piece of it and put it through the window right into its cohort's mouth. The chimpanzees are able to request foods and to comply with one another's food requests with high levels of accuracy (better than 95% correct). The chimpanzees ask first for things that they like best, and when a given type of food is gone, they do not ask for it anymore.

A very close friendship has developed between these two chimpanzees during the course of training, and many of the sharing behaviors that they have learned have now generalized beyond the food request task. For example, on one particular day a coin-operated vending machine had been loaded with small peanut butter and jelly sandwiches. The experimenter then dropped two coins down a long tube. Sherman picked up a magnet (which was tied to a string) and dropped it down the tube and fished with it to get the money. When he heard the "click, click" as the money attached to the magnet he pulled it out, and removed the two pieces of money from the magnet. Then he gave one piece of money to Austin, and kept one for himself. They both went over to the vending machine and Sherman put in *his* money, got a peanut butter and jelly sandwich, which he then gave to Austin. Austin took his money, put it in the machine, got his peanut butter and jelly sandwich, and gave it to Sherman. Thus, money, food, and turns were each shared.

EXTENSION OF THE RESEARCH TO THE INTELLECTUALLY DISADVANTAGED

The whole thrust of this programmed research, since its inception in 1970, has been to develop a technology that would (1) objectify ape-language research and (2) have applied value for language programs aimed at severely disadvantaged children who have not learned to speak.

Figure 2

Following a three-year pilot project with Lana chimpanzee, we initiated a feasibility study with such children at the Georgia Retardation Center.

Figure 2 portrays language training with a young boy. Another of our students was a girl who was severely brain damaged, because of a gas leak in her home, when she was 1 ½ years old. Her mental age is that of a 3 to 3 ½-year-old child. She cannot speak at all well because of neurological deficits. She has, however, learned approximately 150 lexigrams and has spontaneously extended use of them to productive combinations and is now able to engage in rudimentary conversations with others. She also uses lexigrams to gain compliments from her teacher and to negate things that are incorrect on her teacher's part. We have to date worked with nine children quite intensively; five of them have developed substantial language skills. These children are brought to the language lab in groups of two or three. They work together and learn language cooperatively, through various activities, whether it be to make popcorn, to name parts of the body, to name various parts of a dress, or to name activities and people who come into the room. Rather remarkable changes have taken place in the children's behavior, apparently as a result of this kind of experience. The majority of these children displayed a number of behavioral management problems at the onset of the study. These problems have steadily abated, not only within the language training context, but in the cottage situations as well.

PERSPECTIVES

Although we have learned much about the chimpanzee and its language learning capabilities, it is not the case that they begin to challenge humans in their use of language. It is very clear that, on the other hand, the chimpanzees can learn words. They can learn to use those words in symbolic communication between themselves and with humans about them. They can learn to use those skills to identify tools and objects that are important to them jointly in problem situations that they face. Their deficit at this point in time appears to be an inability to exploit language as a skill for learning about their environment at large.

Addendum

Since the presentation of this paper, substantial progress has been made in ape-language research. Chimp–chimp communication (Savage-Rumbaugh et al.,1978) through use of symbols has been reported in detail. That chimpanzees of our project use their symbols representationally, as referents for items not present in time and space, has been demonstrated (Savage-Rumbaugh et al., 1980). That their symbols function thus means that our chimpanzees have at least a rudimentary form of semantics. The interested reader will want to study these and more recent articles to achieve a current perspective of the field.

ACKNOWLEDGMENT

This paper and the research of the authors have been supported by NIH grants HICHD-06015 and RR-00165.

SECTION V

Human Interests, Porcine Interests, and Chipmunk Interests

Introduction

Almost (but not quite) everyone who considers the matter comes quickly to the belief that human beings should assign higher moral priority to the interests of other humans than to the comparable interests of other animals. But a convincing justification for this belief has not been easy to find. Lawrence C. Becker attempts to provide such a justification. For one to be virtuous, morally excellent, it is not sufficient to *do* the right thing, one must *feel* the right way. Virtue entails having and acting on certain 'uncalculated feelings.' (Here Hume and Baier would surely agree.) Many of these feelings, and the traits of character that they partially constitute, decrease in strength with increasing 'social distance.' My family's interests *should* (generally) be more important to me than the interests of my friends, and those more important than the interests of strangers.

In particular, the traits of reciprocity and empathetic identification are parts of human virtue, and entail ordering preferences by social distance. With few exceptions, members of other species are at greater social distances from us than are members of our own species. Thus, if we are to be virtuous, we must give preference to our fellow humans. Some sort of speciesism, Becker concludes, is therefore justified. But what sort? Certainly not an absolute speciesism that would count any human interest as weightier than any sum of animal interests. But at least a weak speciesism, holding that sometimes equivalent animal interests should be sacrificed for human ones, is a necessary corollary of human virtue. Exactly how the virtuous man or woman will deal with animals we cannot now say, for our understanding of virtue is still inadequate.

James Cargile agrees that human interests should be given priority, but he argues that Becker has failed to provide a justification of such priority. Empathy, Cargile claims, is not the point. There are people with whom it is very difficult to feel sympathy, but we are nonetheless obliged to place their interests higher than those of animals to which we may be very attached. There are humans that no one (themselves included) can stand, but we are not justified in disregarding their interests. Parents are, of course, more concerned with *their* children's interests than with the interests of other children, but few parents are so obsessed as to affirm moral priority for the interests of their own children.

221

In fact, asks Cargile, might not a virtuous human decline to give priority to humans, without thereby losing his or her claim to virtue? It will not do to answer that such a moral view would entail great sacrifices. Many highly plausible moral views, if acted upon, would entail great sacrifices even if the moral universe were limited to human beings. It is not unusual for our ideals to make us uncomfortable.

The last paragraph of Cargile's comment provoked more discussion at the conference from which this volume arose than any other piece of comparable length. Cargile buys, raises (quite humanely), and slaughters pigs. In much the same terms as Narveson, he argues that the (short) life of his pigs is better than no life at all, and thus his practice is not only acceptable, but morally praiseworthy. His challenge to vegetarians to do as much for animals as he does did not, however, go without answer. One such answer is offered by Bart Gruzalski.

Gruzalski expounds and defends utilitarianism in regard to raising and killing animals for food. Gruzalski holds that utilitarianism prohibits such uses of animals. The greater part of his paper is concerned with countering three proposed utilitarian defenses of raising animals in order to eat them.

The first such defense considered is that just given by James Cargile. Cargile claims that his pigs live happy, if short, lives, and that the net happiness of pigs is increased by his use of them. Gruzalski's counterargument relies on the great reduction of farmland that would be needed to feed a vegetarian population. Releasing these resources, allowing the land to lie fallow, would greatly increase the opportunities for wild animals. Cargile is right that the quantity of pig pleasure would decrease, but this would be more than balanced by the increase in chipmunk pleasures, deer pleasures, raccoon pleasures, and so on.

The second utilitarian defense of raising and killing animals for food is that suggested by Narveson's paper in Section I. If the pleasures of beings with sophisticated capacities are more important than those of simpler creatures, perhaps the sufferings of the animals *are* outweighed by human carnivorous pleasures. Gruzalski attacks this argument in two ways. Granted that humans have more sophisticated capacities than (most) nonhumans, it does not follow that human pleasures are more important—perhaps they are less important, since humans are so prone to distraction by memory, anticipation, or other sophisticated activities. And perhaps the pains of simple animals, unrelieved by hope or reverie, are more important than human pains. The confinement any domesticated animals must endure, and the terror usually preceding slaughter, are substantial, and the human pleasures gained from eating meat are relatively trivial.

The third utilitarian defense of meat-eating is that, given the size of the meat industry, the chance that any individual's abstention from meat would decrease animal suffering is infinitesimal. Thus, if I obtain *any* pleasure from eating meat, I should continue to do so. Against this Gruzalski urges that (a) one individual's abstention has effects on others, (b) since the cost of becoming a vegetarian is so slight, even a very small probability of reduction of animal suffering suffices to make act-utilitarianism entail vegetarianism, and (c) the meat industry is more responsive to small changes in demand than this defense assumes.

This section concludes with James Cargile's responses to Becker, Gruzalski, and unnamed others at the conference from which this volume arose. His challenge to vegetarians—'what are you doing for animals destined to be slaughtered?'—is reissued. In response to Gruzalski's criticisms he makes several empirical claims about the capacity of old-fashioned farms to support both wild and tame animals, and about the relative hedonic level of wild and domesticated animals. This last consideration bears on (and is borne on by) Gross's paper in Section VIII.

The Priority of Human Interests

Lawrence C. Becker

My purpose here is to put forward an argument in defense of the moral priority, for humans, of human interests over comparable ones in animals.[1] In outline the argument is certainly not original. But I am not aware of any previous attempt to work it out in detail. For that reason, and also because the subject matter itself is somewhat fuzzy, the argument lacks the precision possible these days in discussions of utility functions, duties, rights, and obligations. But it is an important argument nonetheless—even if it turns out to be wrong—and its neglect has contributed a great deal to the suspicion that there is something fundamentally amiss in current discussions of the treatment of animals.

In outline, the argument is simply this: There are certain traits of character that people ought to have—traits constitutive of moral excellence or virtue. Some of these traits order preferences by "social distance"—that is, give priority to the interests of those "closer" to us in social relationships over the interests of those farther away. Animals are typically "farther away" from us than human beings. Thus, to hold that people ought to have the traits constitutive of virtue is to hold, as a consequence, that people ought (typically) to give priority to the interests of members of their own species.

That is the outline, and it will require a great deal of filling in to make it convincing. But I want to make it clear from the outset that no amount of filling in will turn this argument into a defense of the proposition that humans are morally superior to animals (whatever that might mean). Nor will the argument deny consideration to the interests of animals in the making of moral decisions, or deny that those interests can often override human ones. My argument is not a defense of the cruelty to animals found in factory farming and much scientific experimentation.

[1] I shall usually follow the convention of excluding humans from the class denoted by 'animals.'

(But as far as I can tell, the argument is indeterminate with regard to using some sorts of animals for food and for some experiments.)

Basically the argument presented here is about priorities in situations where animals' interests conflict with *comparable* human ones. It operates most directly as a refutation of a line of reasoning sometimes put forward as decisive evidence of the irrationality of our treatment of animals—a clincher, so to speak, designed to show that preference for human interests is at bottom a prejudice (called speciesism) comparable to racism and sexism.[2] The line of reasoning to which I refer goes something like this:

Animals (at least the "higher" ones) have some of the same interests that humans have: avoiding pain, for example, and seeking pleasure. Furthermore, some human beings—such as infants and the severely retarded—have interests only in the sense that the higher mammals do: they lack the self-consciousness, complexity of purpose, memory, imagination, reason, and anticipation characteristic of normal human adults. Yet we treat the animals very differently from the humans. It is customary to raise the animals for food, to subject them to lethal scientific experiments, to treat them as chattels, and so forth. What justifies such differential treatment? It must be some morally relevant difference in the characteristics of humans and animals per se, or in their circumstances vis-a-vis the world at large, or in their rights and our duties to them, or in the consequences (for social welfare) of differential treatment. But in *some* cases it is plain that there is no such morally relevant difference between humans and animals. Hence our preference for the interests of the humans in these cases is just a prejudice.[3]

---

[2]The term 'speciesism' was coined by Richard Ryder (1975, p. 16) and is used also by Peter Singer (1975). 'Humanism,' unfortunately, is already in use for other purposes.

[3]Such reasoning is implicit in many classic and current writings on our treatment of animals. For an explicit use of it, see Singer, 1975, pp. 17–18. Put more fully, and more formally, the argument goes like this:

(1) It is undeniable that many species other than our own have "interests"—at least in the minimal sense that they feel and try to avoid pain, and feel and seek various sorts of pleasure and satisfaction. (Many also appear to be purposive in a stronger sense as well, but that is a more complex issue.)

(2) It is equally undeniable that human infants and some of the profoundly retarded have interests in *only* the sense that members of these other species have them—and not in the sense that normal adult humans have them. That is, human infants and some of the profoundly retarded lack the normal human adult qualities of purposiveness, self-consciousness, memory, imagination and anticipation to the same extent that some other species of animals lack those qualities.

(3) Thus, in terms of the morally relevant characteristic of having interests, some humans must be equated with members of others species rather than with normal adult human beings.

(4) Yet predominent moral judgments about conduct toward these humans are dramatically different from judgments about conduct toward the comparable animals. It is customary to raise the animals for food, to subject them to lethal scientific experiments, to treat

My contention is that this line of reasoning is incorrect—not because it fails to find some patent and morally relevant distinguishing characteristic that it ought to have found, but because it assumes that such a characteristic must be found in order to justify preferential treatment for humans. On the contrary, I shall argue that some differences in treatment that favor our own species are justified because they are the product of moral virtue in human agents.

ASSUMPTIONS ABOUT VIRTUE

I begin with some assumptions about moral virtue. The assumptions are as uncontroversial as I can make them—which does not mean, of course, that I think they can always be used without analysis and justificatory argument. But for present purposes they seem to be unproblematic.

The first is that moral virtue is, at bottom, a matter of character traits. It is defined by a complex of propensities and dispositions to feel, to imagine, to deliberate, to choose, and to act. Being a good person is not just acting on principle, or doing the right thing, for the right reasons, most of the time. To be a good person is to be someone for whom right conduct is "in character." The good person is, in part, one whose responses, impulses, inclinations, and initiatives—*prior* to a reasoned assessment of the alternatives—are typically toward morally good feelings, deliberations, choices, and conduct.

---

them as chattels, and so forth. It is not customary—indeed it is abhorrent to most people even to consider—the same practices for human infants and the retarded.

(5) But absent a finding of some morally relevant characteristic (other than having interests) that distinguishes these humans and animals, we must conclude that the predominant moral judgments about them are inconsistent. To be consistent, and to that extent rational, we must either treat the humans the same way we now treat the animals, or treat the animals the same way we now treat the humans.

(6) And there does not seem to be a morally relevant characteristic that distinguishes all humans from all other animals. Sentience, rationality, personhood, and so forth all fail. The relevant theological doctrines are correctly regarded as unverifiable (at least in this life) and hence unacceptable as a basis for a philosophical morality. The assertion that the difference lies in the *potential* to develop interests analogous to those of normal adult humans is also correctly dismissed. After all, it is easily shown that some humans—whom we nonetheless refuse to treat as animals—lack the relevant potential. In short, the standard candidates for a morally relevant differentiating characteristic can be rejected.

(7) The conclusion is, therefore, that we cannot give a reasoned justification for the differences in ordinary conduct toward some humans as against some animals.

(Further arguments are then given to show that the change required of us is the upgrading of the treatment of some animals rather than the downgrading of the treatment of comparable humans.)

The second assumption about moral virtue, or moral character, is that it sometimes produces spontaneous, uncalculated conduct. Utility theory itself requires that we develop habits of thought, expectations, rules of thumb, reflexive responses, and so on. The alternative is a ludicrous form of paralysis that is self-defeating on rigorously act-utilitarian principles alone. I take it that the other standard types of moral theory come to the same conclusion: that the good person is one who *sometimes* acts without weighing the consequences, or canvassing peoples' rights and duties, or in any other way deliberating about what to do. Sometimes, *as a necessary consequence of being morally virtuous,* a good person just has, and acts on, uncalculated feelings, beliefs, expectations, and preferences.

The third assumption I make about moral excellence is that the character traits that define it form a coherent system constrained both by welfare needs and by obligations. Coherence is assumed to avoid the problems raised by conflicts among traits: Unconditional truth-telling may conflict with tact; but I am assuming that as these things enter into the dispositions that define virtue, a rough balance is struck that in principle permits both tactful and truthful behavior. Constraints imposed by welfare needs and by obligations are assumed to avoid the problems raised by fanaticism. Loyalty may be an element of virtue, but not when it is blind to the consequences for welfare, or to the violation of rights and duties, or to the requirements of justice generally.

The fourth assumption is that the ability to develop and sustain friendships is a necessary part of moral excellence. (I mean to restrict this assumption to situations in which people can meet their survival needs without extreme difficulty, and in which they are dealing with people of good will. Further, as I use the term friendship, it includes intimate and intense love relationships as well as those characterized by mutual respect, admiration, and affection.)

Finally, I assume that the traits that define moral excellence produce "open" but stable and unambivalent feelings, beliefs, expectations, and preferences. The feelings, beliefs, and so on must be open to change in the sense that the moral person must be *persuadable*. Fixed attitudes, as opposed to stable ones, are not part of moral excellence. But the person who lives in an agony of uncertainty about every act, every feeling, every preference, or who is thrown into confusion by every *suggestion* of error, does not exemplify moral excellence either. That is why the traits that make up moral character must be stable and the beliefs, attitudes, and so forth that the traits produce must be unambivalent.

With these few assumptions about moral excellence in the background, then, I want to argue for some favoritism toward members of our own species.

SOCIAL DISTANCE AND PREFERENCES

When hard choices have to be made, I am ordinarily expected to rank the interests of my family above those of my friends, friends' above neighbors', neighbors' above acquaintances', acquaintances' above strangers', and so on. In general, the expected preference ordering follows typical differences in the intimacy, interdependency, and reciprocity in human relationships. Such differences are constitutive of what may be called "social distance"—an imprecise amalgam of relevant facts about tolerable spatial arrangements, the frequency and nature of permissible social interactions, and roles in social structures.[4]

There are exceptions to these expected preferences, of course, along several dimensions. One is obligations: I may have made agreements with strangers that override ordinary commitments to family. Another is proportionality: the trivial interests of a friend do not outweigh the survival needs of an acquaintance, for example. And still another dimension of exceptions has to do with deviations from the typical pattern of relationships: if my familily has abused me and cast me out, whereas some friends have taken me in, I may be expected to reverse the usual preference order. (People sometimes explain this by saying: These friends are my *real* family.)

In addition to the exceptions, there are the well-known conceptual problems raised by any such ordering of preferences. Who is my neighbor? Is it mostly a matter of geography or of social organization? Is a family a biological unit or a sociological one? Where are the lines between friendship and mere acquaintance, and between acquaintance and lack of it?

---

[4]The concept of social distance is a slippery one. As it has been used in social psychology, it mostly has to do with tolerable levels of social "relatedness": Would you marry a ____? Would you accept a ____ as a close relative by marriage? As a roommate? As a neighbor? As a member of your club? The answers to such questions are thought to establish a social distance scale—particularly with respect to race, nationality, social class, and religion. See Sherif, 1976, for an overview of this material. Her references to the work of H. C. and L. M. Triandis are especially worth pursuing. The relation of spatial arrangements to social distance has also been explored. See the discussion and references in Shaver, 1977, pp. 108–111. But I have not been able to find—either in texts or in primary sources—a careful analysis of the *concept* of social distance. And the empirical work so far done in the area has ignored the feature that is of most concern to me here—namely, preferences in the distribution of scarce goods. Would you give the last available food to ____ over ____? is a sort of question that has not been asked in these studies. As a result, I shall have to proceed in terms of what seem to me to be plausible assumptions. Cultural anthropology seems to promise more, but it too (at least to the untrained eye) operates without a detailed analysis of the concept of social distance. See, for example, the interesting material in Bohaman and Bohaman, 1953, pp. 25–30 and Middleton, 1965, Ch. 4.

Finally, the operation of such preference ordering is constrained by principles of justice: Similar cases must still be treated similarly; decisions should be non-arbitrary; and in some highly regularized cases, we require that the decision process not be covert, or manipulative, or involve ex post facto legislation or self-interested adjudication.

I do not mean to minimize the importance of all these matters. But I am concerned here with two other issues: the moral justification, if there is one, that can be given for any preference ordering by social distance; and the consequences of that for our treatment of animals.

Utility, Rights, and Duties

The act-utilitarian justification of preference ordering by social distance is notoriously weak. It depends upon the highly contestable empirical contention that aggregate welfare is best maximized when individuals use such orderings. This is obviously false in many cases where our firmest moral intuitions (as the saying goes) still insist on preference for those closest to us. And more to the point is the wide range of cases in which utility would require reversing or at least ignoring matters of social distance. The interests of children who are statistically unlikely to lead socially productive lives would have to be subordinated to the interests of other children; the interests of the infirm, or the chronically unemployed, or the aged would likewise have to be subordinated. And as long as no countervailing disutility resulted, such subordination would have to occur in families and among friends as well as in public policy.[5] Such cases can easily be multiplied, and they are enough to cast doubt on any act-utilitarian justification for preference ordering by social distance.

Deontological accounts fare little better. It is easy enough to show, of course, that people typically have *more* duties toward those close to them than toward those far away. For one thing, as social distance decreases, the number of contacts—and hence duty-making agreements —between people increases. For another, the number of role relationships in which duties are constitutive parts (e.g., parent–child; teacher–student) varies inversely with social distance. But showing that the *number* of duties varies with social distance is not quite the same as showing that there is a comparable variance in *preference* ordering. Deontological theorists typically insist, after all, that there are human rights (and ''natural'' duties to all) as well as ''special'' ones. And it is hard to

---

[5]As Jan Narveson has pointed out to me, act utilitarians would immediately reply that such cases are far-fetched. Parents do normally love their children more than any friend. Friends put each others' interests ahead of acquaintances', and so on. Consequently, as long as people have such feelings, there will always be disutility in rejecting social distance preference orderings. But that is not enough to satisfy anti-utilitarians. They want some ground for deciding whether such preference orderings are good independently of whether people just happen to have them.

see, on the face of it, how a natural duty to protect human life, for example, by itself, would put any given human life ahead of any other. To hold that such preferences (say for family over friends) are built into the definition of role-related duties is just to beg the question. What *justifies* building them in?

Further, we do not always want to describe social distance preferences as matters of duty (or right). A preference for a hopelessly sick child over a healthy adult (e.g., in terms of distributing scarce food) may be something we approve of—even though we cannot give it a justification *either* in terms of duty or utility.

Virtue and Social Distance

What I want to explore is the notion that some traits of character that are constitutive of moral excellence entail social distance preferences. The traits I have in mind are reciprocity (i.e., the disposition to make a proportional return of good for good), and empathic identification with others. (There are no doubt other traits for which the same argument could be made. I do not propose my list as exhaustive.)

RECIPROCITY.   Reciprocity is a pervasive social phenomenon—and one that appears not only as a mere practice, but as a norm for conduct in virtually every society of record.[6] Returning good in proportion to good received—at least in many common social exchanges—is prescribed, as well as predictable, human behavior.[7] It is evident, by inference, that the *disposition* to reciprocate (leaving aside the issue of proper motives) is quite generally regarded as an element of moral virtue.

Further, it seems clear that one can justify the inclusion of such a disposition in an account of moral virtue. It has obvious social utility that its absence or opposite would lack. It is, for example, necessary for sustaining conviviality, friendships, and certain sorts of cooperative endeavors. For those reasons, and perhaps others, it is also plausible to think that rational contractors would choose a world in which people had such dispositions over one that differed only in lacking them. Rights theory insists on the mutual respect, balanced exchanges, and so on that are characteristic of reciprocity. And reciprocity is obviously embedded in Aristotelian accounts of moral character. In short, if any traits of character can be given a reasoned justification as necessary parts of moral virtue, reciprocity is among them.

EMPATHIC IDENTIFICATION.   A similar case can be made for the ability and the propensity to see situations from other points of view, to understand and indeed to share others' experience empathetically. (I include here also the ability to identify with characters in narrative art and

---

[6]See, for example, Gouldner, 1960.
[7]The return of bad for bad is a much more complex matter.

to have vicarious experience through such identification.) Aside from its utility in settling conflicts, empathy is a prerequisite for applying the utility calculus. How else can we estimate utilities for others?

I assume that other standard moral theories would also list empathy as an element of virtue. Rational contractors would most likely prefer a world in which agents had this trait to one in which they did not. Deontological theory cannot work without the means for deciding what counts as a violation—an injury—to another. And that seems to require in moral agents the ability and propensity to understand the suffering of others. (I assume that right conduct, in deontological terms, is more than a mere mechanical performance of tasks—that it requires proper motives as well.)

RELATION TO SOCIAL DISTANCE.   It is easily seen, I think, that both the disposition to reciprocate and the disposition to empathize ordinarily result in *distributions* ordered by social distance. Given limited resources with which to reciprocate, and limited energy, time, and imaginative ability for empathic identification, those closest to us will inevitably get a disproportionate share—both of the goods we distribute and the attention we pay to them. But do we prefer satisfying the interests of those closer to us? That is, supposing we have the dispositions to reciprocate and empathize, do we, as a consequence of that fact, order *preferences* (as well as actual distributions) by social distance? I think so, for the following reasons.

Take reciprocity first.

(1) The smaller the social distance between people, the more intricate and pervasive are the exchanges between them. Consequently, the difficulty of deciding who is in debt to whom, or when equilibrium has been achieved in a relationship, varies inversely with the distance. Such calculations are virtually impossible within a nuclear family, and extremely difficult even for close friends. In such relationships, it would nearly always be reasonable for everyone involved to feel either in debt or cheated no matter what choices were made—at least, that would be possible if people tried to keep a strict accounting of who owed what to whom. The potential for continuous ill-feeling—and the consequent breakdown of close relationships—is obvious. With good reason, therefore, we do not cultivate "reciprocity accounting" *at all* in close relationships—as long as the relationships remain stable and roughly balanced.

(2) This seems an eminently justifiable position to take with regard to moral excellence. If it is a part of moral excellence to be able to develop and sustain friendships, and if the parts of moral excellence must form a coherent whole (both of which I am assuming here), then the disposition to reciprocate must be compatible with the ability to develop and sustain friendships. Thus the disposition to avoid strict accounting—at least in close relationships—is required.

(3) The required disposition changes as social distance increases, however, partly because the potential for reasonable disagreement over credits and debits decreases. Many exchanges with strangers are discrete and of assessable value. And many of the benefits we receive from strangers are so indirect that reciprocity for these can be equally indirect (e.g., by our being law-abiding, productive citizens). So the stability of relatively distant relationships is not threatened by a more calculative approach.

(4) Finally, we are, typically, *always* more "in debt" to family than to friends, to friends than to acquaintances—if for no other reason than the sheer frequency of exchanges. The more transactions there are in a relationship, the more likely it is that there will be "loose ends." When all of this is put together—the fact that the closer the relationship, the more likely we are to be "in debt," and the fact that the closer the relationship, the less exact is our knowledge of debts—it follows that it is always reasonable for virtuous people to think that anything they have to give is more likely "owed" to those closer than those farther away. Distributional preferences, given the disposition to reciprocate, will therefore be ordered in terms of social distance.

Something similar may be said of empathy. We identify most fully with those closest to us. That is, their interests are "real" to us in a way that the interests of more distant people are not. Empathic identification with the suffering (or pleasure) of people whose very existence we know about only indirectly (through the descriptions of others) cannot help but have an imaginative, dilute, and dubitable quality. In contrast, the interests of those close to us—the interests communicated to us directly—have a vividness, immediacy, and *in*dubitability that imaginatively constructed empathy can never match. It is certainly plausible to suppose that, insofar as empathic identification produces conduct "for" the interests of others, it will produce preferences for those with whom our empathy is strong over those with whom our empathy is weak. The consequence is preferences ordered by social distance.

SOCIAL DISTANCE ACROSS SPECIES LINES

My argument so far has been that the virtuous person—as a consequence of certain traits constitutive of virtue—orders preferences by social distance. I want to argue now that, certain exceptions aside, the social distance from us to members of other species is greater than to members of our own species. The consequence—for virtuous people—is a systematic preference for the interests of humans over the interests of other animals. The argument is fairly straightforward.

First Step

Social distance decreases as the quantity and "immediacy" of social interaction increases. This is just definitional. When I interact *directly* with someone—without intermediaries—and when I do so frequently, the social distance between us (other things being equal) is less than it would be if the interactions were indirect and infrequent.[8] That is part of what is meant by "social distance." (I say "part" because there are other ways in which social distance can increase or decrease.)

Second Step

Dependence, when it is recognized as such by one or more of the parties, is a feature of relationships that typically reduces social distance—by increasing both the quantity and immediacy of interactions. The dependent one struggles to stay "close"; the one depended upon must continually deal with the demands of the other—even if only by rejecting them. Thus, the more dependent a being is on another, the smaller the social distance between the two tends to be.[9]

It is again definitional, at least when the notion of a "relationship" is suitably restricted. *Social* distance concerns interactions in which beings may be said to be acting *toward, with, for,* or *against* each other. It is only those sorts of interactions that I refer to as "relationships." Thus the causal relation (of interdependence) that we have with certain symbiotic microorganisms is not a relation*ship* in this sense. (Or, put another way, it is one in which the social distance between the parties is infinite.) Similarly, our dependence on oxygen is not to be analyzed in terms of social distance, nor are the causal relations between ourselves and vegetables. But we *can* have relationships in the requisite sense with many sorts of animals, and with virtually all human beings. In these relationships, our recognition of the truth about dependence is one of the factors that determines social distance. And the more the dependence, the less the social distance.

Third Step

Animals are typically much less dependent on us, in our relationships with them, than are those humans (infants and so on) to whom the animals are comparable (in terms of their interests, intelligence and so

---

[8]The "other things being equal" clause is crucial here. After all, the interactions in hand-to-hand combat are direct and immediate. And though there is sometimes a bond between enemies that could conceivably be described as "closeness," its relation to social distance as I am using the term is certainly not an easy one to explicate.

[9]It is worth noting that *affection* between the parties is not necessarily involved at all. Affection is one sort of "closeness" in relationships, but not the only sort. See, for example, Hacker, 1951.

forth). Romulus and Remus aside, helpless humans are dependent on other humans for survival, health, and happiness to a degree that the comparable animals are not. The social distance from human adults to human infants is thus typically smaller than the distance to comparable animals.

Final Step

Consequently, given the ordering of preferences by social distance entailed by moral excellence, we will typically prefer the humans. (I say "typically" because in special cases—such as pets, wounded or crippled animals, and those who suffer directly from human actions—the same kind of dependence can exist.)

A much richer account of the increases in social distance across species lines can probably be constructed from social-psychological findings—for example, about the propensity for and limitations of empathic identification. But such complications are not necessary to the argument already made. Similarly, it would be possible to enrich the argument greatly by developing an account of the greater intricacy and potency of reciprocal relationships among normal adult humans compared to that between humans and animals. But that would take the argument well beyond its present purpose.

VARIETIES OF SPECIESISM

To review the argument so far, then, I have argued that certain elements of moral virtue order preferences by social distance, and that social distance typically increases across species lines. The result is the conclusion that moral character disposes us to prefer the interests of humans to those of animals. But to what extent? Here it is worthwhile distinguishing some possible varieties of speciesism to see which sort the argument supports.

Categorizing types or degrees of speciesism is a somewhat arbitrary process. I am not prepared to say that the spectrum from weak to strong versions is continuous, but neither are there indisputable "natural" breaks that justify a unique list of types. The four varieties distinguished below are thus offered more as illustrative of important differences than as definitive of fixed positions.[10]

Absolute Speciesism

To hold that *any human interest outweighs any (sum of) nonhuman interest(s)* is to hold what I shall call the absolute version of speciesism. The reason for the label should be clear. This version refuses to rank any ani-

---

[10]In an unpublished paper, Tom Regan has also argued for distinguishing several varieties of speciesism. I have profited from his discussion.

mal interest (no matter how serious) above any human interest (no matter how trivial). It also refuses to rank any *sum* of animal interests, no matter how large, above even one trivial human interest. An absolute speciesist would hold, for example, that it would be moral for a human being to cause the most extreme suffering imaginable to millions of animals in order to satisfy a whim (as long as doing so did not frustrate any comparable or greater human interest).

As a perfectly general principle, this is a straw man. No one seriously defends it. No defensible moral theory could support it, and my argument in this paper is no exception. The very character traits that entail preference ordering by social distance also entail the subordination of *some* human interests to animal interests. After all, the dispositions to reciprocate and empathize do not operate only with respect to members of our own species. They operate in any "relationship"—as that term was defined earlier. And the dispositions must order preferences by significance level[11] as well as social distance—else empathy would lose its usefulness in assessing priorities for conduct, and reciprocity (as a *proportionate* return of good for good) would be impossible. So the recognition of the need to reciprocate to animals, and the empathic identification with their interests as well as with those of humans, necessarily admits the *possibility* of subordinating human interests. Absolute speciesism is ruled out.

Resolute Speciesism

The absolute position may be weakened in a number of ways. To hold, for example, that any *significant* human interest outweighs any (*sum of*) nonhuman interests, is to hold what I shall call the "resolute" version of speciesism. Here the number of animal interests is not important, for even one "significant" human interest will outweigh any number (no matter how large) of nonhuman interests. (It also follows that a trivial human interest outweighs any sum of trivial animal interests.)

What counts as a significant interest is a central concern here, of course. But significance *level* is not. This resolute position asserts that *any* significant human interest outweighs any sum of nonhuman ones. Just as any number of animals may, on this view, be sacrificed for the survival of one human being, so too they may be sacrificed for health, or happiness, or psychological growth.

---

[11]By a significant interest I mean, roughly, one whose satisfaction is necessary for biological survival, physical health, physical security, physical comfort, pleasure (of a sort that comes from satisfying basic drives), and psychological growth, development, and health. It has been hypothesized (and partially confirmed in sociopsychological studies) that there is a rough hierarchy of significant interests common to all sentient beings we have studied. But finely drawn hierarchies—especially among interests that could be called trivial—are probably impossible.

I cannot find any support for such a position—either from the virtue argument advanced here or from standard moral theories. The dispositions to reciprocate and empathize *must* take account of significance level, as I noted earlier. And given that they operate in all of our relationships with other beings—and not just in our relationships with other humans—the idea that significant human interests can *never* be overwhelmed by significant animal interests is implausible. In fact animal interests do, often, outweigh nontrivial human ones for people we believe to be virtuous. If we have good reasons for accepting this (partial) account of virtue—and I think we do—then we must accept the consequence that virtuous people will reject the "resolute" speciesist position.

Weak Speciesism

The situation is different when the speciesist position is weakened still further, however. The minimal position—what might be called "weak" speciesism—simply holds that when human and animal interests are equivalent (in terms of both significance level and number) the human interests are to prevail. Since this is the minimal version of speciesism, if my arguments support speciesism at all (which I think they do), they must support at least this version.

I noted earlier, however, that there are exceptions—produced by the very traits of character that typically produce preferences for humans. Relationships between humans and animals often develop that reverse the typical preference ordering. (Just as friends can sometimes be closer than families.) But this is not indicative of an inconsistency in moral character, or of a problem that needs to be resolved. On the contrary, it is a perfectly consistent expression of the traits of reciprocity and empathy.

What is a problem, and a serious one, is defining equivalence in significance levels—especially for cases in which the humans are self-conscious and purposive while the animals are not. Is a threat to a human's "life" as used in the sentence "My life was over when I retired" equivalent to a threat to the biological existence of an animal that does not have such a "career"? That is, would a virtuous person necessarily regard it as such? I do not know. Are the pleasures of the table (for a human lifetime) equivalent to the lives of the animals used to supply pleasant eating? Until some answers can be given to such questions, it is hard to say just how much of a speciesist position is authorized by my arguments.

Moderate to Strong Speciesism

Similarly for a whole range of positions—in between the weak and resolute versions of speciesism—which attempt to give rough weights to the *number* of interests involved. How *many* animals may be sacrificed for a human life? Ten? Ten thousand? Here the quest for precision seems ludicrous and offensive. And indeed it is unnecessary *if we are prepared to accept, as morally right,* whatever virtuous people generally agree is right.[12]

VIRTUE RATHER THAN PRINCIPLE

And that is ultimately the recommendation I come to on this matter: that we should rely on the collective judgment of those among us who come closest to exemplifying (what we can defend as) moral excellence. I think that course commits us to some version of speciesism—not the absolute or resolute varieties, but conceivably something in the moderate to strong range.

My reason for advocating this course is simple. The speciesist tendencies I have described are consequences of (parts of) moral excellence. The fact that we can find no reason for speciesism when we consider the consequences, or the morally relevant characteristics of animals viv-a-vis some humans, is irrelevant. If we want people to *be* virtuous—not just to act on principle, but to have the traits characteristic of virtue—then we are going to get some version of speciesism in people's behavior. Since the problem of determining equivalent significance levels is so resistant to analysis (after all, the notorious problem of inter*personal* comparisons of utility is just part of it), it seems reasonable to accept, as moral, whatever behavior follows from the traits that constitute moral excellence. One reason that we cannot be very sure where that will lead us (e.g., with regard to vegetarianism) is that we do not have modern analyses of virtue that are comparable in subtlety and detail to those we have for utility, duty, obligation, and rights. We need such analyses, and I hope that if the argument I have presented here does nothing else it makes that need more apparent.

SOME PROBLEMS WITH THE ARGUMENT

There are a number of important objections that can be raised against the argument I have given. I shall try to answer some of the most pressing ones, at least in enough detail to indicate how an adequate response could be developed.

Preferences Do Not Justify Decisions[13]

Objection: Even if virtue disposes me to order my preferences by social distance, it does not follow that it permits me to act out those preferences

---

[12]An instructive attempt to work out the grounds for accepting, as *just*, whatever decisions are made by "competent moral judges" may be found in Rawls, 1951. Other relevant articles include Thomas, 1980; Wallace, 1974; and Becker, 1975.

[13]This objection, and the one that follows it, were raised by James Cargile in an excellent set of comments on an earlier version of this paper.

to the disadvantage of people (or animals) who happen to be "farther away" from me than family or friends. If a judge, for example, or the captain of a sinking ship, or a physician forced to distribute scarce life-saving resources, is faced with putting the welfare of a friend before that of an equally needy and deserving stranger, we would (?) expect the judge (or captain, or physician), if virtuous, to *prefer* the friend but to *act* impartially. Virtue is complex in this respect, sometimes forcing us to act in ways that go against our sentiments and preferences. So the argument for speciesism (from virtue) is not sound. It may establish that it is virtuous to *prefer* the interests of humans, but it has failed to show that it is virtuous to act out those preferences.

Reply: It is important to notice that the examples that make this objection plausible are all drawn from "public" morality. They are cases in which people, in their roles as "officials" or professionals, are required to be impartial—to suppress their social distance preferences. Such impartiality, when combined with an overriding concern for applying the calculus of utility, is what is typically called ruthlessness.[14]

Ruthlessness (or at least impartiality) may be a virtue in public life—that is, we may have good moral grounds for wanting officials and professionals to develop such dispositions. (Even that is a dangerous doctrine.) But in private life it is surely implausible to think that impartial conduct, contrary to feelings and preferences, is virtuous. The criminal law makes explicit exceptions for the family members of the accused[15]; tort law imposes duties of care on family and friends that it does not impose on strangers[16]; contract law applies different equitable standards depending on whether or not a transaction was at "arm's length." [17] And indeed, "friends" who never *act out* their friendship (i.e., preferences), or family members who never act out their special love for one another would fail to be what we mean by friends or family. In short, in private life at least, the dispositions to reciprocate and empathize—as parts of moral excellence—must produce not only preferences ordered by social distance, but conduct based on those preferences. To prohibit such conduct is to prohibit one aspect of (private) virtue.

Impartiality Is Compatible with Virtue

Objection: Perhaps, though, impartial attitudes and conduct are compatible with private virtue in the way asceticism is thought (by some) to be.

[14]See Nagel, 1978.

[15]I think here, for example, of the rules that give spouses immunity from having to testify against their marriage partners.

[16]See the discussion of the duty to rescue in Prosser, 1971, § 56.

[17]See, for example, the case of *Jackson* vs *Seymour* 71 S.E. 2nd 181 (1952) in which a man was penalized for making a large profit on a business deal with his sister—under conditions that would not have raised an eyebrow if he had had no "special relationship" to the "victim."

That is, it might be *saintly*—the sort of "perfect" virtue that we cannot expect of ordinary folk, but by which we identify the very best among us. If so, then we cannot say that virtue per se entails preferences and conduct ordered by social distance, but only that imperfect or ordinary virtue does so.

Reply: My inclination is to reject this—to argue that saintliness (of the sort under discussion) is not perfect virtue at all, but rather an awe-inspiring amplification of one or a few of the elements of virtue—to the detriment of the others. Ascetics, like the perfectly impartial saint, seem to me to stand in relation to the morally virtuous much as body-builders stand to athletes. It is hard to deny flatly that body-builders are athletes. But there is some question about it, and they are certainly not candidates for athletic perfection. But even if it were true that perfect impartiality were the perfection of virtue, it would not follow that we either could expect or would want that perfection in very many people. If not, then the argument I have given for the priority of human interests—at least for ordinary people—stands.

Racism, Sexism, and Social Distance

Objection: Is it not notorious that social-psychological studies of social distance invariably report that racism (and perhaps sexism) are in part *defined by* increases in social distance across racial (and perhaps sexual) lines? And do these studies not further report that racism and sexism so defined are pervasive in every society so far studied? If so, and if the facts about social distance warrant speciesism, then does the same line of reasoning not support racism and sexism as well?

Reply: The answer is no. The argument I have given is not based *only* on facts about social distance—as found in people's actual attitudes. If it were, the paper would have been very short indeed, for the facts about people's social distance attitudes toward animals are even more obvious and entrenched than their attitudes about race and gender differences. The argument I have given is based instead on the logic of various elements of moral character—namely the dispositions to reciprocate and to empathize. These necessarily yield preferences ordered by social distance that, when combined with some facts about actual differences between humans and animals, result in preferences that can be called speciesist. But surely it is generally agreed that racism and sexism are not entailed by the logic of virtue in a similar way. Quite the contrary, they result from a lack of moral excellence in (among other things) precisely the traits under discussion here: they come in part from a *culpable failure* to reciprocate and to empathize across racial or sexual lines. The failure is culpable in the case of racism and sexism because it is based on false beliefs—negligently or willfully held—about the inappropriateness of reciprocating and the futility of trying to empathize. (Similar analyses apply to other forms of dispositional discrimination that we find objectionable—such as the favoritism shown young and attractive patients in hospital emergency rooms.)

The case is very different with animals, even though much of our cruelty to animals *is* based on a culpable failure to exercise our empathic powers. The argument here condemns such cruelty. But there are also differences—in the powers of normal adults and the dependency (on us) of the others—that entail greater reciprocity and empathy for members of our own species than for others. The consequent "failures" to reciprocate or empathize as fully with animals as with humans are therefore not culpable in the way that the failures of the racist or sexist are. In short, the argument here, if sound, justifies some degree of speciesism, but not racism or sexism.

Virtue versus Principle

Objection: But does the whole argument not rest on simply *defining* elements of virtue in such a way that they yield speciesism? For example, if we can find good reasons for thinking that animals have rights comparable to those possessed by human infants, then is it not *irrelevant* that the social distance to the humans is smaller and that we typically give priority to the human interests? Or put another way, if we were to decide that animals had such rights, would we not then have to redefine the traits of character constitutive of virtue so that they would not yield speciesism? After all, the disposition to be just, e.g., to respect rights and to treat similar cases similarly—is also presumably a part of virtue. If animals have rights similar to human ones, then any trait of character that encouraged us to ignore the equality could not be a part of virtue—because it would not be consistent with the disposition to be just. So what we would need (for virtue) would be dispositions to reciprocate and to empathize that subordinated themselves to the demands of justice. And that seems correct in any case. Feelings of "closeness" should not control our conduct; principles should. And if our moral principles tell us that human interests should not be given priority over comparable animal interests (because there is no morally relevant difference between the two sorts of interests), then that is how we should act—whether this overrides our dispositions or not.

Reply: This is an important objection, and somewhat more difficult to handle than the third one. The general form of an adequate reply seems to me to be the following: The disposition to be just is certainly a part of virtue. And if it turned out that some animals had rights equal to those of some humans—and there were no other relevant differences—then the animals would fall under the similar cases rule. But that still leaves the question of how to decide conflicts. (After all, conflicts among humans who all have equal rights also have to be decided somehow.) The argument here would still support deciding the conflicts along species lines, I think.

The general point about the subordination of what our virtues incline us to do to what our moral principles tell us to do raises a much more fundamental and difficult issue. The only thing I can say about it here is, I am afraid, none too helpful. The conflict between virtue and principle is the stuff of which moral paradox and tragedy are made. We want people to develop traits of character that are stable and that yield immediate, wholehearted (unambivalent) conduct. Even if such traits are "open" to change, they cannot be both stable and at the same time sensitive to every small change in utilities; they cannot produce immediate unambivalent conduct and at the same time wait to feel the impact of applied moral reasoning. The occasional consequence—to take an extreme case—is a tragedy like Oedipus'. Oedipus had the traits that constituted excellence in a king (e.g., decisiveness, honor, the willingness to sacrifice self-interest, trust in those closest to him), and these traits brought him down. The result is tragic, even paradoxical, when it turns out that utility itself recommends the development of such traits. But I do not think that there is any straightforward way of concluding that it is the traits that ought to be abandoned.

ACKNOWLEDGMENTS

I am indebted to Charlotte Becker, Kay Broschart, Lamar Crosby, Art Poskocil and Elinor Sosne for discussion and bibliographical help.

Comments on "The Priority of Human Interests"

James Cargile

There are, apparently, people who think that a dog's life is as valuable as a human's, who would hold that if it is a choice between saving a human being or saving a dog but not both, there is no prima facie basis for preferring the human. One can even imagine someone extending this attitude to insects or plants. And there is no *a priori* reason to assume that such people are unlikely to be converted by argument. It is logically possible that someone might be led reluctantly to such a position by the fallacy of affirming the consequent, and give it up with relief when his mistake is pointed out. But on *a posteriori* grounds I am pessimistic about the prospects for discussions with such thinkers.

Professor Becker argues that morally virtuous people will, and so people generally ought (typically), "to give priority to the interests of members of their own species." I strongly agree with this conclusion and I wholeheartedly endorse his suggestion that the concepts of virtue and excellence of character should not be neglected in favor of those such as utility, duty, obligation, and rights. Furthermore, I cannot confidently predict that having subtler and more detailed analyses of the concept of virtue is not going to help settle arguments over vegetarianism or other questions about animals. Who knows what considerations will lead people, or to what views? However, at our present level of understanding virtue, I do not think Professor Becker's argument for giving priority to human interests is adequate.

Professor Becker argues that morally virtuous people cannot reciprocate and empathize as well with animals as with people, which leads to there "typically" being what he calls a greater "social distance" between people and animals than between people. One objection he considers is that some people also are a greater social distance from blacks or women, so that his argument for discriminating against animals could also be used to justify discriminating against blacks or women. His reply is that virtuous people will be able to reciprocate and empathize with blacks and women.

As Professor Becker points out, the notion of social distance is extremely unclear. It involves giving priority to one creature over another, but in many cases the giving of priority will not be a moral matter. Giving priority to the company of one creature over another does not entail giving the needs of the one priority over the needs of the other, nor does it justify doing so. Even if a good person could not be unable to reciprocate and empathize with blacks and women, a good person surely need not be able to reciprocate and empathize with *everyone* (I agree). There are people who, though not at all evil or lawless, are just impossible to relate to. But that does not justify using them for food, or even rating their interests below those of other people in importance. I am on much better terms with my dogs than I am with some people I know pretty well. If I had to be sent to a desert island, I would rather have the dogs for company than some of these perfectly respectable people. But if I had to save one of these people or both my dogs, and not both, I would have to choose the person even though I would grieve more over the dogs than I would have over the person. Even if the class of animals is harder, in general, to relate to empathetically than the class of people, the class of people that virtually no one can stand is harder to relate to empathetically than some classes of animals. So the principle that if As cannot relate to Bs, but can relate to Cs or can relate better to Cs than Bs, then it is right for As to give priority to the needs of Cs over Bs, does not seem right. The principle is in terms of classes or kinds, and yet the priorities will have to be reflected in the treatment of individuals, and this allows logical conflicts arising from the fact that individuals belong to many different classes.

In this connection, it may be worth mentioning that, while I of course attach more importance to the needs of my children than to the needs of other people's children, I do not claim my children's needs are more important than the needs of other children or that my children's needs have a moral priority over those of other children. We may expect even a virtuous person to give priority to his children. But we (rightly) do not take that as establishing that his children ought to be given priority over others. It is not even true simply that a person morally ought to give priority to the needs and interests of his children. If I am a judge in awarding advancements for merit, I will naturally want my children to do well, but that does not justify my giving them awards in preference to equally qualified children. Or again, if I am put in charge of allocating food during a famine, I may deeply regret having my children get no more than an equal share of food, and my failure to do perfect justice might even be understandable, but there is no clear moral principle suggested by this. I do not say that "Put your own family first" is a false principle, but it is a vague slogan with false interpretations.

It might be replied that other people's children are the same kind of creatures as mine, so that my closer relation to my children than to others is not based on the kind of creatures they are, while my closer relation to my children than to animals is. But this gets back into the difficulty about there being kinds of people to which even virtuous people cannot relate.

If a morally virtuous person is put in charge of dividing scarce food among a group including both people and animals, he may be asked to justify giving people priority over animals. My position would be that it is right to give priority to people, but that I do not know why, or at least cannot think of any nonquestion-begging, sound argument for why. Professor Becker wisely wants to avoid the losing game of trying to find morally relevant differences between humans and animals, such as that people are smarter. But it seems to me he has not successfully avoided that game by appealing to the idea that animals are not as empathetic with people as other people can be. If the morally virtuous judge offered that in response to a request for justification of his distribution in favor of people, he would lose just as surely as if an attempt had been made to claim that people are smarter.

Perhaps Professor Becker intends that the judge would not have to offer reasons. Perhaps the idea is that the judge would just say that it seems obvious that animals rate lower, and it would be Professor Becker's discussion that would let us philosophers in on the true basis for the judge's intuitions. This is reminiscent of the way a rule utilitarian might defend following a rule even in a case where the consequences are bad—having the rule calls for a certain amount of unreflecting acceptance of it that keeps it in force even in cases where its utilitarian basis for the particular case is deficient.

Similarly, the argument might go, a good person will develop, as a result of his or her natural and proper closeness to some people, a feeling of unique value for humans in general compared with other species. The good person may be closer to the family dog than to most people, but he will ideally be so much closer to the family that he will see them as a higher order of beings than the dog. And then he will naturally bring these higher order beings under the concept "person" or "human" rather than the concept "immediate family" by way of identifying the key feature to their unique value. That is, he will do this if he is virtuous and thus not lacking in perspective. And so finally he will have a general feeling for people that will override his loyalty to his dog in hard cases.

I hope I am here understanding Professor Becker's argument. It seems to me forceful and important, but for all that, not adequate as a reply to a request from animal rights advocates for a justification of a policy of giving priority to humans. If our judge allocating goods is a philosopher with time to discuss the issue, he will not adequately answer his animal rights critics by describing, however insightfully, the origin of his tendency to accord higher priority to humans. The critics can point to the humans in the group with whom they are wholly unable to empathize and ask why they should be given priority over some lovable dogs. They can ask why the judge should not discipline his feelings to the contrary. The judge might appeal to the fact that other people in the group would be shocked if such priority were not given, but this is another one of those losing moves that can be ruled out by suitably rearranging (or simply adding to) the details of the case. My opinion is that Professor Becker's argument brings up a valuable idea, but that it is not an adequate answer to the non-priorists.

Furthermore, I doubt that someone possessing the traits required for moral virtue would be forced, by those traits, to give moral priority to humans, though it is not clear whether Professor Becker would make that strong claim. He may mean only that good character traits generally do lead to a preference for humans, rather than that they must.

To judge a claim about what is required by a certain kind of character may call for a very full description, not merely of one character of that kind, but a wide range of such characters. This could call for the skills of a novelist and also for the space of a novel-length book. I do not claim that a virtuous character could refuse to give priority to humans over animals, but I suspect that this is possible.

Consider a character rather like a Hindu holy man, only in this instance one with a wife and children. Could not such a character be outstandingly virtuous while accepting, and preaching, that people should not be given priority over animals? He need not be *ideally* virtuous, in the sense of being the *ideal* person. There are probably a variety of very different characters who could be outstandingly virtuous. Very different kinds of heroes may all make perfectly good heroes for different kinds of people.

One mistaken reason for rejecting my saintly non-priorist would be the idea that he would have to be inhumanly lacking in priorities of all kinds. It might be thought he would have to be indifferent whether the day was spent with his family and friends at a picnic or spent letting hungry mosquitoes bite him. (Unfortunately these activities tend to overlap in real life!) But duty only calls our saint when he is confronted with creatures in need, for their life or well-being. If the mosquitoes are doing all right biting the cows in the field, then our saint can perfectly well give priority to having his vegetarian picnic on a screened porch.

At one extreme, the non-priorist could be on an island with family and friends where every creature is perfectly supplied with its needs. This perfectly possible situation does not allow a full display of his character. But it does allow quite a lot to be shown, including priorities in the choice of company. If we add some bores to the population, our saint could avoid them or even express a good deal of irritation with them without downgrading their needs. It is possible that a terrible bore might desperately need to inflict himself on someone in order to survive. In this possible situation, our saint might face a difficult time. But any system of moral commitments is going to add to the burden of its adherents in some possible situations, over the burden of those who do not accept the system. In real life, most bores can be avoided without in that way depriving them of something they need in order to live and be healthy—and that is the type of need that raises questions of moral priority.

Our saint's position is that when two creatures have needs in order to live and be healthy, and someone can help one or the other but not both, then the claims are of equal strength. This probably does not adequately characterize the position of not giving moral priority to humans. For one thing, it leaves the position in a very vulnerable form by extending it to all creatures, including insects. It would be much easier to defend the position with respect to higher animals. But it seems to me clear that the position, even in its strongest form, allows lots of priority judgments that are not affected by the position, but that are sufficient indicators of a good capacity for reciprocity and empathy.

Conditions under which decent life is impossible can and unfortunately do arise. No style of life can be condemned just because there are situations in which that style of life could not sustain life. Overcrowding can create terrible pressures. Fortunate people are not usually confronted at close range with many needy claimants. But modern communications have forced upon us an awareness of needs and suffering that raises a deep moral problem. Why should I not sell my property when I know that with the proceeds I could save at least a dozen children in India and Africa? I can learn the whereabouts of starving people just by watching television. Why do I not get going right now?

I cannot answer that. A welfare worker who has no time for his own children can be an unattractive character. We must have some relief from our recognition of suffering and our response to it if we are to have time for things essential to a decent life. But I cannot assume that it is inconsistent with a virtuous character to find that certain situations make a decent life morally unacceptable. A saint may regret that circumstances have made it impossible for him to take time off for the pleasures of yoga exercises, piano lessons, and friendly discussions, and that his preoccupation with alleviating suffering makes him boring company. But it is one thing to have a capacity for reciprocity, empathy, and the like, and another to regard your circumstances as such that taking time for the exercise of these capacities with pleasant companions is morally acceptable. If a suffering animal needs help, then that may demand attention, and be regarded as equal in importance to human suffering. Having that view is perfectly compatible with having preferences for humans, and for some humans more than others, in situations where moral priority orderings are not involved.

My saint does not have to tolerate animals in his house, nor does he have to stand idly by while predators devour his children. Whether he will be successful in locking out the man-eating tiger until the beast is accustomed to soybean steaks, depends on the case. But lots of good people might look inadequate with a tiger at the door. It is only fair, in looking at our holy man, to consider him in a case in which his moral priorities do not require self-sacrifice. The hard cases also need considering, but to understand his character we should see how he might flourish. I think he could be extremely impressive, discriminating, intellectually and esthetically critical, and the like, in addition to being generous, kind, and the usual saintly things.

It is true that my holy man would have to take food from his hungry child to feed a more desperately hungry dog, if he were to be true to his principles. (And that's not to mention desperately hungry snakes, rats, and bugs.) But to determine whether this is compatible with his being a virtuous person is not simple. It is here that more details are required than I can present.

For one example, it would be unbearable to have the holy man fight a bloody battle with his desperate eight-year-old daughter to pry the crust of bread from her fingers and give it to the needier dog. We can consistently get rid of this possibility by writing into his character an unwillingness to use that sort of force. We can focus then on a case in which the daughter happily accepts her father's teaching and insists on feeding the dog, or at worst on a case in which the father sadly notes that the daughter is doing wrong, but cannot force her into virtuous conduct.

The holy man is preaching that when there is not enough to go around, when it is either "10 percent eat and let 90 percent starve" or "all starve," that the people who know the truth will starve with style. And similarly with hard choices among all creatures. There will be times when saints live happily, but there may be times when they must die with grace rather than live at the expense of the well-being of other creatures, or even merely by holding back from sacrificing for them.

I do not accept this preaching. But I cannot believe it prescribes attitudes that are incompatible with being a morally virtuous person. Someone who tells me that he cannot go on living a conventional American life while people are starving in Africa, and who then orders his life accordingly, may make me uncomfortable, but I cannot dismiss him as a flawed character lacking sufficient feeling for the camaraderie of us fellow Americans that our style of life can be easily given up. Extending the concern to animals is more extreme, but I cannot agree that any person who refuses to give priority to humans must be lacking in virtue.

It might be replied that even if it were possible to be morally virtuous without giving priority to humans, it is certainly compatible with moral virtue to give such priority. Then the argument might run, not that 'since it is essential to virtue to prefer humans, we ought to,' but rather, that 'since this is compatible with virtue, it is all right to prefer humans.' (Of course another version could keep the conclusion in terms of "ought".)

This certainly appeals to me. Vegetarianism has gotten more adherents than the old Intercourse for Procreation Only League of Bernarr Macfadden, but it is similar in being up against very natural impulses. It is similar too in that there have always been people who worry about sexual impulses to the point of wanting to suppress them, and suppressing them or sublimating them is common among saints. Taking saints as moral ideals and models has been a long-standing source of problems with sexual urges. The idea that one can be perfectly virtuous without being saintly is a good one. Where an impulse is very natural, we ought to hesitate at making a virtue of stifling it.

However, once we admit that there is some value in the life and well-being of animals a question of priorities arises, and the mere fact that even a good person may be inclined to put the needs of animals second does not seem to me enough to settle the question as to what is objectively right. For most people, my imaginary saint is an impossible model, though such a holy person would probably go over much better in India than in America. It would be nice if our ideals did not make us uncomfortable, but I must admit that mine tend to bother me.

The animal rights advocates and vegetarians, though, should not be too comfortable either. Every year I buy several pigs from a neighboring hog farm and raise them to slaughter for food. They are given lots of room and food, everything a pig could want for a good life—but a short one. It would be nice if they could have longer lives. But I believe that their good, short lives are better than no life at all. Thus I think I have done more for the happiness of pigs than most vegetarians. My imaginary saint might gently try to lead me to do better, perhaps wanting to see me buy out the hog farm and raise the inmates to old age. But that would be awfully hard. You do not see the vegetarians doing that either! There are millions of animals in this country that have the life they have, as opposed to no life, only because they will be used for food. Those that are suffering might have been better off not being born. But there are still very many, such as my cattle and pigs, that I think are definitely better off with the lives they have than with none at all. These animals are getting the best deal people are willing to give them, and I do not see the vegetarians as giving them even that much. Let the cruel animal raisers be stopped. But when all raising of animals for food is stopped, the number of animals getting to live at all will be drastically reduced. I do not see any vegetarians out to make the sacrifices that would be required to alter that fact.

The Case Against Raising And Killing Animals For Food

Bart Gruzalski

The important ethical view that one ought to live in such a way that one contributes as little as possible to the total amount of suffering in the world and as much as possible to the world's total happiness is called utilitarianism. In this paper I develop the classical utilitarian argument against raising and killing animals for food. I then examine this position in light of several arguments which have recently been raised to show that utilitarianism permits this use of animals. Throughout the paper I refer to nonhuman animals as animals, and to human animals as humans. Although such usage suggests an elitism that might offend some humans, the substantive arguments in the paper are better expressed if we follow ordinary usage, however unenlightened it may be.

THE UTILITARIAN ARGUMENT AGAINST RAISING AND SLAUGHTERING ANIMALS FOR FOOD

According to the classical utilitarianism of John Stuart Mill, actions are right insofar as they tend to produce the greatest happiness for the greatest number. For our purposes it will be helpful to interpret this general slogan as a specific principle regarding the foreseeable consequences of individual actions. In so doing we want to be responsive to the fact that most acts have several mutually exclusive foreseeable consequences that are of different values (e.g., rolling a die has six foreseeable consequences and we may value some more than others). One way to take these different contingencies into account is to assign a number to each foreseeable consequence to represent its desirability (or lack thereof). If we multiply the desirability of a foreseeable consequence by its probability and then sum these products of the likelihood and the desirability of each of the foreseeable consequences, we have the *expected desirability* of doing the action, which, roughly, tells us the odds that the action will produce consequences of a certain value. According to classical act utilitarianism so interpreted, an action is right if it is the best bet a person has to avoid producing painful consequences and to bring about pleasurable or happy consequences (more technically: if its expected desirability is no less than the expected desirability of any alternative).[1] In applying this view we will be using the standard conception of consequences: an event is a consequence of an action only if there is some other action the agent could have performed that would have prevented the occurrence of the event in question.[2] For example, my glass being full of water is a consequence of my holding it under the tap, since had I placed the glass on the counter, an alternative I could have performed, it would still be empty.

From the utilitarian viewpoint there are strong reasons for thinking that raising and slaughtering animals for food is wrong. When we raise animals they suffer because of confinement, transportation, and slaughter-related activities in ways they would not suffer were we not raising them for food. These actions are therefore wrong on utilitarian grounds unless there are other consequences which outweigh these sufferings inflicted on animals. Of course, there are other consequences of these acts besides the pain the animals experience. The most obvious of these consequences is that the animals become tasty morsels of food. But it is doubtful whether the enjoyment of those who are eating these animals can overcome the pain of captivity and slaughter. Does a family at a Kentucky Fried Chicken experience such pleasure from eating chicken that this pleasure overcomes the frustration, pain, and terror which the chicken had to undergo in order to wind up on a cole slaw garnished paper plate?

The plausibility of a utilitarian justification of raising animals for food is even weaker than the previous rhetorical question suggests. In order that the practice of raising and slaughtering animals for their flesh be justified, the animal's pain must not only be *outweighed* by the omnivore's pleasure, but *there can be no alternative act that would foreseeably result in a better balance of pleasure over pain.* Since eating plants is one alternative, and since this alternative produces the pleasures of taste and health without inflicting pain on animals, it follows that, if one is interested in contributing to the total amount of happiness in the world and not contributing to any unnecessary suffering, then one ought

---

[1] Although Bentham's and Mill's interpretation of utilitarianism in terms of expected or foreseeable consequences has for a time lost favor in the twentieth century, this interpretation, which I adopt in the text, has been recently adopted by Richard Brandt (1979, pp. 271ff) and defended by me in ''Foreseeable Consequence Utilitarianism,'' *Australasian Journal of Philosophy*, **59**, No. 4, June 1981.

[2] Those who discuss this notion of consequences include Lars Bergstrom (1966, p. 91), D. Prawitz (1968, p. 83), and J. Howard Sobel (1970, pp. 398–400).

not to raise, slaughter, or eat animals, for by doing any of these actions one contributes to a kind of suffering that is unnecessary. Although the above argument may seem sound, some philosophers have raised objections to it while accepting the utilitarian point of view. In the following sections I shall examine three utilitarian defenses of using animals for food.

OBJECTION ONE: RAISING ANIMALS BENEVOLENTLY

According to the first utilitarian defense of raising and slaughtering animals, the use of animals for food can be justified on utilitarian grounds even if we take into account only the pleasures and pains of the animals involved. James Cargile states (this volume) this defense of a carnivorous animal husbandry as follows:

> Every year I buy several pigs from a neighboring hog farm and raise them to slaughter for food. They are given lots of room and food, everything a pig could want for a good life but a short one. It would be nice if they could have longer lives. But I believe that their good, short lives are better than no life at all . . . These animals are getting the best deal people are willing to give them, and I do not see the vegetarians as giving them even that much.

Cargile concludes that he has done "more for the happiness of pigs than most vegetarians."

In his book *Animal Liberation,* Peter Singer claims that the argument Cargile raises, which I refer to as the animal husbandry argument,

> . . . could be refuted merely by pointing out that life for an animal in a modern factory farm is so devoid of any pleasure that this kind of existence is *in no sense a benefit* to the animal.[3]

Cargile does not overlook this important and ethically relevant consideration. He agrees that we should stop "cruel animal raisers." His, however, is not a cruel form of animal husbandry. If we raise animals in such a way that their lives are *more* of a pleasure than a burden, as Cargile does, then Cargile can claim that we do what will increase the amount of happiness in the world and so what is right on utilitarian grounds.

The animal husbandry argument rests on an assortment of claims:

(1) The pleasures of the animals we raise would not occur if we did not raise them.

(2) The pleasures of these animals increase the total amount of happiness in the world.

[3]Singer, 1975, p. 241. Those unfamiliar with how animals are turned into meat will find Chapter Three of Singer's book enlightening.

(3) The burdens of these animals are outweighed by their pleasures (and, if not, then that sort of animal husbandry is immoral).

(4) There is no alternative policy that would increase the foreseeable amount of happiness in the world.

It is reasonable for us to assume that (1), (2), and (3) are true. Importantly and realistically, (3) assumes that the animals raised for consumption suffer from their confinement and slaughter. But the core of the argument is that there is no alternative to a humane animal husbandry that would foreseeably produce more pleasure.[4] The alternative in question is vegetarianism. What can be shown is that vegetarianism produces more foreseeable pleasure *on the whole* and that, not unsurprisingly, the animal husbandry argument does not provide a utilitarian justification of raising animals for food.

If we adopted the vegetarian alternative and stopped raising marketable animals we would need to farm much less land to feed the same number of people we feed by raising such animals for food, since plants yield about ten times more protein per acre than meat (Singer, 1975, pp. 170–151).[5] Hence, if our concern is to feed the same number of people we now feed and produce as much animal pleasure as possible, there is an alternative available that would accomplish this better than raising any animals for food. The alternative is to allow 90 percent of the resources currently used to raise livestock to be idle. These resources—lands lying fallow, empty barns—which previously supported market-bound animals, would then support other sorts of animals: chipmunks, rabbits, snakes, deer, and similarly unmarketable animals whose numbers are cur-

[4]On utilitarian grounds the animal husbandry argument can be made even stronger. Peter Singer (1979), modifying some of what he wrote earlier [Singer (1975)], distinguishes between self-conscious beings who have preferences, including the preference to stay alive, and beings which, though sentient, do not have such preferences. When a creature without preferences is killed, the only disutility which occurs is the disutility of the pain of dying and the disutility of the loss of future pleasures. Hence, if we can minimize the pain of dying and prevent the loss of future pleasures by replacing the killed animal with another animal, a humane Cargile-style animal husbandry will produce foreseeable pleasure. As Singer says:

Some of the animals commonly killed for food are not self-conscious—chickens could be an example. Given that an animal belongs to a species incapable of self-consciousness, it follows that it is not wrong to rear and kill it for food, provided it lives a pleasant life and, after being killed, will be replaced by another animal which will lead a similarly pleasant life and would not have existed if the first animal had not been killed (1979, p. 153).

What Singer calls the replaceability argument is, however, objectionable for the same reasons that defeat Cargile's animal husbandry argument.

[5]Or consider the following from Frances Moore Lappé (1975, p. 14):

To imagine what this means in practical, everyday terms simply set your self at a restaurant in front of an eight-ounce steak and then imagine the room filled with 45 to 50 people with empty bowls in front of them. For the "feed cost" of your steak, each of their bowls could be filled with a full cup of cooked cereal grains!

rently restricted by our practices of animal husbandry. These other animals would use the resources we currently direct to the animals we butcher for market, and these other animals would experience the sort of pleasures experienced by Cargile's cattle and pigs without suffering from restricted movement and slaughter. Animals in the wild do not have to experience the frustrations and anxiety of confinement or the terror of waiting passively "in line" to be killed. Although it is true that by failing to raise pigs or chickens, we fail to produce *pig* pleasures and *chicken* pleasures, there is no reason to think that the pleasures of these animals is not on an even par with *chipmunk* pleasures, *rabbit* pleasures, *prairie dog* pleasures, and *snake* pleasures. This observation defeats the central idea behind the animal husbandry argument, the idea that the total amount of animal pleasure is best increased if we raise livestock, something we will not do unless we subsequently slaughter these animals for food.

The policy of allowing 90 percent of the resources we currently use to support livestock to lie idle may, however, be undesirable from a utilitarian point of view. Many of the peoples of the world are suffering and dying from protein deficiencies. In the United States during 1968, we fed to livestock (excluding dairy cows) 20 million tons of plant protein that could have been consumed by humans. Although the livestock provided 2 million tons of protein, the 18 million tons of protein "wasted" by this process would have removed 90 percent of the yearly world protein deficit. Thus, a more humane use of our farming resources would eliminate a great deal of human suffering without imposing any additional suffering on market-bound animals. It is generally thought that a policy of reducing the suffering of beings that do and will exist independently of our choice is a more efficient way of maximizing happiness than a policy that involves the creation of additional beings. If this is true, then using the resources we currently expend on livestock to feed starving peoples would be the policy justified on utilitarian grounds.

More likely than not, a move toward vegetarianism would involve a mix of both policies. The result would be the alleviation of human starvation as well as an increase in the number of wild animals. What is central to this criticism of the animal husbandry argument is that each of these alternative policies increases pleasure and decreases suffering without imposing any additional suffering on animals.

OBJECTION TWO: THE INSIGNIFICANCE OF ANIMAL PLEASURES AND PAINS

Although raising animals for food is not the best way to increase the amount of animal pleasure, it may be argued that it is the best way to

increase the amount of animal *and* human pleasure.[6] In developing the
classical utilitarian position on the use of animals for food, I claimed that
it is implausible to think that the suffering an animal experiences from
confinement, transportation, and slaughter-related activities are outweighed by the pleasures of eating these animals. The carnivorous utilitarian may object by claiming that I have overestimated the disutility animals experience as a result of our using them for food and that once this
exaggeration of animal suffering is corrected, a utilitarian defense of eating meat can plausibly be made out.

One source of miscalculation may be thought to be my implicit assumption that animal suffering has the same objective disutility as human
suffering and animal pleasure has the same objective value as human
pleasure. Narveson has argued that human and animal pleasures are not
on a par. His argument has two steps. The first is that human beings have
higher capacities. These capacities are grounded in our ability to be
"acutely aware of the future stretching out before us, and of the past in
the other direction," (Narveson, this volume) and so, unlike lower animals, a human has a

> . . . capacity to have a conception of oneself, to formulate long-
> range plans, to appreciate general facts about one's environment and
> intelligently employ them in one's plans, and rationally to carry out or
> attempt to carry out one's plans (Narveson, 1977, p. 166).

In light of these "higher capacities," Narveson proceeds to the second
step of his argument:

> Isn't it reasonable to hold that the significance, and thus the quality,
> and so ultimately the utility, of the sufferings of beings with *sophistica-
> ted* capacities is different from that of the sufferings of lesser beings?
> Suppose one of the lower animals to be suffering quite intensely. Well,
> what counts as suffering of like degree in a *sophisticated* animal—one
> like, say, Beethoven or Kierkegaard, or you, gentle reader? If we are
> asked to compare the disutility of a pained cow with that of a pained

---

[6]Any such argument must deal with the objection that human pleasure would be
maximized by feeding starving people the grain that would otherwise be fed to livestock.
One reply to this objection is that feeding starving peoples will only encourage them to
propagate themselves even further, and so feeding some starving peoples now only produces many more starving peoples later. If the objection were accurate, then using livestock grain to feed starving humans would not be a good way of producing the best consequences in the long run, and so the question of maximizing human pleasure by raising
livestock could still arise. (There are alternatives besides eating meat and letting the un-
dernourished starve. For example, a mixed policy of food aid, agricultural support, and
birth control might save lives without condemning future generations to starvation, and so
would be the utilitarian policy of choice.) The question of whether human pleasure
justifies some animal suffering could also be raised under a more optimistic scenario: assuming that the world deficiency of protein for human consumption were alleviated,
would it then be permissible to raise animals to satisfy a human taste preference?

> human, or even a somewhat frustrated one, is it so absurd to think that the latter's is greater? (Narveson, 1977, p. 168.)

Narveson's tentative conclusion is that the pleasures and pains of sophisticated beings are more valuable than those of less sophisticated beings. Hence, if our livestock and poultry are unsophisticated in the crucial sense, then the human pleasures of eating meat, being the *very* significant pleasures of sophisticated beings, may outweigh the requisite suffering of the unsophisticated poultry and livestock.

But Narveson's argument is not successful. The fact that we as humans are able to anticipate the future in ways (we believe) that animals cannot does not show that we have a greater capacity for pleasure. One reason for doubt is that such abilities increase our capacity to fail to appreciate whatever is not present to our senses. The Narveson who is eating filet mignon while anticipating an upcoming philosophical exchange or a Beethoven concert is, precisely because of this future-oriented mental activity, unaware of some of the pleasure eating would have otherwise provided. But even if we assume that humans have some capacity to enjoy life more than animals, the capacity Narveson cites suggests that humans, because distracted by thoughts and fears, may well enjoy life less, and in particular enjoy eating less, which is the critical experience on the human side of this controversy. On the animal side, however, Narveson's considerations not only do not support the claim that animal pains are dim but, ironically, they support the conclusion that the pains animals experience due to confinement, transportation, or slaughter are keenly felt, for there are no future-oriented distractions to mitigate these powerful sensations! In short, as far as the relevant pleasures and pains are concerned, humans and animals are, as far as we can tell, on a par.

OBJECTION THREE: THE SIGNIFICANCE OF HUMAN PREFERENCES GIVEN THAT ANIMALS DIE ANYWAY

Although we cannot justifiably downgrade the value of animal experiences, there is a second reason one may think that I have overestimated the amount of animal suffering that results from our use of animals as food. Whether we raise animals or not, animals must die and experience whatever anguish is involved in dying. Hence, whatever animal suffering is generally associated with the death of an animal cannot be considered a result of our raising animals for food. This reduction in the amount of suffering attributable to our animal husbandry is significant, for the main source of suffering that remains is the suffering caused by the frustrations of confinement. Since this is a frustration of animal preferences, and since not eating meat is a frustration of human preferences, there seems to be no significant difference in terms of total pleasure between satisfying the animal preference for less confinement and satisfying the human preference for the taste of meat. Hence, it may seem that we cannot condemn eating meat on utilitarian grounds, for both eating meat and its alternative lead to a similar amount of pleasure and frustration.

A chief source of animal suffering is the animal frustration caused by various sorts of restrictions. Domestic animals, in order to be profitably raised at all, must be somewhat restricted. The restrictions will be on movement (do Cargile's pigs forage through the Blue Ridge Mountains?), on social intercourse (in packs of ten or twenty?), and diet (eating acorns?). Although domestic animals are selectively bred, it is reasonable to believe with the experts that

> . . . the natural, instinctive urges and behavioral patterns
> . . . appropriate to the high degree of social organization as found in
> the ancestral wild species . . . have been little, if at all, bred out in the
> process of domestication.[7]

Where there are animals being raised, even humanely, it is noncontroversial that there will be a good deal of frustation even under the care of humane animal husbanders. The issue is how much of such frustration is justified by the pleasure of eating meat. Would the frustration experienced by a young boy locked in a room be outweighed by the pleasure of a parent derived from watching an ''adult'' TV show? If that comparison cannot *clearly* be made out in favor of the satisfaction of the parent's preference, it becomes hard to imagine anyone reasonably claiming that satisfying the preference to eat meat outweighs the many months of animal frustration caused by space, diet, and socialization restrictions.

But there are two additional factors which make this allegedly utilitarian defense of eating flesh implausible. The first focuses on the kind and numbers of deaths domestic animals undergo to satisfy the meat eater's taste for flesh. The second raises the issue of whether the pleasures of taste that ''justify'' raising animals for food are not, in the final analysis, *trivial when compared with the animal suffering required to satisfy these tastes.*

Even in those slaughterhouses in which the animals are killed as painlessly as possible, the animal hears, sees, and smells the slaughter and becomes terrified. In terror, in an unfamiliar environment, the animal, physically healthy, is prodded along.[8] At that point, in these ''hu-

---

[7]W. H. Thorpe in the Brambell Report (1965), quoted in Singer, 1975, p. 135.

[8]Richard Rhodes, who regards such killing as ''necessary,'' reports what he felt and observed in a slaughterhouse that was doing its job of slaughtering pigs ''as humanely as possible.'' He writes:

The pen narrows like a funnel; the drivers behind urge the pigs forward, until one at a time they climb onto the moving ramp . . . . Now they scream, never having been on such a ramp, smelling the smells they smell ahead. I do not want to overdramatize because you have read all this before. But it was a frightening experience, seeing their fear. . .

''Watching the Animals,'' *Harper's,* March 1979, quoted in Singer 1975, p. 157.

mane'' slaughterhouses, the animal is stunned by a captive-bolt pistol or an electric current before being killed painlessly. In smaller settings these stunning devices are too expensive and, in addition, skill at killing quickly is not as practiced. One would expect that the small farmer who clubs, slices, or shoots his animals must not infrequently confront an injured, squealing animal that by now is utterly terrified and even harder to kill.

Everything we have learned about animals suggests that in terms of experiencing terror, pain, grief, anxiety and stress these sentient beings are relevantly similar to humans. It is reasonable to believe that our knowledge of the quality of human dying will also tell us something about the dying process of other animals. For humans, the most horrible deaths involve terror. When this factor is not present, and especially when the process of dying is not unexpected for the dying person, dying can be peaceful. From this minimal observation about human dying and the observation that domestic animals are typically slaughtered in circumstances that are unfamiliar and terrifying for the animals, it follows that the experience of being slaughtered is no worse for these animals than the worst deaths experienced in the wild and significantly worse than the deaths of wild animals that die from disease or old age in familiar and unterrifying surroundings. In addition, because the life of an adult animal raised for food is much shorter than the life of a similar animal in the wild, there will be more dyings per total adult population among these animals than among wild animals of similar species. Hence, both in quantity and quality of deaths, rearing animals for food produces a great deal of death-related anguish and terror that is directly a consequence of humans using them for food.

These are some of the foreseeable disutilities that are a consequence even of a ''humane'' animal husbandry.[9] In order to justify our producing this foreseeable animal suffering, we must ask whether it is plausible to believe that these foreseeable bad consequences are outweighed by the foreseeable pleasures of eating meat. But we must first clarify this question for, as stated, it suggests that whatever pleasures we derive from eating meat are to be compared with whatever sufferings ani-

[9]There are two other foreseeable disutilities of an animal industry, even a ''humane'' one. The first is that it is only a small step from the perception of animals as beings we kill to satisfy human tastes to a perception of animals as meat-producing mechanisms, a perception that is part and parcel with the cruel practices of today's meat industries. Hence, among the foreseeable consequences of even a ''humane'' animal industry is the sort of cruelty imposed on animals daily under the guise of meat production. The other foreseeable disutility of the meat industry is the horrible suffering animals experience when they are transported. The following account is typical: ''For an 800-lb. steer to lose seventy pounds, or 9 percent of his weight, on a single trip is not at all unusual'' (Singer, 1975, p. 118). This loss, not only from the fleshly parts of the animal but also from the head and shanks, indicates ''a severe amount of otherwise unmeasurable stress on these animals'' (p. 120). This account is not one of the horror-stories of animal transportation, but is typical and indicates the amount of anxiety and fear animals experience in trucks and trains.

mals experience solely as a result of farming practices. But that is not an accurate interpretation of this crucial question. Rather, we are *only* interested in *the amount of pleasure that would occur were we to eat meat and that would not occur were we to eat tasty vegetable dishes instead.* That is the amount of pleasure which is a consequence of our eating meat (as opposed to eating in general). Since much of the world's population finds that vegetarian meals can be delightfully tasty, there is good reason for thinking that the pleasures many people derive from eating meat can be completely replaced with pleasures from eating vegetables. Hence, the pleasures to be derived from the eating of meat are so minimal as to be insignificant. It follows that any defense of flesh-eating along the above lines is totally unacceptable for the utilitarian.

OBJECTIONS FOUR AND FIVE: THE INCONVENIENCES OF A VEGETARIAN CUISINE

Two additional issues are relevant to any attempt to defend eating meat on utilitarian grounds. The first is raised by Narveson, who points out "that the vegetarian diet is more limited, since every pleasure available to the vegetarian is also available to the carnivore" (Narveson, 1977, p. 14). Although Narveson concedes that this will not be a decisive consideration for most of us, it is worth asking whether it should be decisive even for a person who attaches a high value to greater esthetic variety in diet and who feels that meatless eating would be boring. At issue is whether it is justifiable to cause animal suffering in order to satisfy the tastes of those who prefer a diet with a variety only available by including meat. On the animals' side of the argument there is a concise but powerful reason for thinking that satisfying this preference does not justify the requisite animal suffering. Part of the suffering we inflict on animals is the frustration and intense boredom of a monotonous diet. To claim that *this* intense animal boredom is outweighed by the pleasure of increased variety for humans would be to assume what is unreasonable: that avoiding some human frustration justifies producing a great deal of animal frustration, even when the frustrations are causally identical (viz., lack of variety in diet). When we take into consideration *all* the kinds of animal suffering produced in order to maintain the variety meat adds to a diet, there are *no* reasonable grounds for thinking that satisfying this preference for variety could compensate for the animal suffering involved.

The second additional issue relevant to a utilitarian appraisal of eating meat is that eating meat is, for most omnivores, a deeply engrained habit. Since changing habits is always difficult, it follows that whatever inconvenience and frustration a person experiences as he or she shifts to vegetarianism are a cost that can be avoided by continuing to eat meat.

But there are several reasons for thinking that this cost does not justify continuing to eat meat.

One is that this cost is a relatively short-term affair, whereas the avoidance of unnecessary animal suffering is a very long-range and ongoing consequence. Once the change of eating habits is accomplished, not only will the new vegetarian not contribute to the unnecessary suffering of animals, but most likely neither will any of the vegetarian's progeny, whose vegetarian habits would make eating meat difficult. Hence, although creativity, exploration, and initiative are required to change eating habits in this fundamental way, the payoff is avoiding years and even decades of animal suffering, and such a large good seems to outweigh by far the inconvenience of changing habits. Secondly, it is important to point out that the adventure of such a change of eating habits is, for many people, an exciting and deeply satisfying adventure. Part of the adventure is the discovery of new sources of culinary delight. For many people it is not only liberating to be able to cook without having to rely on meat, but it is also fun to discover that vegetables are more than soggy garnishes. In addition, there is the satisfaction of eating and knowing that one is not contributing to the suffering of animals. Finally, there is the adoption of a way of eating that in many ways is far more healthy than a diet relying on animal flesh.[10]

It is time to bring these observations to bear on our central question: are there any pleasurable consequences of raising and eating animals which would outweigh the frustration, terror, and pain these animals had to undergo in order for us to experience these pleasures? Once we properly focus our attention only on those pleasures that are not replaceable by an alternative style of eating, it is not plausible to believe that there are any.

OBJECTION SIX: THE IMPOTENCY OF THE INDIVIDUAL TO AFFECT ANIMAL SUFFERING

A final attempt to defend eating meat on utilitarian grounds rests on the claim that, for those of us who do not raise our own animals, there is no chance that any one of us will make a difference in the total amount of animal suffering by failing to buy and eat meat. Suppose that I buy meat from a retailer who is supplied by the giant meat industry. Because the meat suppliers are so large, if I stop eating meat, my action will have no effect on the number of animals raised. Hence, if I am concerned to pre-

---

[10]A vegetarian diet tends to be low in cholesterol and animal fats, high in fiber content, and without the dangerous additives that are often used in livestock feed (for example, hormones to stimulate growth and antibiotics to decrease stress-related disease).

vent suffering, there is no animal suffering I prevent by becoming a vegetarian. But if I enjoy eating meat, and if becoming a vegetarian would
cause me to suffer, then I ought to eat meat, for by continuing to indulge
this habit I produce my own pleasure without producing any avoidable
suffering for any other sentient being (Wenz, 1979).

The problem is not that the odds are very small that my action will
have an effect on industry production. For if I only had a small chance of
changing industry production, then one of the foreseeable consequences
of my action would be the prevention of a great amount of animal pain.
Such an action would be a good gamble for preventing animal suffering,
i.e., a small sacrifice for a great gain, and so it would be obligatory on act
utilitarian grounds. The problem, rather, is the much more severe problem of individual impotency in large market situations. The defender of
eating meat claims that only a large number of actions can produce a
change in the production of meat and that no single person's actions are or
would be necessary for this change to take place. Since meat production
will remain constant or will change *regardless* of what I do, any change
in meat production cannot be considered a consequence of *my* action and,
hence, my action has no foreseeable effect on the suffering of domestic
animals being raised for food. It follows, according to this objection, that
if I am an act utilitarian and assess my acts in terms of their own
foreseeable consequences, or if, what amounts to the same thing, I am
concerned to do what will contribute as little as possible to the amount of
suffering in the world, then there is no reason for me to take into account
the suffering inflicted on animals by the meat industry when I am choosing between continuing to eat meat or becoming a vegetarian.

One reply to this utilitarian defense of eating meat begins with the
assumption that my eating meat, or failing to eat meat, will not alter the
consumption behavior of others, i.e., others will not eat either more or
less meat if I change my eating habits. This assumption permits the utilitarian to argue that by becoming a vegetarian I diminish the demand for
meat and so there will be an eventual diminishment in meat production
and, significantly, a diminishment of the corresponding suffering of animals.[11] The assumption that my becoming a vegetarian will not influence
the consumption of others is crucial, for it prevents the defender of eating
meat from claiming that the meat I fail to eat will eventually be eaten by
those who previously ate less meat, or none at all.

But this assumption must be rejected, partly because it denies what
is at the core of the utilitarian defense of eating meat, and—not
irrelevantly—because it is blatantly unrealistic.[12] Profit margins in
various parts of the meat industry are so small, for example, that the
amount of meat "wasted" by even one person becoming a vegetarian

[11] I am indebted to Jan Narveson for this argument.
[12] But note that in 1973 the average American consumed 175 pounds of red meat and
over 50 pounds of chicken (*Information Please Almanac*, 1975)!

would in fact be recycled into other markets, even if only fertilizer markets or pet food markets.[13] Of course, once we abandon the assumption that my failing to eat meat will not alter the meat consumption of others, the act utilitarian can bring a new consideration into the discussion, viz., the ripple effect on others caused by my example of becoming a vegetarian. But the utilitarian defender of eating meat will reply that other markets and other consumers will absorb what any single person or a few persons do not consume, and so my failing to eat meat, even if conjoined with the acts of those who follow my example, will make no difference in the amount of animal suffering produced by the meat industry. A key issue in this controversy is whether one person's example of becoming a vegetarian has a foreseeable possibility of altering the meat consumption of others to such an extent that animal suffering is reduced. Although I think it is true that one person's example has a real chance of being efficacious in this way, even if true, that would hardly close the issue. For instance, should many people become vegetarians, the resulting meat surplus would temporarily depress the price of meat. One result would be that those who previously could not afford meat would enter the market, thereby becoming meat consumers, and some of these new consumers would likely continue purchasing meat after the price rose back to normal. But even if the price of meat were depressed for a long time, that in itself would not guarantee a reduction in animal production and a corresponding reduction in animal suffering. A price reduction might have the undesirable consequence that some meat producers, given the reduced price for meat, would intensify even further their methods of factory farming in order to make a profit, and these methods would cause even more suffering since they would involve additional diet and movement restriction for the animals. Obviously if *enough* people stopped eating meat and the price became depressed *enough,* then the businessman-farmer would cease producing meat and so cease causing suffering to animals. But such a line of argument is sufficiently dependent on long-range and controversial economic probabilities that, on the face of it, the strongest criticism that could be made against the defender of eating meat is that he is willing to take a chance, perhaps an infinitesimal one, of causing intense animal suffering for the sake of convenience, pleasure, or out of habit. Since this criticism packs little force if the odds are truly infinitesimal, it is time we examined two arguments that do not depend on these economic contingencies and yet show that our consumption of meat is causally related to animal suffering.

Each of these arguments depends on the general idea that if a number of acts together produce some group result of value or disvalue, a propor-

---

[13]It follows, as an editor of Humana Press suggested, that pet owners should try to avoid using meat products as far as possible. Note that the three chief ingredients in commercial dry cat food and dry dog food are corn, wheat, and soy. It also follows that people should try to avoid using meat products to feed birds. For example, peanut butter may be substituted for suet in order to feed such birds as woodpeckers.

tion of the value or disvalue of that group result is causally attributable to each of the contributing individual actions. Consider the following example of David Lyons (Lyons, 1965, p. 39):

> If it takes six men to push a car up a hill and, not knowing this, eight lend a hand and do the job, what are we to say? If all pushed, and pushed equally hard, and delivered equal forces, are we to say that only some of them actually contributed to the effects because fewer *could* have done the job?

As long as we cannot distinguish those acts that are necessary for the result from those acts that are not (and this is the hypothesis of interest to us), Lyons' conclusion is that each of these acts does contribute causally to the group effect. Lyons does not attempt to analyze the concept of "contributory causation," and that is a task too large to attempt here. But surely it is reasonable to think that, when a number of actions contribute equally to an effect, we are to causally attribute part of the value of the effect to each individual action. In whatever way the account of contributory causation is worked out, what it would account for is the claim that when many acts contribute to some common result, there is some consequential value to be attributed to each of these actions because each does, in fact, contribute to that result. (More technically, we will assume that if n acts equally contribute to E, and the value of E is V, then the value of the consequences of each act is a function of the value of V/n)

The first "contributory account" of why an individual is not impotent in the marketplace involves two steps. *Step one*: we assume that a large number of persons do become vegetarians, that this results in a reduction in the demand for meat, that no particular action was necessary for this result to take place, and that the reduction in demand reduces animal suffering by curtailing meat production. Since the reduction of animal suffering is of positive value, the cause of this reduction—the reduction in demand for meat—is also of positive value. It is at this point that the contributory causation analysis is operative. Since many acts have contributed to the valuable result of lower demand for meat, there is a consequential value to be attributed to each act of becoming a vegetarian because each does, in fact, contribute to the lower demand for meat. Hence, given our assumptions, the value of the consequences of each act includes part of the positive value of the reduction of animal suffering. That's step one. *Step two* requires that we replace the *unrealistic* assumption that enough people *will in fact* become vegetarians to alter meat production with the *realistic* assumption that there is *some chance* that enough people will become vegetarians to alter meat production. Whatever probability there is that enough people will become vegetarians to reduce meat production is also the probability that an individual person, by becoming a vegetarian, will contribute to the good of a significant reduction of animal suffering. Hence, that is the probability that a person by becoming a vegetarian will be performing actions with positive consequential value. Since by becoming vegetarians we may prevent, or at least help prevent, some of the suffering that animals would otherwise have had to experience because of confinement and the terrors of slaughter, we must do so on act utilitarian grounds.

There is a second "contributory" rebuttal to the objection that one's own actions are impotent to increase or decrease the suffering of animals. This second reply has the virtue of not relying on any probability assessments about people becoming vegetarians. Rather, this second argument focuses squarely on a central aspect of act utilitarianism: an action is right only if *there is no alternative action* that is a better bet to avoid painful consequences or to bring about pleasurable or happy consequences. In the current market situation the person who eats meat is contributing to the demand for meat. But this demand itself is a cause of meat production and, hence, is a cause of the terrible animal suffering involved in meat production. Because causes are evaluated in terms of the value of what they cause—a mudslide that kills twenty people is thereby bad, whereas a rainfall that ends a famine-causing drought is good—the general demand for meat is of great negative value because of the animal suffering it causes. Since individual acts of buying meat contribute to this great evil (the demand for meat), some negative consequential value is attributable to each act of buying meat, for each such act causally contributes to the general demand that causes animal suffering. Significantly, becoming a vegetarian is an alternative that avoids this contribution to the suffering of animals. Hence, one should become a vegetarian if one is trying to do what has the best bet of not bringing about consequences that are pain-producing or of negative value. In short, a person who is trying to live in such a way as to contribute both as little as possible to the total amount of suffering in the world, and as much as possible to the total amount of the world's happiness, will not purchase meat in today's marketplace, for any such act contributes to the brutal and exploitive practice of raising and slaughtering other sentient beings for their flesh.[14]

[14]I am indebted to Henry West for encouraging me to write the hedonistic act utilitarian's account of our duties toward animals; to William DeAngelis, Michael Lipton, Stephen Nathanson, and others who commented on an earlier version I read at a colloquium for the Department of Philosophy and Religion, Northeastern University, December, 1979; and to the editors of Humana Press for many helpful suggestions. I am also grateful to Sharon B. Young for her many helpful comments on earlier versions of this paper as well as for sharing with me her active exploration of vegetarian cuisine. Finally, I am pleased to thank Barbara Jones and Walter Knoppel for initially introducing me to vegetarianism in an intelligent, effective, and tasty manner.

Postscript

James Cargile

In a revised version of his paper, Professor Becker replies to two objections that were raised in my comments. The first objection was that even if having a virtuous character will dispose one to favor people over animals, it does not follow that acting according to such a preference is just, any more than it would follow that a judge is being just when he discriminates against people whom he rightly, and in keeping with a good character, regards as unbearable (say, some outrageously long winded bore who is nonetheless law abiding).

Professor Becker replies that " . . . the examples that make this objection plausible are all drawn from 'public' morality . . ."

" . . . 'friends' who never *act out* their friendship (i.e., preferences) or family members who never act out their special love for one another would fail to be what we mean by friends or family . . . in private life at least, the dispositions to reciprocate and empathize—as part of moral excellence—must produce not only preferences ordered by social distance, but conduct based on those preferences. To prohibit such conduct is to prohibit one aspect of (private) virtue."

The second objection was that it is not necessarily true that having a virtuous character must lead to acting in favor of people over animals in cases where basic welfare is at stake. That is, there could be a virtuous person who refused to give moral priority to people over animals.

Professor Becker replies that a saintly ascetic who lived in this way would represent "not perfect virtue at all, but rather an awe-inspiring amplification of one or a few of the elements of virtue—to the detriment of the others." He also says

> . . . even if it were true that perfect impartiality were the perfection of virtue, it would not follow that we either could expect or would want that perfection in very many people. If not, then the argument I have given for the priority of human interests—at least for ordinary people—stands.

Someone in the audience peremptorily dismissed my suggestion that not
giving priority to people over animals might be compatible with a virtuous character on the grounds that his wife would never have married him
if he had not shown a preference for her and given some priority to her
interests over those of others. This was just the sort of misunderstanding
that I had tried to dispel by stressing that the saintly character I was so
briefly sketching could have all sorts of preferences and priorities as long
as basic welfare was not at stake. To prefer the company of your family,
and to feed your family and not feed others, is not a moral issue if the
others are not starving. There could easily be a society in which issues of
basic welfare did not arise because all creatures had sufficient food and
shelter. Living in such a society would not have to be noticeably different
from present American society for the many people who act as if our society were like that already. At any rate, in such a society, the saintly character I was trying to describe might be hard to distinguish from everybody
else on questions of basic welfare. A consistent generosity and other subtle matters of bearing might give us indications of saintliness, but it is
hard to manifest nobility when life is too easy. But in this kind of society,
there could still be normal friendships. Whether Professor Becker is right
that " 'friends' who never act out their friendship . . . would fail to be
what we mean by friends" seems to me an interesting semantic question
(the corresponding claim for "family" is obviously too strong) but more
importantly, not a relevant one to the issue whether accepting a moral
principle of equality between people and animals is compatible with
virtue.

After all, Professor Becker does not intend to prove that it is incompatible with virtue not to regard the basic welfare of some people as having more objective moral importance than that of other people. He only
wants to show that the basic welfare of people has more objective moral
importance than that of animals, by showing it is incompatible with virtue
not to have this attitude. But if it is compatible with virtue to regard people as of equal moral importance, then it is possible to have this attitude
while having an adequate special commitment to friends and family. And
once that is granted, no basis is given for not being able to consistently
extend this attitude about moral importance to all creatures, while still
having an adequate special commitment to friends and family.

"My family is of special importance, even moral importance, *to me,*
and yet all people are of equal moral importance." How can *I* say that,
without inconsistency, since if I sincerely assert that all people are of
equal moral importance, does it not follow, not from what I assert, but
from the fact that I sincerely assert it, that *for me,* all people are of equal
moral importance? This is a more or less profound problem, which has
often troubled me in various ways. There is certainly a profound truth in
the point that having no priorities among people is inhuman. Furthermore, the explanation may involve distinguishing between true moral

principles and virtuous character. If the true moral principles are utilitarian ones based on equal value for each person, then good human character may essentially involve a tendency not to be perfectly observant of moral principles in all decisions and actions. Professor Becker's ideas on this matter are very interesting. But this profoundly important study does not provide any straightforward basis for finding the true moral principles to discriminate in favor of people over animals. If the true principles are somehow tied to character in a non-utilitarian way so as to allow special treatment of friends on matters of basic welfare (and still be universalizable?) then this accomplishment could be applied for animals too, to make it consistent to claim that whatever objective moral worth that remains for all people equally also holds for animals. We must of course not have it morally wrong for a mother to favor her children. But mother dogs favor their puppies. As far as devotion goes, to people or puppies, mother dogs can do as well as anyone. I do not claim that a mother dog is a moral agent, but if someone is acting as her (unauthorized, I concede) spokesman we may tell that person her puppies are no more important than any other creature's offspring, to justify her having to wait in the food line like everybody else. But justifying excluding her from the food line is another matter. (Not to mention mother cows, where it is a matter of their offspring being included in the food line as the main course.)

Thus I conclude, as before, that Professor Becker's arguments, for all that they are insightful and important, do not establish the objective moral rightness of giving priority to people over animals in matters of basic welfare.

On the other hand, from a less theoretical point of view, Professor Becker's observations about character suggest to me the thought that it is a hard morality that would stamp out such a natural human practice as meat-eating. Although I am ready to defer to a saint, I find the holier-than-thou attitudes of some vegetarians much harder to bear. So I exclaimed as follows: perhaps it is bad of me to slaughter pigs because my family and I crave meat, but I give my pigs a good life while it lasts—we have all got to go sometime—and anyway, if the vegetarians are so concerned, why do they not save some animals from slaughter by buying them at the stockyard?

This intemperate outburst, conceptually quite distinct from the comments on Professor Becker's paper that were my assigned task, was not kindly received. One person got up and asked how I would feel about someone who saved people from execution at concentration camps by purchasing them, giving them several happy years, and then painlessly slaughtering them and eating them. The idea was that my justification for eating my pigs was no better. But I did not think I had proved it was all right for me to eat pork. It may be wrong for all I know, though I hope not, and cannot believe it is. But if some vegetarians really think my killing pigs for food is morally similar to some human monster's killing people for food, then they are not much good themselves to be doing no more about a practice they see as so horrible! If a slaughterhouse were operating on humans in my neighborhood, then besides campaigning to have the practice outlawed, I would certainly buy a few would-be victims (this is assuming the set-up works like the present system for animals) to save their lives. Who knows, I might spend half my salary in such a situation! It makes me uncomfortable to speculate on how far I would be willing to go. But a lot farther, surely, than these vegetarians who are comparing me with the cannibals are going to save animals!

A friendly philosopher at this meeting stood up to say I was not guilty of ad hominem argument. I am sorry to have to have made clear that there is the foregoing strongly ad hominem strain in my remarks.

In his very fair-minded and reasonable paper "The Case Against Raising and Killing Animals For Food," Bart Gruzalski cites my comment about pigs as a case of a utilitarian argument for using animals for food, and he replies that even if it is true that animals can be raised for food without suffering, there is a better arrangement from a utilitarian point of view, one in which land used for livestock is turned back over to wild animals. In Professor Gruzalski's ideal state much less land is needed to grow much more food for people, and the quantity of animal happiness is provided by wild rather than domestic animals.

This is an attractive picture, but I do not believe it has a solid basis in fact. It is possible, especially on an old fashioned farm, for the land to support, in addition to domestic animals, more wild animals than it could have if the land had been left untouched by man. Furthermore, I believe that wild rabbits, for example, lead considerably more fearful and anxious lives than domestic rabbits, do not live much longer, and do not eat any better. A human has as much right to eat meat as a hawk or a fox does.

However, who am I to say? At any rate, even if a certain arrangement would be better than the present one, it certainly does not follow that it is morally wrong to go on participating in the present one. Our food industry should be scrutinized with an eye to increasing the happiness of humans and animals, and methods that cause needless suffering should not be allowed. Anyone who wishes can vote against the system by abstaining from meat. But if everybody suddenly did so vote, millions of animals would become a liability, and the production of eggs and especially milk, which is often popular with vegetarians, would become at best very much restricted. I do not believe that would be the best outcome for animals or people.

SECTION VI

Animal Rights?

Introduction

Can we sensibly say that animals have rights? Should we? Tom Regan and Bernard Rollin have argued (in Sections I and II above) that both questions should be answered in the affirmative. In this section James Rachels presents a fully worked-out argument that at least some animals have at least a right to life. R. G. Frey, in contrast, answers the second question negatively. Talk of rights, he claims, sheds no light on the rights and wrongs of human treatment of animals (or humans).

Why, asks Rachels, do humans have a right to life? After rejecting Kant's position that only self-conscious beings can have rights and only humans are self-conscious, Rachels provides an analysis of 'right.' On this analysis, animals have a right not to be tortured because (a) torturing is objectionable in terms of the animal's own interests, (b) the animals (or someone on their behalf) are entitled to complain about their being tortured, and (c) it is permissible for a third party to compel one not to torture the animals. Whether or not entities of a certain sort have rights of a certain kind cannot be decided once and for all by some conclusion about a single characteristic decisive for the possession of any and all moral rights. There is no such characteristic. Different characteristics are decisive for different rights.

The relevant characteristic for having a right to life is having a life. To 'have a life' is not just to be alive, it is to have a biography, to have hopes and plans and memories. Humans in irreversible comas may be alive, but they no longer have lives. Some animals other than humans appear to have lives in this sense. Apes and monkeys pretty clearly do, and perhaps wolves, dogs, pigs, and others very frequently the objects of human actions. Rachels suggests that even the octopus, though a mollusk and no near kin of ours, *might* have a life. Which animals have a right to life, then? We cannot now answer this with much confidence. Insects and crustaceans pretty clearly do not have lives, and many 'higher' mammals pretty clearly do. It does not follow that all lives are equally valuable, for mental complexity *does* matter (despite Johnson's arguments in Section III above).

R. G. Frey's title "On Why We Would Do Better to Jettison Moral Rights" makes his stand clear. We *can* talk about rights if we wish, but such talk solves nothing and obscures much. It is always more profitable

273

to ask directly whether some item of conduct is right or wrong than it is to ask whether or not it is permitted by, or in violation of, some entity's moral rights. As Frey points out, the most influential single recent work on the ethics of human treatment of animals, Singer's *Animal Liberation* (1976) makes no use of the notion of a moral right.

If we can reach agreement on moral principles, we can move directly from those principles to permitted and impermissible conduct. But if we cannot agree on moral principles, we will not be able to agree on rights. Claims of rights are thus either unnecessary or unhelpful. When claimed rights conflict, as they so often do, how are such conflicts to be resolved? By utilitarian considerations? If the grounds of rights are, on the other hand, self-evident, how are we to argue about them? If compromise is appropriate, how are we to distinguish compromise from domination? If intuition is decisive, what if my intuitions accord ill with yours? In none of these situations does a postulation of rights make an intractable impasse tractable.

We are familiar, Frey points out, with acquired rights, such as the rights I obtain by purchasing a parcel of land. But these acquired rights, he claims, are nothing like, and shed no light upon, unacquired moral rights of the sort thought important in judging human treatment of nonhumans.

Frey thus ends where he began, with the view that claims of rights for animals are useless at best. Rachels, obviously, does not agree.

Do Animals Have a Right to Life?[1]

James Rachels

The question of whether nonhuman animals have a right to life has less practical importance than one might think, for even if they have no such right, the slaughter of nonhumans can still be condemned on other grounds. The principal human activities that involve killing animals—hunting, trapping, meat production, and scientific research—all involve such cruelty that they should be rejected for that reason alone.[2] If, in addition, the animals have a right to life that is violated, the already conclusive case against those practices is simply made stronger. Nevertheless, the question in my title does have some practical importance, for there are some cases of painless killing not covered by the moral prohibition on cruelty. And of course it has considerable theoretical interest, not only for those concerned with animal welfare, but for all those interested more generally in the concept of the right to life.

Humans presumably have a right to life. But why? What is it about humans that gives them this right? If humans have a right to life, but plants, say, do not, then there must be some characteristics possessed by humans but not by plants that qualify the humans for this right. One way to approach our subject is by trying to identify the characteristics that a being must have in order to have a right to life. Then we can ask whether nonhuman animals have those characteristics.

[1]This essay is a companion to an earlier one, "Do Animals Have a Right to Liberty?" which appeared in Regan and Singer, 1976, pp. 205–223.

[2]On hunting and trapping, see Amory, 1974. On meat production see Singer, 1975. On the use of animals in scientific research, see Ryder, 1975. Argus Archives, of New York, publishes a number of useful short books on animal abuse, such as Redding and Stewart, 1977, and Pratt, 1976. Since arguments founded upon opposition to cruelty are bound to be clearer and less controversial than appeals to a "right to life," animal welfare activists are wise to avoid the latter notion when possible.

RIGHTS AND CHARACTERISTICS

Traditionally, philosophers have tried to identify some *one* characteristic, or set of characteristics, that a being must possess before it can have *any rights at all*. Often philosophers have assumed that, since only humans have rights, these must be characteristics that other animals lack.

Kant, for example, held that although we have duties involving nonhuman animals, we can have no duties *to* nonhuman animals—just as we can have duties involving trees, but not duties to trees. On Kant's view we may very well have a duty not to kill an animal—perhaps it is someone's pet; the person owns the animal, and would be unhappy over its death; therefore we should not kill it. The reasons why we should not kill it, however, all concern the person's interests and not the animal's. We can never have a duty not to mistreat an animal founded upon the animal's own interests.

If Kant's view is correct, then animals cannot have rights; for, if a being has rights, then the duties we have in virtue of those rights are duties *to* him and not merely duties involving him. The difference between humans and other animals, which explains this difference in moral status, is that humans are "self-conscious," whereas other animals are not (Kant, 1963, p. 239). Thus, on Kant's view, only self-conscious beings can have rights.

There are three main difficulties for this view.

First, and most obviously, it seems false to say that we can have no duties that are founded on the interests of animals themselves. If making an animal suffer has adverse effects on human beings, that is certainly a reason why making the animal suffer is wrong. But even if there are no adverse effects for humans, wantonly torturing animals still seems wrong. Kant maintains, however, that this is not a real possibility: cruelty to animals always involves at least the possibility of harm to persons, because "He who is cruel to animals becomes hard also in his dealings with men" (Kant, 1963, p. 240). But I doubt that this is true. Is there any reason to believe that people in the meat business, or people who experiment on animals, or people who go hunting, are more cruel to their fellow humans than most people? It may be so, but I know of no reason to believe it is so. Those people are, so far as their fellow humans are concerned, moral in the ordinary ways. What happens is not that they become more cruel toward humans; rather, they develop ways of viewing their behavior toward the animals that deny the cruelty done there. Experimenters, for example, have developed an elaborate terminology for describing what happens in the laboratory that leaves out the element of suffering. Writhing is referred to as "movement response" and screams are "vocalizations"; and of course the infliction of pain is itself merely an "aversive stimulus." These ways of speaking are usually defended on the grounds that they represent the facts more objectively, but the reverse seems true: the use of these terms encourages one to overlook elements of the situation that are objectively present.

Second, if the possession of rights depends on self-consciousness, it is not so clear that all animals are thereby excluded as bearers of rights. The notion of self-consciousness is itself problematic. In ordinary speech "self-conscious" means "awkward and easily embarrassed, ill at ease"—as in, "Johnny went to the party, but was too self-conscious to dance." Obviously this is not the philosopher's meaning. The Kantian thesis is not that only the ill-at-ease have rights. But what, then, does self-consciousness in the philosopher's sense consist in? And exactly how are we to tell whether animals are self-conscious? Perhaps one mark of self-consciousness is the ability to use personal pronouns. A being that can refer to itself as "I" and talk about itself seems fairly describable as self-conscious. There are at least a few nonhuman animals that can now do this: the chimps and gorillas who have been taught sign language, and who have been taught to communicate using keyboard symbols. Still, it is not clear that an achievement of this sophistication is required for self-consciousness, because the philosophical concept of self-consciousness is itself so unclear.

One contemporary philosopher who has tried to clarify the notion is Michael Tooley. Like Kant, Tooley believes that self-consciousness is a morally crucial notion; more specifically, he believes that without this characteristic one cannot have a right to life. In his well-known essay "Abortion and Infanticide" Tooley maintains (1972, p. 64) that

> An organism possesses a serious right to life only if it possesses the concept of a self as a continuing subject of experiences and other mental states, and believes that it is itself such a continuing entity.

Unlike Kant, Tooley leaves open the possibility that animals might have a right to life by this criterion—but it seems only a bare possibility, unrealized by actual animals.

In making the Kantian thesis more explicit, Tooley has not made it any more plausible. I assume that David Hume had a right to life, if anybody has one; yet Hume certainly did not believe that he was a self who was a continuing subject of experiences. In fact, he specifically denied that, and so do many contemporary behavioristic psychologists.[3] It might be suggested that Tooley's thesis could be amended, so as to avoid this objection, by leaving off the last clause. But I do not believe that the difficulty is caused entirely by the reference to belief. It is doubtful that anyone other than philosophers even possesses the concept of such a self—my father, for example, is an intelligent man, but not a philosopher, and I doubt that he "possesses the concept of a self as a continuing subject of experiences and other mental states." If he does have this concept,

[3]Leonard S. Carrier makes this point about Tooley's thesis in Carrier, 1975, pp. 292–293.

he certainly is not aware of it. Hume has argued that this concept is not even intelligible (T:I, iv, 6). Others disagree. I do not know who is right, but I would not like the existence of my right to life to depend on the outcome of that debate.

The third difficulty for the Kantian view is philosophically the most interesting one. The Kantian thesis tries to identify *one* characteristic that is relevant to the possession of *any* right. That is far less plausible than to think that the characteristics one must have in order to have a right vary with the rights themselves. Consider, for example, rights as different as the right to freedom of worship and the right not to be tortured. To have the former right, one must be a creature with a capacity for worship; one must have religious beliefs and the capacity for participation in a form of life in which worship could have some place. For this reason, it seems senseless to think that rabbits could have such a right. At the same time, rabbits might very well have a right not to be tortured, for the capacity to worship has little or nothing to do with that right.

Perhaps, by some convoluted reasoning, it might be shown that only self-conscious beings can have a right to worship. But does having a right not to be tortured depend on being self-conscious? One way to get a grip on this is to ask whether a person's self-consciousness figures into the reasons why we object to torture. Do we object to a person's being tortured because he is a self-conscious being? Or, to use Tooley's explication of the idea, because he possesses a concept of the self as a continuing subject of experiences? I do not believe that this is the reason why torture is wrong; the explanation is much simpler than that. Torture is wrong because it hurts. It is the capacity for suffering, and not the possession of sophisticated concepts, that underlies the wrongness of torture. This is true for both people and rabbits; in both cases, torture is wrong because of the suffering that is caused. The characteristic of humans that qualifies them as bearers of a right not to be tortured is, therefore, a characteristic also possessed by rabbits, and by many other animals as well. Thus I believe that, if humans have a right not to be tortured, so do other animals.

This argument contains an important gap. It might be admitted that we have the same reason for objecting to torturing animals that we have for objecting to torturing humans, but still be maintained that torturing humans violates their rights whereas torturing animals does not. For it may be wrong to treat an animal in a certain way without any question of *rights* being involved at all. To address this issue, we need to ask what is the difference between merely having a duty not to treat people in a certain way, and violating their rights by treating them in that way. Three differences come immediately to mind.

(1) When it is a matter of an individual's rights not to be treated in a certain way, treating that person in such a way is objectionable *for that person's own sake,* and not merely for the sake of someone or something else. Thus if a pet-owner's animal is harmed, and the *only* objection we

have is that the owner's interests are violated, we are not considering the animal to have any rights of its own in the matter.

(2) When rights are involved, the rights-bearer may protest, in a special way, if he or she is not treated properly. Suppose you consider giving money for famine relief to be simply an act of generosity on your part. You think you ought to do it, since you ought to be generous to those in need. However, you do not consider the starving to have a *right* to your money; you do not owe it to them in any sense. Then, you will not think them entitled to complain if you choose not to give; it will not be proper for them to insist that you contribute, or to feel resentment if you do not. Whether you do your moral duty is in this case strictly up to you, and you owe them no explanation if you choose not to do so. On the other hand, if they have a *right* to your money, it is permissible for them to complain, to insist, and to feel resentment if you do not give them what they have coming.

(3) When rights are involved the position of third parties is different. If you are violating someone's rights, it is permissible for a third party to intervene and compel you to stop. But if you are not violating anyone's rights, then even though you are not behaving as you ought, no third party is entitled to coerce you to do otherwise. Since giving for famine relief is widely considered not to involve the rights of the starving, but only to involve "charity" toward them, it is not considered permissible for anyone to compel you to contribute. But if you had contracted to provide food, so that they now had a claim of right on your aid, compulsion would be thought proper if you reneged.[4]

When judged by these measures, it appears that animals do have a right not to be tortured, and not merely that we are wrong to torture them. (1) Torturing animals is objectionable for reasons having to do with the animals' own interests. (2) Someone speaking for the animals could legitimately complain, in their behalf, and insist that you not torture them. For this complaint to be heard the animals would need spokesmen only in the same sense in which retarded humans and infants require spokesmen to assert their rights. (3) Finally, the position of third parties is the same as where rights are involved: it is permissible for third parties to compel you not to do it. The weight of this evidence, then, is that the animals have a *right* not to be tortured—and the only characteristic that they need to qualify for this right is the ability to suffer.

[4]Philosophers suspicious of the concept of rights have wondered exactly what it means, and how it might be rendered in terms of the less puzzling notion of permissibility. The preceding observations suggest this analysis. X has a right to be treated in a certain way by Y if and only if: first, it is not permissible for Y not to treat X in that way, for reasons having to do with X's own interests; second, it is permissible for X to insist that Y treat him/her in that way, and to complain or feel resentment if Y does not; and third, it is permissible for a third party to compel Y to treat X in that way if Y will not do so voluntarily. The *understanding* of rights that goes with this analysis is that rights are correlates of duties the performance of which we are not willing to leave to individual discretion.

The strategy I am criticizing is common among contemporary moral philosophers, who theorize that eligibility for rights depends on possession of certain characteristics, without considering that the relevant characteristics might change as different rights are in question. Robert Nozick, author of the widely-discussed book *Anarchy, State, and Utopia,* holds that a being has rights only if it is a rational, free moral agent with "the ability to regulate and guide its life in accordance with some overall conception it chooses to accept" (Nozick, 1974, p. 49). H. J. McCloskey believes that it is the capacity for moral self-direction that qualifies one for basic moral rights (McCloskey, 1979). Other examples could be given easily. These thinkers are all up to the old Kantian trick: the attempt to divide rights bearers from nonrights bearers on the basis of their possession, or lack, of some one (set of) very general characteristic(s). The primary objection must be to the form, not the content, of these proposals. Why should we believe that the *same* characteristics that make on eligible for one right also apply to the others? Surely the sensible approach is to take up the rights, and the characteristics that make us eligible for them, one at a time.

THE RIGHT TO LIFE

With respect to the characteristics that qualify one for a right to life, I wish to offer the following thesis: an individual has a right to life if that individual has a life. Like many philosophical claims, this one is more complicated than it first appears to be.

"Having a life" is different from merely being alive.[5] The latter is merely a biological notion; to be alive is just to be a functioning biological organism. A life, in the sense in which it concerns us here, is a notion of biography rather than biology. "The Life of Babe Ruth" will be concerned not with the biological facts of Ruth's existence—he had a heart and liver and blood and kidneys—but with facts about his attitudes, beliefs, actions, and relationships. It will say that he was born in Baltimore in 1895; that he was the troubled child of a poor family, sent to live at St. Mary's School when he was eight; that he learned baseball at the school and started pitching for the Red Sox at 19; that he was a fine pitcher for six seasons before switching to the Yankee outfield and going on to become the most idolized slugger in the history of the game; that he was the beer-guzzling friend of Lou Gehrig, and married to Claire; that he died of cancer at age 53; and much more.

The contingencies of human existence determine the general shape of our lives. Because we are born physically weak and without knowl-

---

[5]I have learned so much about this distinction from William Ruddick, in conversations over a period of years, that I can no longer tell which ideas are mine and which are his.

edge or skills, the first part of our lives is a process of growth, learning, and general maturation. Because we will not live much longer than 75 years, and because in the last years we will decline mentally and physically, the projects and activities that will fill our lives cannot be planned for much longer than that. The forms of life within human society are adjusted to these dimensions: in our society, families care for children while they are small and are acquiring a basic understanding of the world; schools continue the educational process; careers last about 40 years; and people retire when they are 65 or 70.

The stages of a life are not isolated or self-contained parts. They bear relations to one another that must be understood if any part of the life is to understood. We cannot understand what a medical student is doing, for example, if we do not appreciate the way in which his or her present activity is preparation for the stages of life that will come later. Moreover, the *evaluation* of one stage of a life may require reference to what came before: to be a doorkeeper, with a small but steady income sufficient to pay the rent on a one-room apartment, might be a laudable achievement for one who previously was a homeless drunk; but for one who was a vice-president of the United States, caught taking bribes, the same existence might be a sign of failure and disgrace. (This is of course a fictitious example, since this is not what happens to vice-presidents caught taking bribes.) Thus the fact that people have memories, and are able to contemplate their futures, in a fairly sophisticated way, is important to explaining why they are able to have lives. Those philosophers who have sought to explain the continuing identity of *persons* by reference to memory may have been barking up the wrong tree since it is more plausible to think that the connections of memory are necessary for the unity of a *life*.

The concept of a life is useful, too, in explaining the kinds of goods and evils to which people are susceptible. Consider the plight of a young concert pianist who loses the use of his hands: why is this a bad thing for him? It is not a sufficient explanation to say that it is bad because it will cause him pain, frustration, worry, and the like. The tragedy consists in the fact that important possibilities for his life have been foreclosed; he could have had a career as a pianist, and now he cannot. If he is frustrated or depressed, he has something to be depressed about. It is not as though the mental anguish associated with his misfortune were only an accidental byproduct; on the contrary, it is a perfectly rational response to a situation that is, independently of his response, bad. We will not have eliminated the evil by eliminating the response to it: we cannot make things all right by getting him to cheer up. It is the loss of the possibility for him, and not simply its effects on his consciousness, that is the evil.

The evil of death is like this. Epicurus, who believed that all good and evil consists in pleasant or unpleasant sensation, thought that death is not a evil because when dead a person experiences no sensations. He failed to appreciate that a person can be the subject of good or evil, not only because he has the capacity for enjoyment or suffering, but also because he may have hopes or aspirations that go unfulfilled, and because he is the subject of a life with possibilities that may go unrealized.[6] When Frank Ramsey died, it was a tragedy, for Ramsey was a young man whose life was only beginning; when Bertrand Russell died, after a full life, things seemed different. Epicurus could not account for the difference. (If death, at even an advanced age, is thought to be an evil, it must be because we are able to view a life as in principle open-ended, always having possibilities that still might be realized, if only it could go on.)

Death is an evil when it puts an end to a life. Some humans, tragically, do not have lives and never will. An infant with Tay-Sachs disease will never develop beyond about six months of age, there may be some regression at that point, and it will die. Suppose such an infant contracts pneumonia: the decision might be made not to treat the pneumonia, and allow the baby to die. The decision seems justified because, in the absence of any possibility of a life in the biographical sense, life in the biological sense has little value. The same sort of consideration explains why it seems so pointless to maintain persons in irreversible coma. The families of such patients are quick to realize that merely being alive is unimportant. The mother of a man who died after six years in a coma told a newspaper reporter, ''My son died at age 34 after having lived for 28 years.''[7] It was a melodramatic remark, and on the surface a paradoxical one—how can one die at 34 and have lived only 28 years?—yet what she meant is clear enough. The man's *life* was over when he entered the coma, even though he *was alive* for six years longer. The temporal boundaries of one's being alive need not be the same as the temporal boundaries of one's life.

Therefore, I believe that it is unwise to insist that any animal, human or nonhuman, has a right to life simply beause it is a living being. The doctrine of the sanctity of life, interpreted as applying merely to biological life, has little to recommend it. My thesis about the right to life is that an individual has a right to life if that being has a life in the biographical sense. By this criterion, at least some nonhuman animals would have such a right. Monkeys, to take the most obvious example, have lives that are quite complex. They are remarkably intelligent, they have families and live together in social groups, and they apparently have forward-looking and backward-looking attitudes. Their lives do not appear to be as emotionally or intellectually complex as the lives of humans; but the

---

[6]See Thomas Nagel, ''Death,'' in Rachels, 1979, pp. 449–459. This essay deserves a place among the classic refutations of hedonism.

[7]*Miami Herald,* September 29, 1972, p. 29A.

more we learn about them, the more impressed we are with the similarities between them and us.[8]

Of course we do not know a great deal about the lives of the members of most other species. To make intelligent judgments about them we need the sort of information that could be gained by observing animals in their natural homes, rather than in the laboratory—although laboratory-acquired information can be helpful. When baboons, dogs, and wolves have been studied in "the wild," it has been found that the lives of individual animals, carried out within pack societies, are surprisingly diverse. But we are only beginning to appreciate the richness of the animal kingdom. Take the octopus, for example. It is a mollusk—like a clam—but it has the most complex central nervous system of all invertebrates, apparently developed originally to control the many gripping-pads along its tentacles. It has a home, stakes out a territory, and fights other octopuses that invade its territory, but not to the death. Observers believe that it shows emotions like fear and anger by changing its skin color (the ability to change color also allows it to blend with its surroundings when natural enemies are nearby). When fighting its prey, it shows no emotion, but when fighting other octopuses it does show emotion. Its intelligence has been tested in such ways as giving it food in a jar to see whether it can figure out how to screw the top off. The animal does figure this out, within a couple of minutes. Now we do not know very much about octopuses, and I am not mentioning all this in order to argue that they have lives and therefore a right to life. I only want to point out the way in which new information can make a difference to our view of animals' lives. Speaking for myself at least, even this meager information makes the octopus seem very different from before.

In our present state of semi-ignorance about other species, the situation seems to be this. When we consider the mammals with which we are most familiar, it is reasonable to believe that they do have lives in the biographical sense. They have emotions and cares and social systems and the rest, although perhaps not in just the way that humans do. Then the further down the phylogenetic scale we go, the less confidence we have that there is anything resembling a life. When we come to bugs, or shrimp, the animals pretty clearly lack the mental capacities necessary for a life, although they certainly are alive. (But being alive is not so important, even for us humans, except in that it is necessary for the continuation of our lives.) Most of us already recognize the importance of this—we think that killing a human is worse than killing a monkey, but we also think that killing a monkey is a more morally serious matter than swatting a fly. And when we come to plants, which are alive, but where

[8]Monkeys even have moral characteristics, such as compassion. On this, see J. Rachels, "Do Animals Have a Right to Liberty?"

the notion of a biographical life has no application whatever, the moral qualms about killing have vanished altogether. If the thesis that I have suggested is correct, these feelings have a rational basis: is so far as we have reason to view other creatures as having lives, as we do, we have reason to view them as having a right to life, if we do.

One final question: If humans have a right to life, and some other animals do also, does this mean that all their lives are equally valuable and have equal moral protection? No. In a situation of forced choice, in which two lives are at stake and only one may be saved, it is reasonable to give preference to the life of the more mentally complex being. That is why human life may rightly be regarded as in general more valuable than nonhuman life (although *some* humans are mentally less complex than some nonhumans, and so should not be given preference).

Philosophers sometimes doubt whether mental complexity has this kind of importance.[9] Some explanation should therefore be provided of why complexity matters. Complexity matters because, when a mentally sophisticated being dies, there are more reasons why the death is a bad thing. A young writer who dies in a car-accident will not get to finish her novel; she will not see her children grow up; her talents will remain forever undeveloped, her aspirations unfulfilled. In the case of a simpler being, not nearly so much of this kind can be said, and so its death, while a bad thing, is not comparably tragic.

---

[9]In his essay "Life, Death, and Animals: Prolegomena," included in this volume, Edward Johnson asks "Why should mental complexity count for anything?" and he concludes that it should not.

On Why We Would Do Better To Jettison Moral Rights

R. G. Frey

I want to approach the issue of animal rights indirectly, by considering a broader issue about moral rights generally.

In *Interests and Rights* (Frey, 1980) I develop two theses, so far as the moral rights of animals are concerned. On the one hand, I suggest that, if there are any moral rights, animals do not possess them; on the other, I suggest that there are no moral rights. The present paper develops a third thesis, to the effect that we will do better in trying to get clear about moral issues such as our treatment of animals (or abortion or euthanasia) by dropping altogether the language of moral rights.

Part of my discussion of this broader issue about moral rights generally finds an analog in earlier discussions of intuitionism; but this should not be surprising, since, as R. M. Hare has remarked, rights are traditionally the 'stamping-ground' of intuitionists (Hare, 1975, p. 203).[1] Intuitionists have been, are, and, I suppose, always will be stern opponents of (act-) utilitarianism; and one of the strongest weapons in their armory, they think, is the use of rights against the utilitarian. I have never been convinced by their case, but my own utilitarian inclinations are not at issue here. For though I am going to be critical of moral rights, I am not going to be so from any peculiarly utilitarian viewpoint.

ARGUING ABOUT RIGHTS

I do not believe any clarity is shed upon the important questions of the rightness and justification of our treatment of animals by trying to discuss these questions in terms of rights. Indeed, I suspect obfuscation is nearly

---

[1]Though only a few sentences in Hare's article are devoted to rights, the present paper owes a debt to them. For a fuller account of Hare's view on rights, see his *Moral Thinking* (Oxford, Clarendon Press), 1982.

always the result of such efforts. For what inevitably ensues is that we turn our attention away from the immediate and important problems of whether our present treatment of animals is right and can be justified and from the necessity of thrashing out principles of rightness and justification of treatment. Instead, we come to focus upon the much less immediate, important, and easily resolvable because wholly speculative questions of whether there really is this or that moral right that some people but not others allege that there is and of what the criteria are—and of how we are to decide what the criteria are—for the possession of this or that right. Speculative questions invite speculative answers, and a vast industry has arisen as a result, both as to the nature and type of moral rights we and other creatures are alleged to possess and as to the grounds in virtue of which we and others are alleged to possess them. So far as I can see, there is no internal limit to this exercise, except that of human ingenuity in being able to think up still different conceptions of moral rights and still further arguments in support of this or that criterion of right-possession.

There are a number of reasons why talk of rights does not shed light on important ethical questions, but I shall here confine myself to a single such reason, namely, the obvious difficulty we all experience whenever we try to argue with each other about moral rights. This difficulty has (at least) three sources.

(1) As Hare has remarked (1975, p. 202), it is not obvious how we are to move back and forth between talk of rights and talk in terms of the moral concepts of right, wrong, and ought. Several views of the matter exist, but none, so far as I know, commands widespread assent. The result is obvious: if we cannot link up the moral concepts and rights, it is not only difficult to see how we could ever argue about moral rights, but also difficult to see quite how we would analyze or explicate the concept of a moral right.

Consider an example: Albert loves fried eggs for breakfast, and his wife knows this; but though she fixes scrambled eggs, poached eggs, boiled eggs, and omelettes, she never fixes fried eggs for breakfast. Albert and his wife are married, there are duties on both sides, and marriage is, we say, a matter of mutual accommodation; yet, though she knows his desires and preferences perfectly, Albert's wife never fixes fried eggs for breakfast. She does not gloat over not doing so; she just does not do so. Now I can imagine a third party saying that it is wrong of Albert's wife not to fix fried eggs occasionally or that she ought to do so or even that, given she is married, knows Albert's desires, and finds fixing fried eggs no more trouble than fixing any other sort of eggs, she has a duty to fix fried eggs occasionally; but does anyone really think we can move from saying these things to the view that Albert has a moral right to fried eggs for breakfast? And if you think we *can* make this move, then how are you going to prevent the complete trivialization of the notion of a moral right, since there appears no end to the possible development of similar examples? Of course, you might say that Albert's right to fried eggs is not a moral but an institutional right, a right arising within the confines of marriage; but many, if not most, people regard this and many other social institutions in a moral light and so regard the institutional rights that arise within them as moral rights as well. I shal not labor the point; but from the fact that you think it wrong of Albert's wife not to fix fried eggs occasionally for breakfast, nothing obviously follows about Albert's having a moral right to fried eggs.

However, even if we cannot move in this direction, can we not move in the other, from talk of rights to talk of, e.g., what is wrong? That is, if we assume that Albert has a moral right to fried eggs for breakfast, can we not conclude that it is wrong of Albert's wife not to fix fried eggs occasionally? But this way of approaching the matter brings out its own shortcoming; for what if I am not prepared to allow you the assumption that Albert has such a right? In that eventuality, you will have to show how you reach this right, on the basis of what is right and wrong or ought to be done in respect of Albert's breakfast; but when you try to do this, the previous argument applies. Likewise, if you assume that squirrels have a moral right to chestnuts on the ground,[2] then you will doubtless conclude that it is wrong of me to deprive them of these nuts; but if you are not granted your initial assumption of such a right, how are you going to reach it merely from your view that it is wrong for me to roast chestnuts in such quantities as to deprive squirrels of this food source? For unless by fiat you simply turn whatever you judge to be wrong into a right on the part of some creature not to have that thing done, you are continually going to run up against the fact that, e.g., it will not follow from its being judged wrong to deprive squirrels of chestnuts on the ground that they have a moral right to these nuts. In short, to link up and tie together moral rights and the concepts of right, wrong and ought requires more than the merely one-way transaction that the present line of argument envisages.

Nor is what makes wrong acts wrong the fact that they violate some alleged right or other. Even in a world without moral rights,[3] acts can still be right and wrong and principles of rightness and justification of treatment can still be presented and argued. Significantly, the book *Animal Liberation* (Singer, 1975), which has done so much to stimulate discussion and re-assessment of our practices in respect of animals, neither prosecutes its case in terms of moral rights nor appeals to moral rights in order to reach substantive moral conclusions about our treatment of animals. Let there be no moral rights whatever: Peter Singer thinks

[2] I owe this example to Stephen Clark.

[3] If one regards social institutions, wherein quasi-contracts are made, in a moral light, then institutional rights may be regarded as moral rights. Quasi-contractual arrangements of this sort are not at issue here, since no one to my knowledge defends an animal's supposed moral right to life on such an institutional basis.

animals can still feel pain and on that basis alone have interests to be weighed by us in deciding upon our treatment of them; and he plainly thinks he has a case for morally condemning certain of our practices for ignoring or undervaluing the interests of animals, though he in no way grounds his condemnation of these practices upon their violation of putative rights.

Of course, it is fashionable today to say that moral rights are the last refuge of the weak and defenseless, the implication being that, in a world, probably utilitarian in character, in which there are no moral rights, the weak and defenseless will go to the wall. But the weak and defenseless *do not* go to the wall in *Animal Liberation*, though Singer studiously avoids moral rights; and Singer *is* an act-utilitarian. Indeed, far from this utilitarian taking advantage of the weak and defenseless, he staunchly defends them, but not, it must be said, on the basis of any alleged rights they may be assigned. I do not agree with a good deal in *Animal Liberation*, as I have made clear in *Interests and Rights* and in my forthcoming book *Rights, Killing, and Suffering*; but it is, I think, a cardinal instance of how we can raise and come to grips with the important issues of the rightness and justification of our treatment of animals without relying upon or invoking moral rights, in a word, without being altogether side-tracked by moral rights.[4]

(2) Another source of difficulty in arguing about moral rights is that what passes as such argument is in fact nearly always argument about the acceptability of the moral principle that is said to be the ground of the alleged right in question. Thus, when someone says that women have a moral right to an abortion on demand, and when we go on to challenge this claim, what we eventually find ourselves disputing is the acceptability (and interpretation) of the moral principle that is alleged to confer this right upon women. The acceptability of this principle is crucial, since you only accept that women have such a moral right if you accept the principle that confers this right upon them.

Three facts about arguments over the acceptability of moral principles are noteworthy. First, we do not agree on moral principles, as everyone knows, particularly if one has followed contemporary discussions on sexual morality, abortion, euthanasia, infanticide, and so on. Second, we do not even agree on the criteria of acceptability or adequacy in moral principles. Are these criteria formal or material? Do they involve reference to 'ordinary moral convictions' and the views of the 'plain man'? Or do they refer to some special subset of these convictions and views? Are they bound up with the achievement of a condition of reflective equilibrium between our moral principles and considered judgments? Or is this

---

[4]What is true here of Singer, (1975) is equally true of Stephen Clark's *The Moral Status of Animals* (1977): as Clark makes very plain, its positive theses can all be raised and discussed outside the trappings of moral rights. I have taken Singer as my example here primarily because he is an act-utilitarian.

type of criterion merely a sophisticated intuitionistic one and so objectionable on grounds similar to those urged against W. D. Ross? In order to ground a moral right, in short, it will not do to assume the acceptability or adequacy of some moral principle; when you then try to show that this principle *is* acceptable or adequate, however, you find that the criteria of acceptability in moral principles are themselves up in the air. Third, as we have seen, we can argue about the acceptability or adequacy of moral principles even if no moral rights are alleged to be grounded upon them, even if, that is, there are no moral rights whatever. Thus we can argue about the acceptability of a moral principle enjoining respect for life that encompasses fetuses and animals under its protection, whether or not we think that fetuses and animals have a moral right to life; and we have to argue about the acceptability of this principle in any event, if you maintain fetuses and animals have a right to life on the basis of it.

In the light of these facts, I think we can produce an argument that shows that moral rights are superfluous.[5] If we cannot reach agreement on the criteria of acceptability or adequacy in moral principles, and if we cannot, therefore, reach agreement on our moral principles, then we are not going to agree over whether there is this or that right, conferred by this or that principle. If, however, we do reach agreement on the criteria of acceptability or adequacy in moral principles, and if we do then reach agreement on moral principles, then there is no longer any need to posit the existence of a right. For if we take morality seriously and so try to live up to our principles, we shall behave in the way the right's proponent wants us to, without that proponent having to postulate the right's existence. In other words, if moral rights are put forward on the basis of unagreed moral principles, we will not agree on whether there are such rights, whereas if they are put forward on the basis of agreed moral principles, they appear unnecessary, since our principles will now lead us to behave in the way the right's proponent wants us to behave.

One might argue, of course, that postulating moral rights upon the basis of agreed moral principles serves the function of an insurance policy, in case people do not take morality seriously or fail to live up to their principles or succumb to temptation. Perhaps this is so; but the important point is that this line of defense no longer can sustain a view of moral rights as the central concern, the very heart and soul of a theory of morality. They become mere appendages, and often not even that, to agreed moral principles.

(3) A third source of difficulty in arguing about rights is the problem of conflicts and their resolution. Advocates of rights do well to be sensitive to this problem, and I have never read one yet who did not concede that the problem of conflicts posed serious difficulties for him. Certainly, it is clear that inability to resolve conflicts is a severe hindrance to arguing about rights.

[5]See Frey, 1980, Chapter 1, for a fuller statement of this argument.

It is to a discussion of conflicts and of several features in respect of them that I now want to turn. But, first, it is important to notice just why such a discussion, which might appear tangential to the issue of moral rights for animals, is in fact germane.

Advocates of moral rights invariably hold that there are a good many such rights, and in cases where animals are ascribed some moral right or other, some conflicting moral right of human beings can normally be ascribed as well.[6] Thus, if squirrels are held to have a moral right to the chestnuts in my backyard, it is also the case that I am held to have a moral (and legal) right to the use and enjoyment of my property. If chickens are held to have a moral right to life, it is also the case that they are the property of chicken farmers, who are held to have not only a moral (and legal) right to the use of their property, but also the additional moral right to earn a livelihood and to support their families. It seems obvious that, if we are ever going to make anything of arguing moral questions such as our treatment of animals in terms of rights, we simply have to have some way of resolving these conflicts.

ARGUING ABOUT RIGHTS IN THE FACE OF CONFLICTS

The picture on conflicts we carry away from the discussion of rights among earlier intuitionists this century is this: there are a number of *prima facie* rights, and these not infrequently come into conflict with one another; we resolve this conflict by means of moral intuition, either by intuiting which is the more stringent and compelling *prima facie* right on this occasion or by intuiting which is the more stringent and compelling *tout court*. Now moral intuition, whether as an explanatory device or as a piece of moral epistemology, will not do, for reasons much too familiar to detain us here. So what do we do about conflicts?

Whatever we do, as is well-known, would appear to involve the abandonment of strict intuitionism. What we require for resolving conflicts are ways of measuring the relative stringencies of rights; but either these measures themselves rest upon moral intuition, in which case they are as tainted as the earlier position, or they admit of non-intuitionistic justification, in which case strict intuitionism has been abandoned. Notice that an appeal to some hierarchy of rights in terms of relative stringencies gets nowhere, since we can only compile such a hierarchy in the first place if we are in possession of the very measures we are seeking. Nor am

---

[6]A similar point is made by Hare (1975, p. 203), though not in these terms. For an extended discussion by Hare of moral conflicts, see his 'Moral Conflicts', in S. M. McMurrin (ed.), *The Tanner Lectures on Human Value, 1980* (Cambridge, Cambridge University Press, 1980), pp. 169–193.

I aware of ever having seen an ordered and argued hierarchical list of these non-intuitionistic based measures in the writings of advocates of rights. In short, strict intuitionism is objectionable, a hierarchy of rights cannot be fashioned, and non-intuitionistic measures of stringency are not immediately or even apparently in the offing.

Moreover, in this situation, how can one exclude *a priori* the possibility of a principle of utility serving among the measures of stringency? After all, it is not obviously wrong or silly to say that we determine relative stringency between two conflicting *prima facie* rights by appeal to the consequences of keeping and breaking each of them, which, when generalized, amounts to saying that what it would be right to do is to comply with that *prima facie* right the consequences of which compliance are productive of best consequences. In this eventuality, the very core of one's rights-based theory of morality, almost certainly adopted in the first place as an antidote to (act-) utilitarianism, is infected with a utilitarian principle. Accordingly, if one's rights-based theory of morality is to avoid this fate, one must provide measures of stringency that preclude the admission of a principle of utility among them, but that in no way rely upon a strict intuitionism, a hierarchy of relative stringencies or merely presumed non-intuitionistic measures of stringency in order to do so.

These are not, however, the only boundaries to our problem; there is at least one other important one. In the earlier intuitionists, some of the *prima facie* rights claimed were themselves embedded in intuition. This was the case because duties and rights were seen as correlative, and many *prima facie* duties, such as those of beneficence, self-improvement, and nonmaleficence, did not themselves stem from any voluntary undertaking on our part. In other words, the *prima facie* right had its ground in the correlative *prima facie* duty, and some *prima facie* duties had their ground, not in the quasi-contractual arrangements that ensue as the result of voluntary acts on our part, such as making a promise or entering a marriage, but in intuition. It is within social, contractual arrangements of the above sorts that the correlative thesis of rights and duties finds a home; outside such arrangements, this thesis is of doubtful truth. But whether it is true or false is of no consequence here: if the correlative thesis is true, then our attention turns from alleged *prima facie* rights to alleged *prima facie* duties and to the fact that many such alleged duties, including, I submit, a number of allegedly important ones, such as duties to respect life and prevent avoidable suffering, do not find their ground in any voluntary undertakings on our part but, in the earlier intuitionists, in moral intuition; if, however, the correlative thesis is false, then some of these alleged *prima facie* rights must themselves be directly intuited. Whichever it is, my point is this: to the extent that *prima facie* rights are either directly or indirectly embedded in moral intuition, then, given our repudiation of moral intuition, to that extent *prima facie* rights are repudiated. Plainly, therefore, there is an additional boundary to our problem of how to argue about rights in the face of conflicts. In order to resolve conflicts between *prima facie* rights, one must provide measures of stringency that preclude a principle of utility, but do not rely upon a strict intuitionism, on a hierarchy of relative stringencies, or on merely presumed non-intuitionistic measures of stringency in order to do so; but in order to prevent our resolving the conflict by repudiating one or both of the conflicting rights, one must sustain these rights—must sustain, that is, the claim that there really are these rights—without recourse to moral intuition.

Now these boundaries markedly circumscribe possible answers to our problem of how to argue about rights in the face of conflicts. In order to illustrate how limited room for maneuver is, I want very briefly to compare and contrast the answers that three recent advocates of rights provide to this problem. In none of their cases is arguing about rights an easy matter.

MORAL RIGHTS AS SELF-EVIDENT, RECOMMENDED, OR AXIOMATIC

In his recent and interesting paper 'Moral Rights and Animals' (1979),[7] H. J. McCloskey affirms the existence of moral rights and goes on to consider whether animals can possess such rights. He approaches the case of animals through a discussion of moral rights generally and of possible criteria for the possession of moral rights. He distinguishes between accorded moral rights, such as a right to the use of one's property, and basic intrinsic moral rights, such as a right to autonomy and integrity, and it seems reasonable to suppose that basic moral rights are both more important and their violation more serious than is the case with less basic ones.[8] If I understand his paper correctly, McCloskey thinks that the capacity for moral self-direction and self-determination is the characteristic in virtue of which a creature possesses basic moral rights, and it plainly follows that animals are only going to be conceded basic moral rights to the extent that they possess or can be construed as in some wise sharing in this capacity. But this is about who can possess intrinsic moral rights; what about conflicts among these rights and the argument by which McCloskey sustains the claim that there really are such rights, in order to prevent us repudiating one or both of the conflicting rights?

He approaches our problem this way: he allows that there can be conflicts between basic intrinsic moral rights and remarks that we are to resolve these conflicts 'by weighing up the conflicting rights in terms of their various stringencies' (McCloskey, 1979, p. 52). But how do we determine the stringencies of these rights? McCloskey says this (p. 52):

[7]Anyone interested in the subject of animal rights is going to have to study this paper.

[8]It does not follow, of course, that less basic moral rights are not important and their violation a serious affair to McCloskey.

> It is impossible to set out priority rules for resolving such conflicts.
> They are to be resolved only by reference to the relative stringencies of
> the rights involved; and this is to be determined by reference to the
> ground of the right, and the basis on which the individuals concerned
> come to possess the rights. In many cases, the conflicts seem rationally
> to be irresoluble.

My perplexity here is this: if many conflicts of moral rights cannot rationally be resolved, just how is light supposed to be shed upon the moral
questions being discussed in these terms? For example, should it turn out
that conflicts between the rights of animals and human beings cannot in
many cases be rationally resolved, what reason have I to think that we are
really more likely to obtain a clearer view of the moral issues involved in
our treatment of animals by discussing them in this way? This is particularly true, I should have thought, since this impasse over conflicting
rights is reached, not after long bouts of torturous argument, but, if moral
questions are discussed and prosecuted in terms of basic moral rights, at
the very outset. Furthermore, what am I *to do* in these irresoluble cases?
If these conflicts cannot be rationally resolved, then I as a moral being
seem compelled either simply to break down—where conflicts cannot be
resolved I *just do not know what to do* in such situations—or to act irra-
tionally, neither of which is a particularly appealing alternative to moral
beings.

My initial perplexity here is deepened, moreover, when, following
McCloskey's directions, I look to the ground and basis of these basic
moral rights in order to ascertain their stringency. McCloskey says this
(1979, pp. 43–44):

> A persistent claim more common in the eighteenth century than to-
> day, has been that certain rights are self-evidently so. It is an appeal
> which I suggest is appropriate in respect of certain basic rights, if 'self-
> evident' is understood correctly, as by W. D. Ross in the context of his
> discussion of our knowledge of the principles of prima facie obligation.
> Rights such as rights to life as persons, to moral autonomy and integ-
> rity, to respect as persons, appear to be rights which are self-evident,
> and in respect of which, argument other than that directed at clarifica-
> tion of what is involved in acknowledging the moral right is neither
> necessary nor possible.

If conflicts between basic moral rights are to be resolved by appeals to
stringency, if stringency is to be determined by the ground of these rights,
and if their ground is held to be self-evidence, then how am I any further
along towards resolving conflicts than before? For self-evidence either
provides me with no measures of stringency at all or it provides me with
measures argument in support of which is, in McCloskey's words, 'neither necessary nor possible.' And without argument, how are we going to
resolve, having come this far, those disputes that will arise between us
not only as to whether this or that is self-evident, but also as to which of

two equally self-evident, basic moral rights is the more stringent? Appeal
to the basis of such rights, to be found in our capacity for moral self-
direction and self-determination, does not assist us, since both of the conflicting, equally self-evident, basic moral rights can have this same basis.

My point about McCloskey, then, is not merely that few of us are
prepared to hold a right to life to be self-evident and to require no argument whatever in order to substantiate and sustain it; it is also that, if self-
evidence is held to be the ground of such a right, and if the ground of this
right is the arbiter of its stringency, and if other basic rights have the same
ground, then we are effectively deprived of any means of *arguing* about
whether there is a right to life, about the stringency of such a right, and
about conflicts between it and other basic rights. To affirm that there are
such rights, but to leave us no room for arguing about them in the above
ways, does not strike me as the most profitable way of discussing moral
issues; and to affirm that there are conflicts among these rights, but to
allow that, in many cases, such conflicts elude rational resolution, does
not strike me as the best way of trying to provide principles for the direction or re-direction of our behavior towards other creatures.

J. L. Mackie is another recent advocate of rights. His 'Can There Be
A Rights-Based Moral Theory?' (1978) attempts to sketch the bare bones
of a rights-based theory of morality. Essentially, the view that emerges
there is one of a multiplicity of moral rights that are derived from one or a
small number of fundamental moral rights. Three points are noteworthy
about this distinction. First, this fundamental right appears at first to be
the right to choose how one will live (Mackie, 1978, p. 355), but the class
of such rights seems later to be expanded to include Jefferson's rights to
life, liberty, and the pursuit of happiness (p. 356). Second, only persons
possess these fundamental rights, though any and all persons possess
them equally. One implication of this view, as Mackie realizes, is that,
since animals are not obviously persons, they in turn are not obviously the
subject of these fundamental rights. Third, the distinction between less
fundamental and fundamental moral rights is not that between *prima facie*
and absolute rights; in Mackie's view, because even fundamental moral
rights can clash, they. too, are only *prima facie* rights.

How, then, does Mackie approach our problem of how to argue
about rights in the face of conflicts? He claims that people's final rights
result from compromises between their initial conflicting rights,
including their conflicting fundamental rights, and these compromises, he
says, 'will have to be worked out in practice, but will be morally defensible only insofar as they reflect the equality of the *prima facie* rights'
(Mackie, 1978, p. 356). But how are these compromises to be effected in
practice? Mackie suggests a model (p. 356):

> . . . we might think in terms of a model in which each person is rep-
> resented by a point-center of force, and the forces (representing *prima
> facie* rights) obey an inverse square law, so that a right decreases in

weight with the remoteness of the matter on which it bears from the person whose right it is.

The difficulties with this model are just those that Mackie himself brings out, viz., that cases where equally vital rights clash will arise, that cases in which the parties do not know how vital some right or other is to them will arise, that the parties involved will sometimes be mistaken, other times deceived on how vital some right is to them, and so on. Indeed, I think this model may well underestimate the degree of conflicts; for the rights each of the parties in conflict insists is vital to them almost certainly truly represents an interest of theirs that they at least *take* to be vital to them, and clashes of this sort among persons is a common phenomenon. The courts are full of them. So, if one's theory is to avoid this fate, one must have a so-far unspecified means of telling even among numerous and frequent *sincere* avowals as to what one takes to be a vital interest that one is mistaken or deceived or whatever. In fact, though Mackie does say that it is not a reasonable requirement to demand of a theory such as his that it resolve all conflicts (1978, p. 359), he nevertheless acknowledges that resolving conflicts is a serious problem for him; and this very difficulty implies, he says, that it will not be easy 'to determine, in concrete cases, what the implications of our theory will be' (p. 357). If this is so, then it is not readily apparent how we are better off arguing moral issues in these terms. It is not clear how we are to argue clashes in fundamental moral rights, and if fundamental moral rights represent important interests of people, it seems imperative that we have some idea of how to go about adjudicating these clashes, at least if we seek an harmonious society. Being told that we will have to thrash out these clashes in practice, without some decision procedure, is not terribly informative about how we are to set about this task. And what confidence have we that such clashes *can* be thrashed out in practice? So far as I can see, nothing in Mackie's paper provides this confidence.

How, then, does Mackie sustain the claim that there are these fundamental rights? What is the ground he ascribes them? He has this to say on the subject:

> The fundamental right is put forward as universal. On the other hand I am not claiming that it is objectively valid, or that its validity can be found out by reason: I am merely adopting it and recommending it for general adoption as a moral principle (p. 357).

But why should we accept or follow Mackie's recommendation? Judging from Mackie's article, it is because he thinks doing so can best capture our moral intuitions (which do not, of course, have to rest upon moral intuition, as a thesis in moral epistemology). Thus, Mackie berates all forms of utilitarianism, because 'they not merely allow but positively require, in certain circumstances, that the well-being of one individual should be sacrificed, without limits, for the well-being of others' (p.

352). It is this anti-utilitarian moral intuition or judgment that recurs throughout Mackie's essay and provides the backdrop against which his recommendation of moral rights that will not allow the vital interests of one person to be sacrificed for the well-being of others must be seen. Put simply, the basis of Mackie's recommendation of fundamental moral rights is this and doubtless other anti-utilitarian intuitions or judgments of his, and one has no reason to accept or follow Mackie's recommendation on this score unless one happens to share these intuitions.

My own intuitions run in the direction of utilitarianism; to take only the example of sacrificing interests, I support compulsory conscription in time of war and so am prepared to see the interests of this or that individual sacrificed for the well-being of an aggregate of others. But my point here is not just that our intuitions do not agree; it is also that, *unless they agree,* we shall not agree that there are any moral rights or fundamental moral rights. For unless our intuitions run in the direction of Mackie's, we have no reason, in Mackie's terms (p. 352), to 'posit' the existence of such rights. Thus, the case for positing rights must inevitably take the form of pointing to one's moral intuitions and saying that these constitute the reasons for positing some scheme of fundamental moral rights. But why give our moral intuitions this degree of weight? Mackie does not say in his essay; and though he may think that our moral intuitions are what moral theories must account for, enough has recently been written on this issue in methodology to indicate that it is a minefield of controversy.

In short, one is only going to think there are moral rights (or reasons for postulating moral rights) if one's intuitions run this way as opposed to that; if they run differently, then there are no moral rights (or no reasons for postulating moral rights). To a great extent, this places the moral rights we are alleged to have at the mercy of our intuitions; and though one can nurse along and perhaps even cultivate one's anti-utilitarian intuitions, so, too, can one nurse along and cultivate contrary intuitions, with obvious results. Oddly enough, then, on Mackie's view, argument about moral rights and their impact on and implications for us in concrete situations can only take place between people with similar intuitions; for to people with different intuitions, the case for positing such rights in the first place is not made. But surely this gets things in reverse? Those among you whose intuition it is that eating meat is wrong or that vivisection should be completely abolished do not want to argue with someone possessed of similar intuitions; you want to argue with me, with someone who does not share your intuitions. If Mackie is correct, however, you will not be able to do so. Or, more accurately, you will not be able to do so on the basis of rights, since rights ultimately appear to depend upon intuitions and our intuitions are such that we are not going to agree on the postulation of fundamental moral rights in the first place. (With this the case, we never reach the subsidiary issue of whether animals can be the subject of such rights.) Thus, however you decide to argue these moral issues with me, *if* you decide to argue them with me, it will be outside the context of Mackie's rights-based theory of morality.

Since Mackie mentions Ronald Dworkin in his essay, perhaps a very few words on *Taking Rights Seriously* (Dworkin, 1977) may be appended. Dworkin finds himself in opposition to the prevalent theory of law, which is the legal positivism that derives from the work of Austin and Bentham; and one significant aspect to Dworkin's discontent with positivism is his disenchantment with utilitarianism (or what he calls 'economic' utilitarianism), which he describes in his Introduction as the normative underpinning of positivism. His well-known attack on utilitarianism is not so much direct, by rehearsing all its many alleged difficulties in all its forms, as indirect, by showing that a theory of rights that emerges from giving consideration and weight to people's individual rights yields results antithetical to utilitarianism.

Very roughly, Dworkin's view is that persons have individual, political, or moral rights that they hold against encroachment from other people and from the community on their vital concerns; and, if I understand his position correctly, unless these political or moral rights are taken seriously, i.e., are recognized, accepted, and rigorously observed by other individuals and especially by the community, in the form of its political leaders, legislators and magistrates, we shall fail fully to recognize, accept, and observe people's absolutely fundamental right to equal concern and respect. (Since it is questionable whether animals are persons, it is questionable whether they possess these rights.) These individual rights and, in particular, this absolutely fundamental right are not brought into being by community decision, social practice, or public legislation; they survive contrary decisions, practices, legislation, and adjudication; and, in fact, they prescribe the bounds beyond which both individuals and the community may not go in the pursuit of their aggregate goals. As in the case of Mackie, then, the underlying motif is that these individual rights and this absolutely fundamental right may not be infringed in the name of increasing the collective well-being of an aggregate of others; if anything, these rights stand as bulwarks impeding the pursuit of this ideal.

Having examined the cases of McCloskey and Mackie on this question, let us ask first, therefore, how Dworkin sustains the claim that there is this absolutely fundamental right to equal concern and respect? He has this to say on the matter (1977, p. xii):

> . . . our intuitions about justice presuppose not only that people have rights but that one right among these is fundamental and even axiomatic. This most fundamental of rights is a distinct conception of the right to equality, which I call the right to equal concern and respect.

And a few lines later he says that his book 'suggests one favored form of argument for political rights, which is the derivation of particular rights from the abstract right to concern and respect taken to be fundamental and axiomatic' (pp. xiv–xv). Here, too, then, we have a case of positing a fundamental right into existence and then attempting to derive still other rights from that initial positing. Nor does saying that people's individual rights depend upon or can be derived from their absolutely fundamental right to equal concern and respect amount to an argument for that right. Also, it is not entirely clear to me whether my pursuit of my absolutely fundamental right can conflict with your pursuit of yours. Mackie says of this right in his article that 'one person's possession or enjoyment of it does not conflict with another's' (1978, p. 356), but I do not see how one can determine this *a priori*. I can well imagine, for example, that those who campaign for a right to life on behalf of fetuses may indeed hold in certain situations that the mother's right to concern and respect clashes with the fetus's right to concern and respect, which leads to the problem of how this conflict in fundamental rights is to be resolved.

The important point to notice about Dworkin's view, however, is that which links it to Mackie's. Why develop and then adopt this theory of rights? It is not that one finds oneself with a theory and then discovers that it yields anti-utilitarian results; rather, as in the case of Mackie and, I suspect, McCloskey as well, one finds oneself with anti-utilitarian intuitions and so one tries to develop a theoretical framework into which these intuitions can fit and in which they are taken seriously. As Neil MacCormick has recently stressed in his review of *Taking Rights Seriously* (1978), Dworkin's adoption of Rawls's intuitionistic method of seeking a reflective equilibrium between one's intuitive judgments and one's principles, a moral methodology recently slated by Hare (1973) and attacked by others, is the bedrock upon which Dworkin's constructive model of morality rests, and Dworkin's intuitive judgments, particularly as regards the sacrifice of one person's interests for the well-being of an aggregate of others, are anti-utilitarian. Hence, the positing of an axiomatic and absolutely fundamental moral right to equal concern and respect. Ultimately, therefore, the existence of this right depends upon the way our intuitions run, and what I should say on this score in respect of Dworkin I have already said in respect of Mackie.

In sum, not much headway has been made with the problem of arguing rights in the face of conflicts; and in the case of one of the boundaries to this problem, serious difficulties arise. For in order to get basic or fundamental moral rights into existence and to sustain them, so that their observance can, e.g., serve to impede the utilitarian pursuit of the collective good at some individual's expense, McCloskey appeals to self-evidence, Mackie makes a recommendation based upon his moral intuitions, and Dworkin appears to stipulate in accordance with his intuitive judgments about what it would and would not be right to do to individuals on behalf of the well-being of a number of others. It is not easy to see how argument is possible about moral rights put into existence and then sustained on these bases; indeed, as we have seen, all three writers considered, for one reason or another, do not think argument even to be appropriate. Certainly, it is not easy to see how we could argue with any degree of finality about moral rights on either an intuitionism grounded in self-evidence, if we differ over whether something is self-evident and so whether it requires argument in order to substantiate it, or, in the cases of Mackie and Dworkin, an intuitionism grounded in moral intuitions or intuitive judgments, if these vary among us, as they assuredly do. And if our moral intuitions do vary, and we cannot agree as a result on the positing of this or that right, how are we supposed to gain a clearer understanding of the moral problems involved by going ahead and discussing them in terms of rights? Surely the only result of doing so is to divert attention from the task of working out principles of rightness and justification of treatment to the wholly secondary—and, if our moral intuitions vary, wholly unnecessary—task of trying to figure out whether there are such rights in the first place and of what to make of the intuitionistic theses we encounter as a result? I realize, of course, that just as people have utilitarian intuitions, so they have anti-utilitarian intuitions; buy why go on to posit the existence of moral rights or a scheme of such rights on the basis of one's anti-utilitarian intuitions? Could any light possibly be shed on moral questions by doing so? For it is inevitably one's positing of the rights, not the questions, on which attention will be lavished, and when we come across the intuitionistic theses underlying the positing of the rights, with their attendant difficulty of being almost impossible to argue about if we disagree in intuitions, we are even further removed from the prospect of obtaining a surer grip on the issues. Would we not do better, therefore, to jettison moral rights in order to forego being sidetracked by them? We can still ask and debate whether our present treatment of animals is right and/or justified, only now the answer cannot be found in an appeal to moral rights. I do not claim we shall have an easy time of it, even here; but at least the diversions rights create can be avoided, and, if I am right, without any loss to the debate on how animals should be treated.

ACQUIRED AND UNACQUIRED RIGHTS

I want to end with a few remarks on a distinction in respect of rights that I think points to another reason they are of little assistance in discussions, e.g., of our treatment of animals. I shall not father this distinction on any particular advocate of rights, if only because it is probably too crude to serve the purpose of developing a taxonomy of rights; but I think each of the writers earlier considered at least implicitly draws the distinction. The distinction in question is that between acquired and unacquired rights.

Typically, acquired rights are acquired through a person's voluntary act. Property is typically acquired through purchase or exchange or through accepting an inheritance or gift, and, once acquired, its owner acquires rights in respect of it. Once a right to property has been acquired, it can be enjoyed, waived, possibly forfeited, recognized, relinquished, transferred, upheld, infringed, violated, contested, and so on.[9] In my view, we not only understand the right, but also understand what it is to have such a right both by looking to its manner of acquisition and by locating it within the framework of those things that can be done with and to it. Something that has been voluntarily acquired can be voluntarily relinquished, as when I sell my shares in the firm; and subsidiary rights that I enjoy because of my initial voluntary act, such as the right to vote at the annual shareholders meeting, can be waived, eventually relinquished, and so on. It is by arguing within a framework of activities of this sort that we gain an understanding of what it is to have acquired rights.

Now McCloskey thinks there are intrinsic moral rights, Mackie that there are fundamental moral rights, and Dworkin that there is an absolutely fundamental moral right, and what is significant about these rights, I think, is that they are not acquired through any voluntary act of persons, such that failure to perform it results in not having the right or rights in question. (Animal rightists will certainly believe this, since they will not want an animal's supposed right to life to turn upon its 'voluntarily' doing this or that.) Unacquired moral rights we have, presumably, merely through having come into existence and through being the sorts of creatures we are. But what sorts of creatures are we? Rational? Language users? Sentient? Persons? And once we pick and choose among the host of speculations that exist on the subject, we have to go further, into the endless speculations about the criteria of rationality or possession of language or sentiency or being a person. My point here, however, is not this very obvious one, that problems arise over what it is about us in virtue of which we have unacquired moral rights; it is something deeper.

What is important here is that we cannot use the case of acquired rights to shed any light upon unacquired moral rights, not only because acquired rights are typically acquired through voluntary acts whereas unacquired moral ones are not, but also because those very activities that enable us to understand what it is to have an acquired right to, say, property, do not perform a similar function in the case of an unacquired moral right to, say, equal concern and respect. For, in a word, it is not clear these activities can occur (or, at least, are allowed to occur, by the proponents of unacquired moral rights). For instance, can a fundamental right to equal concern and respect be forfeited or relinquished or transferred or

---

[9]Put differently, I take these and other activities to constitute what is involved in having, using, and disposing of acquired rights; and in the present context, I am emphasizing disposing of them. So far as acquired rights are concerned, I think McCloskey would agree with this characterization, though not perhaps with this emphasis; see McCloskey, 1979, p. 28.

waived? What does one have to do before one can be said to have relinquished all right to equal concern and respect or to choosing how one shall live or to life? These questions are sources of difficulty in arguing with advocates of rights; but this is not surprising, since the whole theoretical enterprise hangs by each advocate's initial positing of a scheme of unacquired moral rights in the first place, and each advocate is as free to posit that these rights cannot be relinquished, forfeited and so on, as not.[10]

Moreover, argument about unacquired moral rights is severely constrained *just because* these activities within which we locate acquired rights and what can be done with and to such rights—activities and their upshots that are often argued in the courts—are themselves very much in doubt in the case of unacquired moral rights. If I say a person may give up or forfeit his or her absolutely fundamental right to equal concern and respect, and if Dworkin denies this, what sort of argument could we have? So far as I can see, if Dworkin's theory does not allow that this right may be given up or forfeited, then that is the end of the matter; for no number of murders that the person in question may have performed will show this theoretical point false. Put differently, if an advocate of basic or fundamental moral rights posits such rights, and if the advocate also posits that these rights are possessed by all persons equally, and further posits that they cannot be relinquished, given up, or forfeited, and cannot be ignored, infringed, or violated with impunity, I just do not see how we can argue with that advocate. And if we cannot argue, I do not see how light is shed on moral problems by positings of this sort.

In fine, I think the more we diverge from the cases of acquired rights, the more we lose our grip both on the nature and ground of rights and on our understanding of what it is to have a right. In the case of intrinsic or fundamental or absolutely fundamental unacquired, moral rights, this divergence reaches a radical degree, and I think what grip we had on rights has been lost. Rather, we are at sea in a tide of theoretical claims and counterclaims, with no fixed point by which to steer. Such, it seems to me, is, by implication, the case with the alleged moral rights of animals.[11]

---

[10]See, e.g., McCloskey 1979, p. 28.

[11]I am grateful to R. M. Hare and H. J. McCloskey for their comments on this paper. An amended version of this paper forms part of my forthcoming book *Rights, Killing, and Suffering* (Oxford, Basil Blackwell).

SECTION VII

Breadth of Vision

Introduction

The two papers making up this section are joined by their author's claims that the predominant orientation of this volume is too narrow. Both argue that the general frame of reference so far has been too restricted to permit us to see ourselves, and thus our relation to other animals, aright.

The conceptual blinders of most modern philosophy, Michael W. Fox charges, prevent it from perceiving the true nature of questions about the relations of humans with other animals. Traditional occidental 'anthropocentric' ('homocentric' is Wenz's term) thinking is unable to grasp the structures of ecological interdependence that set the framework both for the problems and for the possible solutions. Excessive abstraction, especially when combined with insufficient attention to empirical studies of animals, results in irrelevant discourse. Fox calls for a more holistic approach, incorporating significant elements of Eastern thought and viewing the entire biosphere as a single moral community. The excessive abstraction of philosophy is both the result of and a contributor to human alienation from the natural world. We must move from an anthropocentric worldview to an ecocentric one.

Fox argues for according rights to animals on grounds similar to those offered by Regan, Rollin, and Rachels, but the primary right he considers is a right to develop one's natural potential in a suitable environment. The correlative duty is one of harmonized restraint, not just in regard to the sentient, but toward the entire biotic community. This is the key to human stewardship and a genuinely human ethics.

T. Nicolaus Tideman agrees with Fox that the question is not simply our relation to other animals, but rather our relation to the whole of nature. And like Scanlon (in Section IV) he urges that we cannot live without competing, or compete without killing. Even if we are vegetarians, we will cause plants to be killed, or at least consume seeds. Tideman thinks it quite plausible that plants are conscious beings, and in fact proposes a panpsychic view—''that we share the quality of consciousness with all creation.'' We cannot, then, avoid killing the conscious (if we avoided killing any others, we would kill ourselves). Tideman finds an escape from the resulting ethical dilemma in a changed view of death. If we, and the rest of nature, existed before and will persist after, our bodily

305

lives, that is if death is not an end, but instead a transition to a different state, then death need not be an evil. One may kill if the killing is harmonious with the purposes of both killer and killed. Determining whether or not that is the case will often not be easy.

Philosophy, Ecology, Animal Welfare, and the 'Rights' Question

Michael W. Fox

PHILOSOPHICAL RELEVANCE: TOWARD A PHILOSOPHY OF ECOETHICS

With few exceptions, most philosophers who present papers debating the moral status of animals in society and the pros and cons of giving animals certain rights, reflect, as one would expect, a classically occidental world view: an anthropocentric view that, as I see it, is not equipped to deal with the complex interfaces between human and nonhuman animals and the environment. Some knowledge of animal behavior and ecology is a prerequisite for anyone who wishes to discuss objectively and accurately, the question of animal rights. Also some knowledge of Eastern philosophy (see later) and the phenomenological approach of Merleau-Ponty (as reviewed by Dallery, 1978) would considerably reduce such anthropocentrism and help close the philosophical gap between reality, humankind, and nature.

The 'techniques' generally employed in Western philosophy too often reflect an overly abstract and theoretical human-centered world view and tend to exaggerate, rather than resolve, the apparent and real conflicts between humans, animals, and nature by focusing upon abstract, contrived, and often irrelevant questions and not upon the real issues of ecology and animal nature and human obligations toward nonhuman entities. Arguments such as: there would be fewer farm animals alive if vegetarians had their way because fewer would be born, thus being a meat eater is in the animals' best interests, are quaintly academic and intellectually stimulating. But in reality, they are fatuous and vacuous, reflecting ignorance of social, economic, ecological factors (80% of all grain produced in the US is fed to animals, for example) and of very real animal welfare concerns (over $1 billion is lost from livestock injuries and stress as a result of improper handling and care while being transported by road and rail for example, Fox, 1980b).

Although there may be a close analogy between such ignorance with comparable academic debates over the rights and intelligence of slaves three generations ago, a more accurate analogy may be of Nero fiddling while Rome burns.

Progress in developing a moral philosophy that can deal with such complexities is further limited by what I would call circular-semantic thinking with its overemphasis upon hypothetical issues and formal logic. The dialectic of contrived polemical questions that ignore ecological factors and real life situations can make certain schools of philosophy seem like a naive form of reductionistic gibberish. The challenge surely is to develop a more appropriate school of philosophy (as attempted by Passmore, 1974, and Midgley, 1978, for example) capable of considering humanity within the context of nature and embracing all the various relationships with wild and domestic animals.

Perhaps philosophy has become increasingly irrelevant to worldly concerns because of its preoccupation with the mind (logic, meaning of meaning, etc.) and contrived dualities rather than with the totality and nonduality (Capra, 1977) of mind and body, consciousness and matter (energy), humankind and nature. Attitudes, values, and morals stem not from the mind, but from interactions between mind, culture, and environment. One cannot therefore study any component in isolation from the whole. Surely the challenge to philosophy is to define those social, psychohistorical, and ecological determinants of the mind's acquired values, attitudes, perceptions, conceptions, and moral codes. Although social ethics maintain social order (by protecting the rights and interests of individuals and groups), ecological and humane ethics are needed to maintain the ecological community of humankind on earth (by protecting the rights and interests of other creatures and the environment we all share) (Stone, 1974). To see and treat animals as objects of property has been relevant to maintaining social order within human societies in the past. But today, with the 'peopleing' of earth, to continue to treat animals and things of nature as objects of property and as resources for humankind's exclusive use, could upset both the social order and the ecological community. Competition for resources, species extinctions, and serious ecological imbalances are already adversely affecting both social order and the ecological community.

Notably lacking in most philosophical discussions on the moral status of animals is any reference to our obligations and duties toward them and to the environment. Philosophers would add much to the validity of their arguments by acquiring a better understanding of animal behavior, ecology, and the many uses and abuses that wild and domesticated animals are subjected to. Although it is an established biological and ecological fact that humans, other animals, and nature are inseparable, it is clear that both culturally and philosophically humans are very separate, if not alienated, from the rest of creation. Those philosophers who argue in support of human dominion and superiority over animals, who are opposed to accepting animal rights, or who do not believe that we have moral obligations toward them, only increase this alienation: an alienation that many (e.g., Roszak, 1975; Fox, 1976) believe underlies the present indifference toward and inhumane treatment of animals in contemporary society, and the exploitation and destruction of the environment. Deanimalization, 'denaturation,' and dehumanization are pathognomonic of such alienation. The challenge, surely, for moral philosophers, is to develop the conceptual tools to analyze and resolve this serious problem: in essence, to resolve conflicts of 'interests,' to heal the earth. Utilitarian, contractual, and libertarian philosophies alone are inadequate to accomplish this task. No single philosophy (utilitarian, libertarian, and so on) is adequate in itself to deal with all the different kinds of relationships and influences between humans, animals, and nature. Some of the seeds may be found in Hume's empiricism, but because much of Western philosophy has grown within (if not out of) the framework of increasing human alienation from animals and nature, I doubt that a more embracing and effective philosophical school of thought can be developed without a radical transformation in consciousness. By this, I imply a more reality-oriented natural philosophy that deals with some of the actual, rather than hypothetical, relationships, ethical questions, and moral obligations vis-a-vis humans, animals, and the environment. The sciences, particularly those of ethology and ecology, could provide philosophy with the factual material for developing a new philosophy of eco-ethics (Fox, 1980a).

What is needed, I believe, is a transformation of consciousness from an egocentric (or humanocentric) world view to an ecocentric one, where humans are seen not as separate from nature and superior to animals, but as an inseparable part of the whole of life, and where we think and act accordingly, and with conscience (Fox, 1980c; Krishnamurti, 1970; Maslow, 1968; Schweitzer, 1965). Though Hume's empiricism, Merleau-Ponty's phenomenology, and Schweitzer's reverence for all life philosophy are close to this broader and more integrated world view, Western philosophers who wish to discuss the issues of conservation and humane treatment of animals would certainly be more effective and relevant if they were to see the connection between animal rights and 'eco-ethics' with the nondualistic philosophies of Shintoism, Taoism, and Buddhism (Watts, 1973). Time, indeed lifetimes, could then be spent in more constructive contemplation within the infinitely more realistic and holistic dimension of humans as animals in nature. This is not a plea for a more mystical and emotionally nonrational philosophy, but rather, a demand for a more realistic, holistic, and worldly moral philosophy that includes, rather than excludes, humanity from nature and the animal kingdom, since after all such inclusion is an existential and biological fact.

The works of one western philosopher, Martin Heidegger, perhaps comes closest to the essence of Taoism and of a nature philosophy and eco-ethics. For example, Heidegger's "*logos* or ordering cosmic wisdom anchored in nature herself" is synonymous with Lao-tze's *Tao,* both concepts, according to La Chapelle (1978) being independent and coming directly from nature (see also Richardson, 1963).

In summary, philosophy could play a vital role in formulating an 'ecoethical' and moral framework to guide human thought and action toward a more stable and harmonious relationship with the rest of the biosphere. A consonance of animal and human interests is needed to temper human dominionism and self-centered utility. Without some knowledge about animals, ecology, history, and politics, philosophy will remain outside the realm of world reality, wherein we all abide and gain our sustenance, inspiration, and wisdom. Our species will fall short of being and becoming fully human until all cultures have developed and assimilated a *natural* philosophy to serve all creation, as well as our own continued evolution.

ANIMAL RIGHTS—A REALITY OR FICTION?

Opponents of animal rights seem to believe that animals simply cannot have rights because, for example, they are not part of the moral community or cannot reciprocally respect our rights. Yet human beings belong to more than a moral community of people. We belong to an ecological community that includes all animals. To deny animals equal consideration of their interests/rights because they cannot reciprocate and respect ours, as philosopher M.A. Fox has suggested, is untenable unless we agree that we only live in a moral community and not in an ecological one, and that all idiots, pre-verbal children, and others who cannot or do not respect other's rights should be cast out into the wilderness.

Many philosophers do not accept the notion of animal rights and present carefully thought out arguments in favor of their point of view. Such academic games overlook the important point that there are political and social dimensions to animal rights advocacy that are far more important than the purely theoretical/philosophical dimension.

If animals cannot have rights, then why can I, since I am also an animal. For me to set myself above animals is surely hubristic speciesism. Such a conceptual separation of humankind from animals and nature is at the root of most of our contemporary social, interpersonal, and environmental problems (Leiss, 1972; Berry, 1978; Fox, 1976 and 1980a; Stone, 1974). Since animals have certain needs, and therefore interests, they have certain rights that we should respect. There is no need for a logical premise to validate such a moral obligation toward nonhuman creations, though many occidental philosophers seem to believe so. Morality need not be based upon logic or sentimentality, but upon ecological sensibility (Stone, 1974) and empathetic sensitivity (Fox, 1980a; Dubos, 1972). We should all show unconditional benevolence toward sentient creatures simply because we are human, endeavoring at all times to prevent any conflicts of 'interest' between animals and humans, and at the same time to minimize our entropy-accelerating impact upon the biosphere.

To contend that animals do not have any rights also overlooks the fact that there are laws to protect certain animal rights. The right to life is protected by the Endangered Species Act; the right to humane treatment is upheld by the Humane Slaughter Act and by state anticruelty statutes and the Federal Animal Welfare Act.

The animal rights question often evokes such comments as: animals in the wild do not give each other any 'rights' and since animals cannot make claim to any 'right,' how can they ever possess 'rights'?

Although a preverbal child, or mentally retarded or comatose adult, cannot voice claim to rights, they can have legal representation and their rights can be defended. Animals are in an analogous category, but that is not to say that their rights are necessarily equal to those of people. As Singer (1975) urges, we have a moral duty to give animals equal consideration of their basic rights and interests. If animals cannot possess or be given rights, they why can people? It is clearly illogical, therefore, for humans to have rights that are upheld by society and for nonhuman animals to have none, unless we reject the fact that we live in an ecological community that includes more than humans and human interests.

Wild animals are biologically constrained (for the benefit of the greater good of others and of their habitat) by natural 'laws,' those social and ecological influences that regulate populations, prey–predator relations, and so on. Within groups of highly social animals, such as baboons and wolves, various rituals and social codes help maintain social stability, cooperation, and ecological fitness (Fox, 1974a). We find similar intraspecies factors at work in the human species together with culturally evolved and perpetuated rules of social conduct. Some of these rules are moral or ethical injunctions, a few of which are now written into law to protect the rights and interests of people, institutions, corporations, and property (including land and livestock).

The point has now come at which it is socially and ecologically imperative for humankind to develop moral codes and laws to protect plant and animal species and the environment as a whole from further human depredation and destruction (Scheffer, 1974; Fox, 1980a).

Since animals, like people, have certain needs and interests (as well as awareness; Griffin, 1976) independent of any external value placed upon them by us, then it is logical to infer that they too are entitled to equal consideration of (but not necessarily equality of) certain rights to protect their interests that could otherwise be jeopardized by those rights that we might claim over them.

The more 'worldly' we become, the more the need there is for ethical restraints at the interfaces between people, animals, and nature: and the more civilized we and the world will become in the process. At one time, as mere objects of property, women, children, prisoners, and slaves had no rights (Stone, 1974). In most 'civilized' countries, they now have rights and as we become more civilized and worldly, animals and all things of nature will fall, not under our dominion, but within the scope of our moral concern. This will facilitate development of a true symbiosis between humans and nature.

It is we, as responsible citizens, who are morally, and as yet to a lesser extent legally, obliged to respect and uphold those rights—the birth rights if you wish—of our fellow humans. Similarly it is the purpose of humane educators, conservationists, lawyers, philosophers, and other animal rights advocates to insure that animals are accorded rights and that those rights are upheld (Clark, 1977; Linzey, 1976; Morris and Fox, 1978, Regan and Singer, 1976; Paterson and Ryder, 1979).

An alternative semantic is to state that we have moral obligations toward animals that impose certain restraints (ethical and legal) upon what we may or may not do to them.

From a strictly utilitarian view, respect for human rights can mean a better society for humankind. Recognition of animal rights can help expand this traditionally human-centered view, and could mean a better world for all people and animals alike. Animal rights, which are only now being explored and defined by philosophers, lawyers, and others could, when incorporated into the moral and legal fabric of society, provide the necesssary restraints and directives for the benefit of both society and the environment as a whole.

Mahatma Gandhi (1972) observed that "The greatness of a nation and its moral progress can be judged by the way its animals are treated." Today, many animal rights and welfare concerns are flatly opposed by economic cost/benefit justifications to an extent and consistency that seem to indicate that we think only in terms of economics and this now takes precedence over ethics. Yet in the final analysis surely the greater concern is poverty, not of the pocket, but of the spirit.

The time to recognize and respect animal rights has come. Veterinarians, ethologists, and humane educators especially who can speak for the health and husbandry requirements of animals define their basic needs (and correlated interests and rights), may do much to improve the 'greatness' of all nations by promoting a deeper understanding of animals and responsible compassion in our relationships with them.

ANIMAL RIGHTS AND THE LAWS OF ECOLOGY: TOWARD A HUMANE STEWARDSHIP

The rights accorded to any animal will vary in relation to the values placed upon it by people. Rights will therefore be different for those animals that are wild or domesticated, for those that are companion or pet animals, and for those that are used for work, food, or research studies (Fox, 1980a). Rights will also vary within and between any given culture, time, and place.

The urgent need to exercise humane compassion and to protect the rights of animals are clear indicators that the rights of animals today are neither understood nor respected and upheld. The extrinsic values that determine the way in which people regard and treat animals should complement and not conflict with or oppose the basic, intrinsic right of any life form.

This basic right may be defined operationally as the freedom for an animal to develop (and actualize) its natural potentials in the environment for which it is best suited or pre-adapted. It should not be subjected to unnatural physical, psychological, social, or environmental stresses or be treated in such a way that causes it to suffer (if it is sentient and capable of suffering, unless it is essential for human survival, and there are no humane alternatives) or for the total well-being or integrity of the biosphere. Nor should any species be treated such that its relationship with others and its natural balance within the ecosystem is disrupted.

This right to 'freedom' to develop natural potentials is conditional in that, through evolved co-adaptation between species, each is constrained. The rights of one are relative and complementary to the intrinsic rights of others. Such reciprocity is manifest as harmonized restraint: it is reflected in physical health, social unity, symbiosis, the balance of nature, and the order of the universe. It is the law of ecosystems that man has frequently violated as a global mega-predator and extractive parasite, rather than respecting as a symbiote and careful steward.

If one creature violates the rights of another, the disharmony that it creates violates this basic law; for a rational creature (a human being), it is to be judged a sin against creation. This natural law of harmonized restraint is applicable to all creatures and, for humans as stewards, becomes an ecological injunction that should take precedence over all other human deeds and priorities. Humans should not create disharmony for their own short-term 'good' because ultimately it will not be in their best interests, for what is good for humankind is, by virtue of the law of ecosystems, good for all life.

For example, the acceptance of, or indifference towards, unjustified pain or suffering in animals may lead to or be associated with a more generalized inhumane indifference, a growing, dehumanizing lack of compassion in society as a whole. Cruelty towards animals has been linked with sociopathic behavior and child abuse. Such correlations warrant our concern and illustrate the point of similarity in relationships between one human and another and between humans and animals. Similarly, the injunction not to create disharmony is as relevant to the way we treat our bodies as it is the way we treat nature and ecosystems. To abuse is to pollute, deplete, and ultimately to exhaust and destroy.

To respect and uphold the intrinsic rights of all creatures entails a radical change in human regard for and relationship toward all nonhuman life. Extrinsic values and uses accorded by humans must become secondary to these intrinsic rights that are mandatory to uphold by virtue of humankind's relationship to nature and implicit by virtue of the very existence of all creatures. To negate or ignore such intrinsic rights is to create disharmony and to violate the basic law of creation. And since what is not good for all life is also not good for humankind, this may be judged by future generations (if they become more enlightened) as a crime against society since it is a sin against creation.

Although the humane movement primarily aims at preventing the infliction of unnecessary pain and suffering in animals, such an ethic can become 'speciesist' for not all things of creation (plants, lakes, deserts, and so on) can suffer or experience pain. Even Singer's (1975) valiant effort to promote a humane philosophy for animal rights is limited because he bases it only upon the argument of sentience: creatures should be accorded rights (to humane treatment, and so on) not because they can feel pain and can suffer; they should, together with all natural creations, be treated with respect because we, like they, exist as part of the biotic community. Thus the humane ethic that is concerned almost exclusively with suffering must be enlarged to incorporate nonsentient creations (plants, rivers, and so on) into an all embracing biospiritual or ecologically humane ethic.

An ecologically sound humane ethic constrains those human actions that may cause: unnecessary pain and suffering; privation of basic social and environmental needs for normal growth and fulfillment; disharmony within and between species (both sentient and nonsentient); and disharmony between interdependent animate or inanimate microcosms and macrocosms of the biosphere's ecosystems, i.e., have adverse environmental consequences.

In summary, we must obey this ecologically humane ethic for it is the key to humane stewardship (Fox, 1980a). We must treat all creatures humanely because we are human and because they exist. And we must relate and act to foster harmony socially and ecologically for the sake of all life on earth and learn to maximize human interests and at the same time minimize the rate of entropy and our impact on the environment and other creatures. And we might consider that the unethical exploitation and suffering of animals and destruction of the environment are symptoms of an unbalanced, if not disordered, state of mind, and of a growing atrophy of the human spirit. Awakening to the realization that animals have intrinsic worth and are beings worthy of moral concern, may help overcome these mental and spiritual disorders, encouraging empathy, compassionate understanding and enlightened self-interest.

Deciding What to Kill

T. Nicolaus Tideman

It is not hard to feel kinship with a dog. A dog expresses itself with its face, posture, and body movement. It is not hard to believe that a dog can understand one's feelings, and vice versa. These empathetic feelings make many people uncomfortable with the idea that there are people who kill dogs for food. How can they be so insensitive to the bonds between man and dog?

But what makes dogs so different from cattle? Cattle do not have ranges of facial expression and body movement as wide as those of dogs, so they do not appear to transmit as many specific feelings. And yet when one is standing near a cow, looking directly at her face, it is not hard to believe that she understands one's feelings. Many people have experienced this and still eat beef without compunction. Others find the sense of kinship with cattle too strong to be comfortable with the knowledge that cattle are killed for their meals.

Similar feelings of kinship occur with other mammals and, to a lesser extent, with lower animals, leading many people to become vegetarians. Sensing the anguish of the slaughtered animals, they want no part of causing it.

Some people are satisfied with a delegation of the dirty work to others. As long as meat appears in neat packages in the supermarket or as cooked slabs on their plates, with its biological origin rarely discernible, they are content to eat it without worrying about what must be done to produce it. And yet, from an economic or a systems perspective, the buyers of meat are generating the slaughter of animals just as surely as if they were to decide that they wanted particular animals butchered for their freezers. One cannot consume meat without inducing others to kill more animals to restock the meat counters.

If one is responsible for the deaths of the creatures that one eats, and if bonds of kinship with those creatures are often perceived, where should one draw the line as to what one will eat? Should one's empathy extend to shrimp and crabs, which are hardly more than overgrown cockroaches?

Do cockroaches also deserve our respect and protection? If all animal life is to be respected, is plant life beyond the pale? Do the amoebae come under our shield and not the algae, because the latter contain a bit of chlorophyll, making them plants? What makes us so sure that trees are less sentient than toads? Plants expend energy more slowly for their size than animals, so it may be more difficult to perceive them as expressing feelings. And they lack the kind of central nervous system that we have come to think of as the *sine qua non* of consciousness. But plants do move purposefully, within the limits of being rooted, and many people regard them as possessing consciousness.

For a discussion of the basis for believing that plants possess consciousness, I would recommend *The Secret Life of Plants* (Tompkins and Bird, 1974) and works cited therein. None of the evidence and arguments that I have seen are so compelling as to admit no other interpretation than that plants possess consciousness, but I accept that interpretation because the alternative, a boundary beyond which there is no consciousness, is so unappealing. For any proposed location of the boundary of consciousness there are borderline cases that make it implausible that a disappearance of consciousness occurs at that point.

The alternative to a boundary for consciousness is a belief that we share the quality of consciousness with all creation, and that nothing can be excluded from the community of beings to whom we own respect by virtue of sharing with them the quality of consciousness.

The possibility of reaching such a conclusion can cause anxiety. The necessity of eating in order to survive implies that one must choose between killing other conscious beings and allowing oneself to die. The prospect of having to deal with that dilemma may lead some to decide that it simply is not possible that all things possess consciousness.

One solution to the dilemma of eating is offered by fructarians, who eat only fruit and seeds. But it is not clear that the destruction of seeds is less catastrophic than the destrution of more mature plants. The mature plant has had at least some life experience, while the potential life experience of the eaten seed is totally unfulfilled. The ingestion of any living or potentially living thing could be avoided by eating only milk and honey, but consuming milk requires that cows eat living things. And no matter what one eats, some of the living cells that line one's stomach will be killed in the process of digestion.

There is no hope of avoiding killing. And to allow oneself to die instead would certainly be irreverent of life. Is there any solution to this moral dilemma? A critical factor in the dilemma is the negativity that is generally associated with death, which is often viewed as an extinction of consciousness. If death is not an extinction of consciousness, the nature of the problem is altered considerably.

Numerous individuals have reported what they regard as communications with persons who have died, but the unshared quality of these ex-

periences and the lack of procedure by which any investigator can repli-
cate them at will have generally made claims of such communications,
and hence of existence beyond death, uninteresting to persons who look
to science as a guide to their beliefs.

Two recent books, however, offer different approaches to verifying
the continuation of consciousness beyond death that may meet the criteria
of science. Neither approach could be said to establish the continuation of
consciousness in an indisputable way, but both are suggestive enough to
warrant serious evaluation by anyone who is concerned with the morality
of killing.

The first approach to verifying the continuation of consciousness beyond death comes from "near death" experiences. Raymond Moody reports the results of an investigation of such experiences in *Life After Life*
(Moody, 1976). After hearing several similar anecdotes, he made a systematic collection of accounts of the experiences of persons who had
come close to death. He found a remarkable recurrence of certain elements in these experiences, including (1) greetings from deceased friends
and relatives, (2) interviews with "beings of light," where one's life
would be reviewed, not in a spirit of judgment, but rather loving illumination, with one deciding for one's self whether the intended lessons had
been learned, and (3) a loss of all fear of death.

In the Introduction to *Life After Life,* Elisabeth Kubler-Ross reports
that in her years of caring for terminally ill patients she found her patients
reporting to her the same patterns of experiences that Moody collected. It
should be possible for other investigators to interview people who came
close to death and see whether they find the same patterns. Determined
skeptics will not be without resources, however. They can say first that
any future reports, since they are subjective, may be those of persons who
have become familiar with the reports already circulating (which may be
fraudulent for all we know), so that future reports may be false reports
motivated by a desire for attention. Furthermore, why should we place
any credence in the delusions experienced by persons who are trauma-
tized by coming close to death?

Such skepticism is an example of the general phenomenon that, for
any collection of observations, there are an infinite number of potential
theories that would explain those observations. In science there is a general preference, in deference to Occam's razor, for simpler theories over
more complex ones, but there is no way of knowing that the simpler
theory is the correct one, or even if identifying unambiguously the simpler one. To my mind, the simplest, most compelling explanation of the
reports of near-death experiences is that death is a transition rather than a
termination. This view is reinforced by inquiries of a second sort.

The second approach to verifying the continuation of consciousness
beyond death is to collect reports from random samples of physicians and
nurses who have observed people as they die. This is the approach of

Karlis Osis and Erlendur Haraldsson, who report the results in their book *At the Hour of Death* (Osis and Haraldsson, 1977). The suggestive pattern that Osis and Haraldsson observed was that about 4% of dying persons, both in the United States and in India, experienced apparitions. The apparitions of dying persons were predominantly of friends and relatives already deceased or of religious figures. The predominantly reported purpose of the apparitions was to convey the dying person to the next world. These phenomena occurred more frequently in persons who were more conscious and less drugged when they died. The work of Osis and Haraldsson would seem to be replicable, and indeed the books of Moody and of Osis and Haraldsson each report instances of the phenomena researched primarily by the other. *Death-Bed Visions* by William Barrett (1926) provides a collection of reports like those of Osis and Haraldsson. *Afterlife,* by Archie Matson (1977) reports both near-death and deathbed experiences of both of the sorts described.

If the work of Osis and Haraldsson can be replicated at will, it will be hard to argue that the phenomenon is deliberate fraud by those who purport to have the experiences, for it seems implausible that a person at death would spend those last breaths in a sudden effort to perpetuate a fraud. The argument that the phenomena are simple delusions could be used, but the lack of an association between such experiences and conditions that are generally hallucenogenic (Osis and Haraldsson, 1977, pp. 70–73) would mitigate against such an argument. As I said before, the simplest and most coherent explanation of the reported phenomena, to me, is that death is a transition rather than a termination. If this is true for humans, and if other creatures share with humans the characteristic of consciousness, which seems to be the characteristic that survives bodily death, then it is plausible to suppose that death is a transition rather than a termination of existence for animals and plants as well as for humans.

What would this mean for the morality of killing? Would it mean that there is never any moral opprobrium attached to killing, since consciousness is not destroyed, only changed in its focus? That depends on whether life has a purpose. If life is without ultimate purpose and death is merely a change of focus, then painless death, at any rate, might be morally unobjectionable.

But I believe that life has discernible purposes. Moody (1976, p. 65) reports that the luminous beings who assisted people in reviewing their lives during near-death experiences seemed to "stress the importance of two things in life: Learning to love other people and acquiring knowledge." Another approach to discovering the purpose of life is taken by Helen Wambach in her book *Life Before Life* (1979). She hypnotizes large numbers of people simultaneously, asks them to go back in their lives to when they were born and before, and try to recall whether their lives came as a result of conscious choices, or were they compelled to be born. About 48% of her participants report that they are able to obtain answers to such questions, and 81% of these reported choosing to be born. When Wambach asked her participants the purpose of their lives (1979, pp. 80–97) they gave answers broadly compatible with the "loving and learning" purpose identified by Moody.

The picture that emerges from these inquiries is one of human life as a transitory state with definite purposes. Suppose the same applies to nonhuman life. What does that imply about how we should feel about killing? The only way that I have been able to answer that question to my own satisfaction is through the idea of the unity of all creation. This is the idea that every consciousness is a component of the Force that created all. Killer and killed are not truly separate, but rather different facets of a unified structure.

How should I feel about killing, if each time I kill I am killing a part of myself, and if death marks an end, not of existence but of a chapter of experience with definite purposes? There is something of a parallel with the way my brain may decide that some of the lining of my stomach will be sacrificed to digest what is consumed to nourish the rest of me. In the same way, I can imagine that it may be consistent with the purpose of creation that one element of creation terminate the life—cause the transition from physical existence to nonphysical existence—of another element of creation. The most plausible criterion for appropriate killing would be the compatibility of the killing with the overall purpose of creation—loving and learning, harmony and advancement of all consciousness.

Among Eskimos, in a time of food shortage, an older member of the group may go out into the cold and freeze to death so that there will not be as many mouths to feed. For myself, I can imagine the further step in finding satisfaction in knowing that my body would provide food for others, especially if my death were imminent in any case. If can therefore imagine that an animal or plant could find fulfillment in being killed and eaten. For the killer, the guide to action would be to ask, "Is this impending death plausibly in harmony with the combined life purposes of killer and killed? Can the life be taken in loving recognition of our common bonds?" There can be no guarantee that adopting this perspective will always produce good decisions, but that is the way that I would recommend deciding what to kill.

As an example of killing in such a spirit, I would offer the following, contributed to *The Last Whole Earth Catalog* (Portola Institute, 1971, p. 268) by Peter Rabbit:

> I have a blind overlooking a big block of salt in a creek bottom the deer
> come there every morning & evening
> I watch them
> I call them with a Herter's Deer Call
> I ask them if any among them is ready to die

I tell them that we will use the energy from eating their flesh
    in a way that would please them
I don't forget those words
almost always it is a doe without a fawn or a lone buck that
    tells me he will join us
I shoot them from no further than 50 yards
they die almost instantly
I feel their spirit enters me
everytime
it feels good
that's all I know about killing deer.

SECTION VIII

Facts and Acts

Introduction

The concluding section of this collection contains papers on scientific studies of nonhuman animals, the epistemological value of such studies, and on the actions of human 'animal liberationists.'

W. B. Gross reports on his research on the effect of genetic endowment and environmental changes on chickens. His work puts into focus a number of perplexing and philosophically important questions. How do we determine the welfare of an animal? Survival probability is pretty clearly one factor. But the survival of the species requires variability of individuals and thus ensures that some individuals will be ill-adapted to the environments into which they are born. Another partial measure of welfare is rate of growth.

The relation of stress to welfare is rather complex. Reaction to particular sorts and intensities of stress is partially genetically determined, partially a matter of prior experience. Very low levels of stress are not conducive to welfare, nor are very high levels. By most available measures, modern poultry farming methods are less stressful than those of the past. The type and level of stress most beneficial for the continued survival of an animal depends upon the environment. In chickens, short-term stresses tend to increase resistance to bacterial, and reduce resistance to viral, diseases. Chickens free of the social stress generated by cage mates appear to be highly contented, but other measures indicate decreased welfare.

Gross emphasizes the attitude of the human handlers of the chickens as one of the most important determinants of welfare. Those who view their animals as 'machines' in Benson's sense (Section I) are likely to be much less effective than Fox's 'humane stewards' (Section VII).

The battery-cage chicken has been one of the most widely denounced of the developments of modern farming, and there is no doubt that some current poultry-raising practices are inhumane. Gross's paper, however, challenges the belief that all modern poultry practices are more deleterious to the welfare of the birds than traditional methods or life in the wild. (This coheres with VanDeVeer's suggestion [in Section III] that there may be gains for the animal in domestication.) Fox's paper in Section VII would give central place, when evaluating our treatment of chickens, to the question 'what are the natural potentials of a chicken?.'

325

Deborah G. Mayo points out that discussions of the ethics of human use of animals frequently focus on farming and hunting, with much less attention paid to the experimental and instructional use of animals. This is true, in general, of the papers in this volume so far, with the obvious exception of Gross's. The research of Gross and his colleagues is concerned with the effects of certain factors on chickens, does not entail extrapolation from one species to another, and has clear practical and theoretic goals. Gross thus escapes many of the charges Mayo makes against the animal research industry as a whole.

Research involving animals is usually justified on the grounds of benefits accruing to humans (and, less frequently mentioned, to other animals) or of contribution to the sum of knowledge. But both these justifications, Mayo argues, are spurious when applied to much of animal research. Many research projects are pointless, gratuitously repetitive, fatally flawed, or concerned with the obvious or the trivial. If animal suffering is of any significance whatever, its infliction can only be justified for substantial, clearly-defined goals. But many experiments are ill-conceived or actually unconceived in that no hypothesis has been formulated before conducting the experiment. One or a very few repetitions of an experiment may be needed to confirm the results, but some experiments are repeated time and again with no epistemological justification.

Even experiments with goals that are neither repetitive nor trivial may well still be invalid. Mayo describes several possible causes of such invalidity. The analogy of artificially induced conditions (e.g., cancers) to naturally occurring ones is often a very dubious one, but is rarely questioned. When this is combined with extrapolation from one species to another so that the analogy is between an induced condition in rats and a naturally occurring condition in humans, the inference may very well be so weak as to be useless.

Especially in toxicological research the differences between species are of such magnitude as to render suspect *any* inference from one species to another. Standard testing methods such as LD_{50} and the Draize test show highly variable results, even when the same species is used. Guinea pigs can eat strychnine, but penicillin is highly poisonous for them. Sheep can eat arsenic, but almonds kill foxes. Thalidomide causes deformity in almost no other species than man.

Yet other sources of experimental invalidity result from background variables not explicitly considered. (Rollin notes such a case in Section II—the starving deer smelling food.) If animals come from varied sources, they may have any number of pre-existing conditions, and if they are specially bred for the laboratory they are likely to resemble little else on earth. In either case the reliability of inference is seriously undermined.

Some of these impediments to valid experimentation on animals might possibly be overcome, but most cannot and thus, concludes Mayo, much animal experimentation cannot be made morally or scientifically defensible.

Le Vasseur and Sipman, the humans who released the dolphins in Hawaii, do not appear to have been significantly concerned with the scientific validity of the experimental work they observed. What they considered intolerable was the solitary confinement of highly intelligent animals in small tanks for life. As Gavan Daws tells the tale, they saw themselves as liberators. But of course they were not charged with being liberators, but with being thieves. Of the several possible lines of defense, the most promising and that preferred by Le Vasseur (Sipman has not yet been tried) was that dolphins are persons, persons are not property, and thus there was no theft.

This line of defense was prohibited by the trial judge. The judge ruled that dolphins could not be considered persons under the law, and that no testimony to support such a claim would be in order. Thus prohibited from raising the moral status of the dolphins as an issue in court, and with the elements of theft clearly established, Le Vasseur was convicted. His appeal to the Hawaii Supreme Court is pending.

As Daws points out, Le Vasseur's conviction has not obviously deterred other animal rights activists. One such activist is Henry Spira. His short paper is a call to action with a sketch of a how-to-do-it manual. Spira's views are not necesssarily those of the editors (the same is true for every other contributor to this volume).

Chicken–Environment
Interactions

W. B. Gross

For over a decade our research program has dealt with the effects of genetics and environmental parameters on the productivity and disease resistance of chickens. The results of this research indicate that rather mild genetic and environmental changes can alter the parameters that we measure in very complex ways. This should be helpful for evaluating their welfare under the various ways in which they are managed.

An animal can be considered to be the creature called for by the translation and implementation of the plan recorded in its genes. It is this genetic plan that is continually evolving so that its product can survive in its environment and exploit new ecological niches. Genes specify a great variety of structural, behavioral, defense, and life strategy characteristics. It is easy to imagine the millions of features that are included in an animal such as the chicken. When one realizes the wide range of values for each feature, it is awe-inspiring to contemplate the variety of combinations possible in a population. Thus a population is a group of unique individuals.

The genetic plan survives only if the specified creature carrying it can continually pass it on from generation to generation. In essence, if the plan is plastic it can have everlasting life. Although natural selection works at the level of the individual the survival of the population is essential. Since the genetic package does not know what the future holds, it must be continually adaptive. In each generation it sends forth a wide variety of similar creatures some of which have a better chance than others to survive and pass their genetic material on to the next generation.

Features that do not favor survival are not entirely lost, but appear less frequently during the next generation. The genetic package must continually expect the environment to change, because if it became overspecialized for a temporary and/or specific environmental niche extinction could occur. Thus variability must be maintained to avoid extinction. Many extra individuals are produced so that the failure of some to survive does not jeopardize the species. For the individual this does not seem to be fair, but all have a chance to survive in some environment. The researcher and the husbander hope to raise all the animals; nature does not.

Animals must continually monitor the use of their limited supply of resources. There are not enough resources for an animal to do all things well all of the time. Allocation of resources can only be made on the basis of the animal's perception of its needs in an everchanging environment. Although the priority ranking of various activities is unknown, our research suggests that adaptation to the environment has a very high priority, while growth and reproduction seem to have lower priorities.

One of the best tools for measuring an animal's reaction to a disease or environmental change is the scale. It often reveals reduced growth rates even though the animal otherwise appears to be outwardly unaffected.

Animals have an adaptive mechanism that continually monitors the allocation of resources and the biochemistry of cells. The environment is monitored by the senses such as sight, smell, touch, and hearing. In addition there are also internal body sensors that measure disease, toxins, and nutrition. These sensory inputs are evaluated in the cerebrum and continuing signals are relayed to the hypothalamus, which is in the brain stem. In response to these inputs, the hypothalamus causes the pituitary gland to produce more or less of a hormone called ACTH (adrenalcorticotropic hormone), which is carried by the blood to the adrenal glands. In response to the level of ACTH in the blood, the adrenal glands produce more or less of a hormone called corticosterone. Corticosterone is carried by the blood to all of the cells of the body where it is transferred to the nucleus. In the nucleus it affects the translation of the operating genes of the cell, which in turn modifies the production of proteins and enzymes by the cell. Increasing levels of corticosterone inhibit some responses and stimulate others. Thus sensory perception can ultimately affect cellular biochemistry and the allocation of resources. Chickens differ in their corticosterone response to the same stress. We have, through genetic selection, developed lines of chickens for either a high (HPC) or low (LPC) plasma corticosterone response to being with unfamiliar pen mates. This demonstrates that there is genetic variation in the response of chickens to stressors.

Some of the environmental factors that may be stressful to chickens have been identified. Among these are: unusual temperatures, unfamiliar pen mates, unfamiliar sounds, unfamiliar or uncaring handlers, lack of food or water, the injection of foreign proteins, toxins, or any disease. The animal's perception of the stressfulness of an event depends on its prior experiences. Taking an animal out of its cage to be weighed can be stressful to it. If done gently, the chickens become habituated and each subsequent weighing appears to be less stressful.

Another example is our finding that a bird fasted for 24 hours has a reduced ability to produce antibody in response to vaccines. A second fasting later in life results in no decrease in the ability to produce antibody in response to vaccines.

Short-term stresses from a variety of environmental events produce similar changes. Compared to unstressed individuals, stressed birds have an increased defense against bacterial and some parasitic diseases. They have decreased defense against viral disease and tumors, and decreased antibody response to vaccines. Their cell-mediated (tissue, T cell, non-antibody) immunity to diseases is reduced after a stressful event.

Many diseases with complex causes follow a stressful event. Among this type of disease are distemper in dogs, shipping fever in cattle, and respiratory disease complex in chickens and in humans. Shortly after stressing, the animal is exposed to a viral infection that by itself might be quite mild. It is easier for viruses to infect an animal after a stress. Since its defense against viruses is weakened, the animal must commit more resources to the defense against the invading virus. This results in fewer resources being available for defense against bacterial infections. Bacteria that normally are found in the air, on the skin, or within the respiratory tract are then able to invade. The host is caught unprepared and the invading bacteria overwhelm the weakened antibacterial defense, which results in a disease that could terminate in death.

From our observations of stressed birds, the HPC and LPC lines of chickens, and the results of feeding corticosterone we are able to determine the characteristics of chickens with increased levels of corticosterone in the blood.

They tend to be smaller, more active, and lay smaller eggs. Their sensory descrimination is good; that is, they can distinguish easily between different sounds. Birds fed corticosterone tend to put on fat and have increased appetite. When placed into unfamiliar cages, the high corticosterone birds tend to adapt easily. Their defense against bacterial diseases is increased, while their defense against virus infections and tumors is decreased. Antibody response to vaccines is decreased. Increased corticosterone levels seemingly adapt chickens for harsher environments.

Chickens with a relatively low level of corticosterone in the blood tend to be larger, less active, and lay larger eggs. Their sensitivity to sensory stimuli is extremely high, but their discrimination between stimuli is poor. For example, chickens housed under a fairly low light intensity become very calm and are low stressed. They can easily hear sounds from outside the building, but cannot evaluate them and thus become agitated. When they are placed in unfamiliar cages, they do not adapt well. Their resistance to bacterial infection is decreased, while their defense against viral infection and tumors is increased. They tend to produce high levels of antibody in response to vaccines. It appears that low levels of plasma corticosterone tend to adapt birds to a relatively favorable environment.

Large birds have an advantage in contests for social position and re-
sultant reproductive privileges. They must however reach reproductive
age before size becomes an important factor. The best-adapted birds perceive less stress from their environment, become larger, and have a better
chance to pass their genes on to the next generation. From the standpoint
of the hens and the population, most roosters are expendable in order to
help maintain adaptability to environment changes.

In general, animals adapt fairly well to continuing mild chronic
stress, but tend to be smaller. An example of this is the results of our
research with males kept in one of three environments for three months.
One group was kept in flocks of five in each of seven cages. Ample food
and water were available so that there was no nutritional advantage for
dominance. Each day one bird in each cage was shifted into another cage
according to a plan that kept contact with familiar individuals to a minimum. For about an hour there was considerable antagonistic behavior in
each cage as the new peck orders were set up. As a result, the birds
tended to be smaller with a slight reduction in feed consumption. Their
combs tended to have blemishes associated with social strife and sounds
associated with agression could be heard occasionally throughout the day.
Resistance to mites, bacterial, and viral diseases was very good and their
antibody response to vaccines was excellent. Their muscular tone was
good and they were quite active and well preened. Apparently their resources allocation favored defense both against disease and agression at
the expense of growth.

Another group of males were kept singly in solid walled cages. They
could hear but not see other birds. They were thus under a very low level
of social stress. Feed and water were always available. They became lethargic and did not seem to be well preened. Feed consumption was reduced, resulting in reduced weight gain. Their vocalization suggested
very contented birds. Defense against mites and bacterial infections was
much reduced. Their defense against viruses and their antibody response
to vaccines was good. When placed together in a floor pen, they showed
aggressive behavior, but had very little skill at fighting. Apparently these
birds had so little stress that they had difficulty maintaining appetite and
bacterial defense.

The last group was placed in flocks of three in stable social environments. Water and feed were always available. Feed consumption and
growth rates were greatest in this environment. Vocalization suggested
that they were contented (but not as contented as the singly caged birds).
Muscle tone was good. The birds seemed to be well-preened and the
combs had few blemishes associated with aggression. Defense against
diseases and antibody response to vaccine seemed to follow genetic tendencies.

Clearly too much or too little social stress was unfavorable for chickens. Optimal stress seems to be characterized by optimal growth rate and

utilization of feed, a stable social structure, and variable (depending on genetic background) disease defense characteristics.

Other experiments have illustrated the effects of too little stress. For example, we have a population of chickens that is frequently kept in floor pens and provided with over four times the space allocated to similar commercial birds. Under these circumstances they grow well and are especially resistant to Marek's disease tumors. When adults from this line are placed in cages, at least 1/3 die from infections caused by bacteria that are always present in the environment. They become lethargic, have reduced feed consumption and may die of starvation. Feed consumption and bacterial defense can be returned to normal by adding corticosterone to their feed. Clearly cages are an unsatisfactory environment for these birds because the stress level is too low. Other stocks, however, do well in the same type of cages.

Most stressors affect birds for only a few days before their responses return to normal. However, stresses early in life can have long-lasting effects. Chilling, overheating, or six hours of water deprivation resulted in changes in growth rate, feed efficiency, and antibody response to vaccines, disease resistance, behavior, and response to subsequent stress. The response to each change (increase or decrease) was variable, depending on the genetic background of the chickens and their subsequent environments. This suggests that the genetic potential can be modified by events early in life, while many systems of the chick are still developing. The variability in responses suggests that strategy for change in response to an early stress may vary with the genetic background of the population.

When chickens are adapted to their handlers and to the experimental procedures, they become calmer and easier to handle. Genetic differences are easier to separate and experimental results may vary depending on whether or not adapted birds are employed. That is, some experiments have one outcome with adapted birds and a different outcome with those that are unadapted. Responses to adaptation include changes in disease resistance, antibody response to vaccines, and blood chemistry values. Responses to adaptation to handlers and experimental procedures are modified by the stock being observed and the presence or absence of a stress early in life.

The response of an animal to any test is modified by many variables in its environment. Thus an animal mirrors its genetic legacy as modified by its environment.

Several methods have been used for evaluating the bird's relationship to its environment, which has often been referred to here as the level of stress. A commonly employed method is to measure the level of the hormone, corticosterone, in the blood. Determination of the corticosterone level is a good test for separating groups of stressed from groups of unstressed birds. Chickens differ in the rate of removal of corticosterone from the blood. Thus, chickens with the same plasma corticosterone level may differ in their rate of utilization of the hormone, and thus their response to various tests.

In our experiments, growth rate and feed efficiency seem to be good indicators of the chicken's response to its environment since they are reduced by many stressful factors and are easy to measure.

Behavioral changes indicate the chicken's perception of present and past environments. Researchers should be keen observers of the many behaviors of their experimental animals. This should result in more consistent and better-evaluated experimental findings, better care, better understanding, and reduced trauma to the animals.

Chickens have a wide variety of vocalizations in response to environmental changes. Vocalizations reflect the chicken's current evaluation of the environment. It seems possible to measure contentedness, some need (perhaps for water, heat, or light), illness, and relationship to the handlers.

In summary, our experimental observations suggest that the chicken's behavior, vocalizations, growth rate, and disease resistance are influenced by its genetic heritage, early environment, long-term environment, recent environment, and the attitude and behavior of its handlers. One of the most important variables is the attitude of the handlers.

Our experimental findings also provide insights into understanding the development and nature of the poultry industry.

The purpose of that industry is to produce disease-free chicken products of the highest quality for the consumer. Under the free enterprise system, the industry has been very successful in meeting this objective.

Years ago poultry was kept in small groups with occasional supplemental feedings of grain. Under these circumstances productivity was low and the price of poultry products was relatively high. Productivity was low because the birds were subjected to the stresses of variable food supply, climatic changes, predators, diseases, and social stress. In addition they were subjected to the attitudes of the people with whom they were associated, which could be quite variable. Clearly this method of raising poultry was far too stressful. Progress in the raising of poultry could be made if methods were found for reducing these stresses.

As a result of research and development, improvements were made in nutrition, genetics, disease control, and housing. When a method for reducing stress was developed, the first producers to use it obtained reduced stress among their birds, which resulted in increased productivity and ultimately in increased profits. Following their example, the remainder of the industry then began to utilize the development, which reduced production costs. Competition thus reduced the cost of the products to consumers, who became the prime beneficiary of this research.

Today's highly productive poultry industry has resulted from the incremental utilization of many methods for reducing the stresses to which chickens were subjected. Progress can be measured by noting improvements in growth rate and egg production. Indeed in our research, we found that maximal growth rates were indicators of optimal stress levels.

Progress began when poultry were placed in small naturally lighted houses that can still be seen (without chickens) in many areas. These birds were protected form severe stresses, predator stress was reduced, and there was better availability of food.

Larger insulated and fan-ventilated houses became necessary for increasing the efficiency of production by creating a stable environment. More efficient disease control methods were developed before the improved housing could be utilized. Even with these improvements, social stress and population pressure remained. Under these conditions most disease losses were from viral diseases, Marek's tumors, and mixed viral–bacterial infections (air sac disease). These are the same diseases that our research found to be associated with high stress environments.

Broiler chickens are most active in setting up their social orders at almost 6–7 weeks of age. Stress-related diseases had a peak incidence at about 7–8 weeks of age, which is also the age of marketing. Inspectors condemn any chicken with Marek's tumors and active or healing lesions of "air sac disease." Therefore, there was a strong incentive for reducing social stresses.

It was found that at reduced levels of light intensity, chickens failed to recognize social and population stresses. Removal of these social and disease stresses released resources that could then be used for increased growth. Another benefit of reduced light levels is reduced feather picking, so that debeaking is no longer needed.

Chickens that have not been exposed to a wide range of experiences interpret any unusual occurrence as being serious. Examples are unusual sounds, such as a squeaky fan, or the presence of a stranger in the house. The latter results in extreme avoidance behavior, accompanied by apprehensive agitated vocalizations that may cause injury to the birds. Because of the problem with strangers, poultry handlers are reluctant to allow visitors into their houses. Another reason is more complicated. Drugs are being used less and less for the control of chicken diseases. Reliance is being placed on vaccination, eradication and, most effectively, on isolation rearing. Since most disease agents are carried to the farms on visitors' shoes, clothing, and hands, any visitor is now a potential source of disease organisms. Obviously, then, this type of housing requires a very careful management.

It has been found that keeping layers in cages reduced the number of their social contacts and resulted in a more stable social order than is generally observed on the floor. Population pressure and the problems of feather pulling and "pick outs" can be almost eliminated by reducing light intensity, but, under these circumstances, northern fowl mites can become a problem. This is one of the diseases that we found to be associated with low levels of stress.

Low-stressed environments and automation make the husbanders more important than ever. The birds are at peak productivity and are unfamiliar with stressful situations. Any little thing that goes wrong results in decreased productivity. A good husbander spends much more than the standard 8 hour day walking slowly and cheerfully among the birds constantly checking the birds and equipment and familiarizing the birds with his presence. Again, I wish to emphasize that the attitude of the husbander is critical to the welfare of the birds. Poor husbandry, regardless of the management system, should not be tolerated and is not tolerated economically.

Along with improvements in housing, nutrition, and disease control there have been changes of a genetic background of chickens. Sophisticated genetic procedures enable selection for adaptability for certain environments. As a result, some stocks do better in cages while others do better on the floor, resulting in genotype–environment interactions. The presence of such interactions precludes the making of general inferences from stock to stock.

On the basis of our understanding of chickens raised under experimental conditions, I will attempt to evaluate the relative stressfulness of various commercial environments. In order of increasing stressfulness they appear to be:

1. A bird by itself in a cage
2. A group of birds in a cage
3. Laying birds on the floor
4. Broilers on the floor
5. The most stressful commercial environment is the raising of turkeys on range.

Environments 1–4 are reduced in stressfulness with increased climate control by means of insulation and mechanical ventilation. Stressfulness can be further reduced by reducing the light intensity.

Although, at first, I was uneasy about raising birds under reduced light intensity, an experiment several years ago modified my bias. The chickens were placed on the floor of an isolation room that had completely occluded windows and a dark entry room. After a week for adaptation to the location of feeders and water the lights were turned off and the birds were raised in total darkness. My daily inspection of the birds and their feed and water was made by touch. The birds in the dark gained 5% more weight and sounded more contented than the controls in well lit cages. In another experiment with birds in dark cages, the birds in the dark became higher stressed than the high stress controls because of a squeaky fan.

Our studies have only begun to explore the interrelationship between the chickens, genetics, behavior, vocalizations, environments, disease resistance, and handlers. Much more research by those who are able (singly or as a group) to evaluate all of these variables is needed before we can have an adequate understanding of the needs of chickens. At this time I believe that studying the vocalizations of chickens can give us a good understanding of their appraisal of the current environment. Behavioral studies can often indicate problems in management, sometimes reflecting events that occurred early in life.

I must reemphasize that attitude is the most important factor in working with animals. Excellent facilities can be good or bad for the animals, depending on the attitude of the caretakers. Lesser facilities may be quite adequate when the attitude of the handlers is good. Attitudes are difficult to modify by rules and regulations. These frequently cause confusion, are sometimes resented, and are occasionally circumvented. They also require personnel for enforcement.

It is my experience that people who understand animals seldom subject them to inhumane treatment. My suggestion is that the best method for improving the welfare of experimental, commercial, or pet animals is increased understanding of their behavior and needs. Students who, after graduation, may work with animals should take a series of courses on animal behavior. Particular emphasis should be made on animals the student is likely to work with. Examples are students of animal agriculture, biology, medical sciences, psychology, and veterinary medicine.

Programs of education through the extension service, humane societies, and other sources could help to inform those already working with animals of our current knowledge of their behavioral needs.

There has been a tendency to place too much emphasis on the physical and nutritional needs of animals with too little regard to their social and behavioral needs. It is easy to employ our human bias while evaluating the animal's environment. Evaluations should be made on the basis of careful observations by well-trained individuals. One of the most important variables in the animal's environment is the attitude of its human associates.

Against a Scientific Justification of Animal Experiments

Deborah G. Mayo

INTRODUCTION

Discussions of the treatment of animals typically focus on their use as food and clothing, omitting the widespread use of animals in laboratory research. Animals serve as experimental subjects in teaching surgical operations; in testing the efficiency and safety of drugs, food, cars, household cleaners, and makeup; in psychological studies of pain, stress, and depression; and in satisfying the curiosity and desire of humans to learn more about biological processes. In so doing they are subjected to shocks, burns, lesions, crashes, stresses, diseases, mutilations, and the general array of slings and arrows of the laboratory environment.

If it is agreed that killing and torturing animals is prima facie wrong, then additional justification is necessary in order to defend the sacrifice of millions of animals each year to research. The justification most frequently offered is that animal research provides increases both in scientific knowledge and in the health, safety, and comfort of humans. As Lowrance (1976) remarks:

> For most people, any qualms over jeopardizing the animals are more than offset by the desire to gain knowledge useful in alleviating human suffering (p. 52) . . . . Few people would engage in such work were it not so essential (p. 54).

So closely is scientific experimentation associated with animal experimentation that those who oppose or criticize animal experiments are often taken to be opposing or criticizing science. One finds advocates of humane experimental methods labeled as anti-science and referred to as "those whose love of animals leads them into a hatred of science and even humanity. . . ." (Lane-Petter, 1963, p. 472). The classic volume,

Experimental Surgery (Markowitz et al., 1959) introduces the student to the Antivivisection Movement with the following remarks (emphasis added):

> It must be apparent . . . that ordinary antivivisectionists are im-
> mune to the usual methods of exposition by reasoned argument. *They*
> *strain at a dog and swallow a baby.* . . . They are an unfortunate evil
> in our midst, and we must accustom ourselves to their presence as we
> do to bad weather, and to disease.

The error in depicting critics of animal experiments as anti-science becomes clear when one begins to question the extent to which the purported scientific aims of these experiments are actually accomplished.
For then it turns out that the experiments and not their critics are
unscientific. However, humanists concerned with the treatment of animals too rarely question the scientific basis of animal experiments and fail
to uncover dissent within the scientific community itself. To the philosopher's arguments that animal experimentation is morally indefensible, the
animal researcher responds that they benefit humanity. But if the most
common uses of animals in research can be shown to be neither significantly beneficial to humans nor scientifically sound, then any appeal to
such benefits in justifying these uses is undermined. It is the purpose of
this paper to undermine the justification of common types of animal experiments by questioning their practical and scientific relevance and validity. Those experiments that cannot be justified on scientific grounds
can be no more justified than the frivolous killing and torturing of animals. In fact they are even less capable of justification, since such experiments block more fruitful uses of scientific resources.

IRRELEVANT EXPERIMENTS

I shall first consider the relevance of animal experiments and then discuss
various problems leading to their invalidity. In an important sense, invalidity is not separate from irrelevance, since invalid experiments are
surely irrelevant ones. However, in this section I shall focus on experiments that are irrelevant because of the triviality or obviousness of the
question they ask. Indeed, many experiments do not even have a specific
question in mind at the outset. They are often carried out simply to see
what will happen and after the results are in some sort of hypothesis is
formulated. Whether the hypothesis has been formulated before or after
the experiment is not reported in the description of the experiment. Yet,
formulating the hypothesis on the basis of the experiment can be shown to
lead researchers to conclude, wrongly, that something of relevance has
been observed (see Mayo, 1981).

    I must emphasize that the examples I here consider are not at all ex-
ceptional or unusual. On the contrary, each represents a basic type of ex-

periment that is performed with minor variations on millions of animals each year. An examination of the *Psychopharmaceutical Abstracts,* in which summaries of published experimental results are reported, will attest to the triviality and repetitiveness of the great majority of inquiries. What is particularly disturbing about the irrelevant experiments mentioned here is the amount of pain and suffering they involve. That such irrelevant painful experiments are not rare even at present is made plain in Jeff Diner's (1979) *Physical and Mental Suffering of Experimental Animals,* in which research from 1975–1978 is reviewed. It must also be kept in mind that these experiments are examples of ones considered important enough to publish. It is fair to assume that, in reality, many more experiments with even less relevance are performed. I limit myself to considering only recent experiments, to make it clear that these are not atrocities of the past.

(i) Infant monkeys were blinded at the University of Chicago in order to assess whether blindness inhibited social interactions as measured by facial expressions. The result: blind monkeys showed all normal facial expressions, except threat (Berkson and Becker, 1975).

(ii) Pigeons were starved to 70% of their weight in the City University of New York. It was concluded that following starvation pigeons ate more than usual (*Journal of Comparative and Phsyiological Psychology,* Sept., 1971).

(iii) The Department of Psychology at the University of Iowa studied the effects of brain lesions on the grooming behavior of cats. Cats underwent surgery to produce various types of brain lesions and films were taken of their subsequent grooming behavior. It was reported that:

> Statistical analyses of the grooming behavior shown on the films indicated that cats with pontile lesions and cats with tectal lesions spent less time grooming . . . . Other studies revealed that cats with pontile or tectal lesions were deficient in removing tapes stuck on their fur (Swenson and Randall, 1977).

(iv) At the Downstate Medical Center rats were surgically brain damaged and then stimulated by pinching their tails. They were then offered substances to drink. It was reported that (Mufson et al., 1976):

> Brain-damaged animals during tail pinch-induced drinking trials are responsive to the sensory properties of the test liquid. Chocolate milk is consumed, but tap water is actively rejected. Tail pinch to sham-operated control rats failed to induce such behavior; instead, it induced rage behavior towards the hand that pinched the tail.

(Is it to be concluded from this that brain damage decreases rage at painful stimuli and increases the desire for chocolate milk?)

(v) The following experiment carried out at the George Washington University Medical Center is a typical example of radiation research. Non-anesthetized rabbits had their heads irradiated while being restrained in "a Lucite restraining device." It is reported that "The developing skin, mucosal and eye lesions were recorded and often photographed, but no treatment was offered" (Bradley et al., 1977). The report continues to describe in detail the monstrous radiation-induced damage without drawing any conclusions.

(vi) An extremely widespread sort of experiment involves assessing the effects of various drugs on "punished responding." Punished responding typically involves first teaching an animal to perform some task such as pressing a key by rewarding it with food, and later changing these rewards to punishments, such as electric shocks. A number of such experiments have been carried out by Dr. J. E. Barrett using pigeons. Here, the "punishments" consist of electric shocks administered through electrodes implanted around the pubis bone. The results are rather inconclusive. It is reported that (Barrett and Witkin, 1976):

> The broad range of effects obtained in the present experiment make it difficult to readily characterize the effects of drug interactions on behavior.

(vii) At Emory University cats were used to study how two different kinds of painful stimuli, foot shock and tooth shock, influence behavior. To administer tooth shock, electrodes were implanted in the upper canine teeth of the cats. Foot shock was administered by means of stainless steel rods that formed the grid floor of the shuttle box in which the cats were placed. The cats were trained to escape the shocks by jumping across a barrier. However, when the cats were also subjected to the tooth shock, they were unable to escape the foot shock. It was concluded that tooth shock exerted a stronger influence than foot shock on behavior. The report regrets that (Anderson et al., 1976):

> Since 14 mA was the maximum amount of current that could be generated by our apparatus, it was not possible to determine if foot shock levels greater than that would have led to escape responding.

(viii) The *British Journal of Ophthalmology* published the experiments of Dr. Zauberman, which measured the number of grams of force needed to strip the retinas from the eyes of cats. There was nothing said about how this or similar experiments that were carried out could be relevant to the problem of detached retinas in humans.

(ix) A good deal of research has as its goal the determination of the effects of various operations on the sexual behavior of animals. For example, for a number of years the American Museum of Natural History in New York has conducted research on the effect of surgical mutilation on the sexual behavior of cats. In 1969 cats raised in isolation had penis nerves severed. The results of years of sex testing on these cats were overwhelmingly unsurprising: genital desensitization together with sexual inexperience inhibits the normal sexual behavior in cats (cited in Pratt, 1976, p. 72).

Numerous other experiments by the same researchers were conducted to determine the effect of surgically destroying the olfactory area on the brain on sexual behavior in cats, monkeys, hamsters, rats, and mice. The conclusions from all of these experiments were reported to be "contradictory."

(x) Some experiments are rendered trivial or useless because they do no more than repeat an experiment already performed numerous times. Even worse is the continuous repetition of experiments whose relevance is dubious in the first place. For example, there is the experiment that has been carried out since the time of Claude Bernard, one of the founders of modern vivisection methods. This experiment involves sewing up the ends of the intestines of dogs rendering them unable to defecate. Death has been observed to follow in some cases between 5 and 11 days, in other cases between 8 and 34 days. To what use is such information to be put?

One of the reasons for continually repeating an experiment that has already produced a result is that by using enough animals a result that is of a sufficient degree of statistical significance can be obtained. It is thought that the more observations the greater the evidence. However, this is based upon a statistical fallacy. The more experiments needed in order to observe an effect that is statistically significant, the smaller and more trivial the effect is. With enough experiments, even a chance occurrence is rendered overwhelmingly significant, statistically speaking.

INVALIDITY OF EXPERIMENTS

Experimental investigations are multi-staged affairs involving a host of background variables, the gathering, modeling, and analyses of data and inferences based upon the data. At each stage a variety of flaws can arise to render the experiment and inferences based upon it invalid. I shall consider some of the most pronounced flaws that arise in carrying out animal experiments and making inferences from them in medical and pharmacological research. These flaws stem from the disparity between experimentally induced conditions and conditions in humans, from within- and between-species differences, and from confounding variables before, during, and after experimental treatment. I consider each of these in turn.

Artificial Induction of Disease

One type of medical research involves ascertaining whether certain pathological conditions in humans can be alleviated or cured by certain drugs. Animals are used as "models" upon which to test these treatments. To do this it is necessary for the animal subject to have the condition in question, and in order to bring this about healthy animals are made sick. To this end they surgically have organs removed or damaged; they are injected with pathogenic organisms and cancer cells; they have irritants applied to their eyes and shaved skins; are forced to inhale various substances, and consume deficient diets. To produce such conditions as fear, anxiety, ulcers, heart diseases, and shock, animals are stressed by electric shocks or subjected to specially made pain devices, such as the Blalock Press and the Noble-Collip Drum.

It turns out, however, that the conditions artificially induced have little in common with the naturally occurring diseases in animals (when these exist) and much less in common with the diseases in man. This renders any conclusions drawn on the basis of treating these induced conditions of little relevance for treating humans (or even animals in cases where the condition naturally occurs). Extrapolating results from animal research to humans also frequently fails because of an absence of comparative examples of diseases in animals (particularly with hereditary diseases). For example, ulcers do not occur naturally in animals, and cancer in animals is quite different from cancer in man.

It is for this reason that the usefulness of animals in cancer research has been questioned. Most of the anticancer agents in use today have been tested in animals, most commonly rodents who have had tumors transplanted into them. This method, however, is of questionable validity. As noted in a review of testing anticancer drugs (*The Lancet,* April 15, 1972):

> Since no animal tumor is closely related to a cancer in human beings, an agent which is active in the laboratory may well prove to be useless clinically.

The situation would not be so serious if these agents were merely useless. In fact they are quite harmful, often causing side effects that may themselves precipitate further ills or even death. When researchers announce that a substance has been found to be effective in treating animal cancers, it is not revealed that the cell kinetics in animal cancers are vastly different than in humans. As the review above states:

> Animal tumours favored as test models have short doubling-times and a large proportion of cells in cycle with short generation-times. Probably, in many human cancers, intermitotic times are much longer and many cells are out of cycle. . . .

Since all anticancer agents in use have had their effectiveness assessed by animal tests, virtually all of them act only on rapidly dividing cells. This is one reason that treating human cancers (which typically do not divide rapidly) with these agents fails. The possibility of transplanting actual human cancers to animals is a suggestion which has been tried, but with poor results. As *The Lancet* review remarks, "This is hardly surprising, in view of the vastly different biochemical make-up of the animal model and of the human tumour which responds."

Hence, people are given drug after drug in the hope of arresting cancer when in fact these drugs have been evaluated upon cancers and organisms "vastly different" from their own. It may be argued that no better method is available for treating cancer, and so for the time being it is the best that can be done. Such, however, is not the case. In the last 30 years or so new techniques have been developed which hold much promise. These new techniques involve testing anticancer agents on cultures of human cancer cells. This has the advantage of permitting the sensitivity of individual cancers to chemotherapy to be estimated, providing each patient with treatment custom-tailored to the type of cancer involved. This would prevent individuals from having to suffer the agonies of numerous trial agents that may be entirely ill-suited for treating their particular strain of cancer. If these newer techniques are to be developed sufficiently, some of the attention presently given to animal testing will have to be channeled into these alternatives. Unfortunately, researchers have been reluctant to do so.

The need to induce pathological conditions in animals in order to test some treatment upon them gives rise to an additional area of medical research. This area involves experiments that have as their sole aim the determination of how various pathological conditions can best be brought about in animals. Although such research has provided means for inducing a number of conditions, cures for these conditions have not been forthcoming.

For instance, research has repeatedly been carried out in order to find ways of inducing peritonitis in dogs. Peritonitis is the painful condition suffered by humans after rupturing their appendices. Even after a standard method for producing this disease in dogs was available (i.e., surgically tying off the appendix and feeding the dog castor oil), further experiments to find improved methods were made. One such experiment is reported by Hans Ruesch in *Slaughter of the Innocent* (Ruesch, 1978, pp. 105–106):

> With each dog strapped down and his belly laid open, the 'surgeons'—subsidized by the American taxpayers who of course had never been asked for their consent—tied off and crushed the appendix, then cut out part of the intestinal tract and the spleen. With the intestinal system thus mutilated and unable to function normally, the dog was made to swallow a large dose of castor oil. The authors stated that thus 'a fatal, fulminating, diffuse peritonitis of appendical origin may be uniformly produced in dogs.'

There was no attempt to cure peritonitis, the aim having been merely to cause it. It was reported that the average survival time of the agonized dogs in this experiment was 39 hours.

Arguments to the effect that it is wrong to cause pain and suffering to animals are often rejected by claiming that animals simply do not suffer.

Descartes, for example, asserted that the cries of an animal are no more significant than the creaking of a wheel. Ironically, it is precisely upon the assumption that animals *do* suffer from stress, fear, and pain in a manner similar to humans that the validity of much of animal experimentation rests. Few, if any, conditions are studied as widely in animals as are pain, stress, ulcers, fear, and anxiety. An enormous amount of data has been compiled about how to produce such conditions, a good deal of which arose from the research of Hans Selye (1956). To obtain this data, millions of animals, primarily rats, mice, rabbits, and cats were and continue to be subjected to burns, poisons, shocks, frustrations, muscle and bone crushing, exposure, and gland removal. However, as is generally the case with artificially induced conditions in animals, laboratory-induced stress and stress-related conditions have little in common with stress and stress-related diseases in humans. To support this claim, it is necessary to consider something about how these conditions are induced in the laboratory animal.

One of the tools developed in 1942 to aid Selye in his research on stress, is the stress producing Noble-Collip Drum, named after its inventors, R. L. Noble and J. B. Collip. The animals, locked and strapped (paws taped) inside the revolving metal drum, are tossed about and in so doing are thrust against iron projections in the drum. This treatment (often involving thousands of tosses at a rate of 40 tosses a minute) crushes bones, tissues, and teeth, and ruptures and scrambles organs. In assessing the considerable work of Selye, the *British Medical Journal* (May 22, 1954, p. 1195) concluded that experimentally induced stress had little in common with conditions that humans develop. Ulcers brought about in animals subjected to a rotating drum differ from ulcers in humans, not only because of a difference in species, but because ulcers in humans stem from rather different origins (e.g., long-term psychological stresses) than do those created in the laboratory through physical torture. Unsurprisingly, treatments developed from Selye's stress research (e.g., administering a hormone excreted by tortured animals, ACTH) are of very dubious value. As one surgeon points out (Ogilvie, 1935):

> They [gastric and duodenal ulcers] never occur naturally in animals, and they are hard to reproduce experimentally. They have been so produced, but usually by methods of gross damage that have no relation to any possible causative factor in man; moreover, these experimental ulcers are superficial and heal rapidly, and bear little resemblance to the indurated chronic ulcers we see in our patients.

The Noble-Collip drum has been criticized not only as being inhumane, but as being too crude to be scientifically useful. In what is considered a definitive review of experimentally produced shock, H. B. Stoner made this remark about it. "It is impossible to describe the effect of the injury and study the injured tissue quantitatively . . . . The method seems altogether too crude for modern purposes'' (Stoner, 1961). Despite this, the Noble-Collip Drum is still used in experiments on stress and shock. Typically, these experiments attempt to test the effect of various drugs on the ability of unanesthetized animals to withstand the trauma of the drum.

The most widely used means for bringing about stress, terror, anxiety, and shock is administering electric shocks. A number of researchers are fond of experiments in which animals are trained to avoid electric shocks by performing some task, such as pressing a lever, and thousands of such experiments are performed yearly. After the animal has learned this "shock avoidance," frequently the experimental condition will be changed so that what previously permitted the animal to avoid shock now delivers shock. When such an experiment was performed on rhesus monkeys, it was found to produce "conflict" followed by gastroduodenal lesions. Countless experiments of this sort are repeatedly performed simply to bring about stressful conditions—without any attempt to treat these. When there is an attempt to treat the induced stress-related condition, the results are often useless or meaningless because of the disparity between natural and artificially induced conditions.

Electric shocks are also commonly used to induce aggression in animals, mainly rats. When the restrained animals are given a sufficient number of shocks, they will bite, box, or strike each other. Then the effects of a number of drugs on shock-induced aggression are assessed. In one case, which is typical of such research, the effect of mescaline on rats in a shock-induced aggression situation was tested (Sbordone and Garcia, 1977). Although in some cases the drug appeared to increase aggression, it was also found that the same aggression was shown by some nontreated rats. Hence, attributing the increased aggression to the drug is of questionable validity. Pratt (1976) describes the work of a prominent researcher in aggression studies, Dr. Roger Ulrich:

> Ulrich's work since 1962 . . . has consisted largely in causing pain to rats and observing the resulting aggressive behavior. This investigator would give painful foot-shocks to the rats through an electrified grid floor . . . (p. 61). Ulrich then introduced other distressing stimuli . . . . Bursts of intense noise (135 db, sustained for more than 1 min.) were introduced. The effects of castration were tried; . . . and, finally one pair had their whiskers cut off and were blinded by removal of their eyes (pp. 61–62).

All of this aggression research has done little to control human aggression, which is rather different from the shock-induced aggression in rats. Ulrich himself has very recently come to question the usefulness of his past research. In a letter appearing in the *Monitor* he confesses (Ulrich, 1978):

> Initially my research was prompted by the desire to understand and help solve the problem of human aggression but I later discovered that

the results of my work did not seem to justify its continuance. Instead I
began to wonder if perhaps financial rewards, professional prestige,
the opportunity to travel, etc. were the maintaining factors.

Another condition that electric shocks are employed to induce is epilepsy.
Monkeys are given electric shocks that produce convulsions similar to
those caused by epilepsy and eventually drive them insane. The insane
monkey is then given a variety of drugs with the hope of curing or
controlling epilepsy. However, while the monkeys display behavior that
appears similar to epilepsy in man, their shock-induced fits have little
bearing on human epilepsy, which has rather different origins. Hence, inferences from such experiments to treating human epilepsy are of very
questionable validity. Unsurprisingly, the animal-tested drugs have failed
to cure or control epilepsy.

The development of drugs to prevent brain hemorrhages proceeds in
a similar manner. To evaluate the efficacy of drugs on animals, it is first
required to create blood clots in the brains of test animals. To this end,
their skulls are cracked by hammer blows, causing the brain to form
blood clots. Drugs are then given to the animals to determine which seem
to improve their wretched state. But blood clots from hammer blows are
rather different from those arising in humans, which are a gradual result
of circulation problems or from long-term unhealthy eating and living
habits.

In the interest of studying the effects of certain drugs on overeating
(hyperphagia) researchers tested the drugs on rats that had overeaten
(Wallach et al., 1977). However, in order to induce overeating in these
rats they were forced to eat by painfully pinching their tails. The procedure was described as follows: "Tail pinch was applied for 10 minutes.
Pressure was gradually increased until the animal either ate or became
frantic." However, humans do not overeat because they are under pain of
torture to do so. As such, the effects of drugs on the overeating of animals
forced to overeat are irrelevant for assessing their effect on humans who
overeat.

Underlying all these cases of induction of diseases is the assumption
that by artificially creating a condition in animals that appears to resemble
a pathological state in man, one can make inferences about the latter on
the basis of treating the former. This assumption is often false because of
the disparity between experimentally induced conditions in animals and
the corresponding conditions in humans. Indeed, there is also likely to be
a disparity between artificially produced disease in animals and its natural
occurrence (if it exists) in that animal. An example of this is seen in the
case of inducing a deficiency of vitamin E in mice. Because this deficiency brought about a syndrome similar to muscular dystrophy, it was
theorized that vitamin E would be effective in treating muscular dystrophy in man. In 1961, researchers showed (Loosli, 1967) that hereditary
muscular dystrophy in mice and the dystrophy brought about by a vitamin

E deficiency in mice were fundamentally different. Hence, the inference that vitamin E is useful in treating muscular dystrophy (as opposed to vitamin E deficiency) is invalid not only for humans, but for mice as well. The researchers concluded that it is never certain that experimentally induced disease sufficiently copies an inborn error.

In addition to producing misleading inferences, the techniques of animal research are seen to be detrimental in that they take attention away from more fruitful methods, such as clinical observation of humans. In a 1978 address, Dr. Alice Heim, chairperson of the psychological section of the British Association for the Advancement of Science, remarked (*The Times*, September, 1978):

> Surely it is more valuable to work with disturbed human beings who seek help than to render cats and other animals 'experimentally neurotic'; then try to 'cure' them; and then try to draw an analogy between these animals and the immensely more complex *homo sapiens*.

Differences Between and Within Species

In addition to using animals as "models" for disease, animals are widely used to test the efficiency and safety of drugs and environmental substances by toxicologists and pharmacologists. The problem of differences between species is perhaps greatest in toxicological research. This problem often prevents the valid extrapolation of the results of animal tests to humans.

In experiments carried out to determine how poisonous various chemicals are, the classic measure of toxicity used is called the *median lethal dose*, abbreviated as LD_{50}. It is defined as the dose of a substance needed to kill 50% of a given species of animals. Each year, millions of animals, usually mice, are force-fed drugs, insecticides, floor polishes, food additives, lipsticks, and other chemical substances, and the dosage required to kill about half of them (within 14 days) is calculated. But the significance of the LD_{50} measure is far from clear.

The purpose of the LD_{50} test is to determine the degree to which substances are poisonous to humans. However, for a number of reasons it fails to provide such a determination. For one thing, many of the test substances are relatively innocuous and hence enormous quantities must be forced down the animals' throats to cause them to die. In such cases, the death is often caused simply by the damage done by the massive quantities and not the test substance itself. The calculated LD_{50}s for these substances have no bearing on the manner in which the substance is to be used by humans.

Having found the LD_{50}, experiments with increasingly lower dosages are made to ascertain the supposedly safe dosage of the drug (the LD_0). The next step is to extrapolate this safe dosage to humans. This extrapolation is made simply by multiplying the weight of the animal proportionately to the weight of humans. However, this safe level applies only to the test animal and it may differ radically for humans. This is particularly true when, as is often the case, the animal's death is attributable to the sheer volume of the substance. Extrapolating in this manner is also based on the assumption that the drug acts in a linear fashion—that twice the dosage means twice the effect. In fact, there is typically a threshold below which substances may have no real effect.

It might be thought that despite the differences between test animals and humans, that the LD_{50} may still provide a rank order of the degree of toxicity of substances. Such is not the case. One problem is that a substance with a low LD_{50} may be extremely poisonous over a long period of time, such as lead and asbestos. Hence, the LD_{50} is not useful for ascertaining the result of chronic exposure. In addition, it has been found that in different laboratories not only do the LD_{50} values differ, but the orderings differ as well. This arises from interspecies differences and a number of environmental factors that I shall take up later.

Scientists themselves have come to see the LD_{50} tests as clumsy and crude and lacking in reliability. In one study on four widely used household chemicals, it was found that the LD_{50}s in six laboratories differed both absolutely and relatively. It was concluded that "neither a particular method nor a single value may be regarded as a correct one" (Loosli, 1967, p. 120). Still, in the US, as in most countries, health agencies require that the LD_{50} be calculated for each of the thousands of new substances introduced each year.

Another test promoted by the FDA involves the use of rabbits' eyes to determine how dangerous various substances are to human eyes. In addition to cosmetics, detergents and pesticides are also tested in this way. In 1973, Revlon alone used 1500 rabbits for eye and skin irritancy tests. As is usual for toxicity tests, no anesthesia is used since it is claimed to interfere with the results of the tests. The rabbits are immobilized in restraining devices for weeks, their eyes (which lack tear glands) being held open with clamps. In assessing the severity of eye irritants, the measure is not numerical, as with the LD_{50}. Rather, the eyes of the restrained rabbits are observed after several hours of having the irritant applied, and are described in terms of such categories as ulcerated cornea, inflamed iris, and gross destruction. Such crude determinations yield results that are unreliable and unmeaningful. In 1971, 25 of the best known laboratories jointly conducted a comprehensive evaluation of irritancy tests on rabbits. It was noted that "extreme variation" existed in the way the laboratories assessed the effects of irritants on rabbits.

The study (Weil and Scala, 1971) concluded that:

> The rabbit eye and skin procedure currently recommended by the Federal agencies . . . should not be recommended as standard procedures in any new regulations. Without careful reeducation these tests result in unreliable results.

As this was reported by the very laboratories that carry out such tests, it may be regarded as an understatement. Despite the acknowledged crudity and unreliability of these toxicity tests and measures, they are still routinely carried out, primarily as a means by which manufacturers can protect themselves and obtain the right to market new substances.

Underlying the use of animal models for pharmacological tests is the presumption that it will be possible to extrapolate to humans. However, the vast differences that exist between species make the results from one species an unreliable indicator for another. There are a number of ways that substances can produce different effects in different animals. Chemicals act upon living things in five main stages: absorption, distribution, excretion, metabolism, and mechanism of action, and interspecies differences may arise during any of these. Even if the difference is quite small at each stage, they may accumulate to yield a large total interspecies discrepancy. These interspecies differences make inferences from animal experiments very much dependent upon which animal is used as the research model.

Richard Ryder (1975, p. 150) illustrates the gross differences between species in the effectiveness of drugs by citing the following results. The effect of a 'Product X' was found to vary as follows:

Species	Body weight, mg/kg
Man	1
Sheep	10
Rabbit	200
Monkey	15.2

From these results it is clear that testing the substance on rabbits will be a poor indicator of its effectiveness in humans. Still, rabbits are a popular animal for this sort of testing.

Interspecies differences may lead to concluding that substances that are innocuous or beneficial in humans are harmful, and that substances that have insidious effects on humans are harmless. For example, penicillin is extremely poisonous to guinea pigs. Had penicillin been subjected to the routine animal tests, as new drugs presently are, it would never have been tried on humans. On the other hand, what is a deadly poison to humans, strychnine, may be safely consumed by guinea pigs. Similarly, a dose of belladonna that is fatal to humans is harmless for a rabbit, the often used laboratory animal. Morphine, while sedating most species, incites frantic excitement in cats, dogs, and mice. Arsenic, deadly for humans, can be safely consumed in enormous quantities by sheep. Still, foxes and chickens die from almonds, and parrots are poisoned by parsley. Tuberculin, which, because it cured TB in guinea pigs, was thought to also be a cure in humans, turned out to cause TB in humans.

Digitalis, which because it was seen to produce severely high blood pressure in dogs was thought to be dangerous for humans, turned out to be a major treatment for humans with heart disease.

In certain species, aspirin is highly toxic and it has been seen to produce malformations in the fetuses of rats (*Newsweek,* Nov. 20, 1972). Other substances that are not teratogenic for humans, but are known to cause malformations in the offspring of laboratory animals, are adrenaline, insulin, and certain antibiotics. The converse is the case with drugs such as thalidomide, which was tested extensively in animals (e.g., the popular laboratory Wistar rat) without any adverse effects to the fetus, but which turned out to produce monstrous human babies. The thalidomide tragedy drove people to conclude that more animal testing was needed to protect the safety of humans, when, ironically, the tragedy was a product of animal testing. More specifically, it resulted from the invalidating factor of interspecies differences. After thalidomide was found to produce deformed humans, researchers tried repeatedly to produce similar effects on animals. One hundred fifty different strains and substrains of rabbits were tested, but with no malformations. Finally, it was found that such malformations could be produced in New Zealand white rabbits. One might ask what the point was in carrying out these experiments after thalidomide was already seen to produce malformed human fetuses. The teratogenic effects of a number of other drugs have also been shown not only to vary with species, but also with different strains of the same species.

It is interesting to note that the German drug company that marketed thalidomide, was, after being tried for two and a half years, acquitted on charges of having marketed a dangerous drug. The acquittal was based on the testimonies of numerous medical authorities who claimed that animal tests are never conclusive for making inferences about humans. Despite the failure of animal tests to reveal the danger of thalidomide, Turkish professor S. T. Aygun was able to discover its teratogenic properties through the use of chick embryos, and prevented its being marketed in Turkey. Thalidomide's dangers have also been revealed by testing it on sea-urchin eggs (Krieg, 1964). Much more on alternatives to animal testing may be found in Ruesch, 1978. Unfortunately, even those medical authorities who have come to see animal testing as scientifically unsound are reluctant to voice their views. The following remarks made by two doctors in 1976 express this point (Stiller and Stiller, 1976):

> In praxis all animal experiments are scientifically indefensible, as they lack any scientific validity and reliability in regard to humans. They only serve as an alibi for the drug manufacturers, who hope to protect themselves thereby . . . . But who dares to express doubts of our much-vaunted technological medicine, or even just to ask questions, without meeting the solid opposition from the vested interests of science, business, and also of politics and news media?

Confusion of Background Variables

Invalid inferences arise when what is attributed to the experimental treatment actually arises from the influences of nontreatment variables. Animals even of the same species react differently under the same treatment as a result of nontreatment variables, both of the genetic and environmental kind. Different responses result from differences in the health of the animal, its age, sex, litter, and strain, its living conditions, stressful or painful stimuli, and even odors and the time of day. These variables arise before, during, and after the experimental treatment. I will now consider some ways in which these nontreatment variables may be confused with treatment variables and hence give rise to faulty inferences.

Animals that ultimately become subjects for research may start out with very different characteristics. An experimental response may be the effect not of the experimental treatment in question, but of some former condition of the animal quite unrelated to the treatment. Hence, the background of experimental animals is of major relevance to the reliability of conclusions based upon them—a fact that is often ignored. How are research animals obtained, and how does their background influence the reliability of experiments made upon them?

Animals that find themselves in research laboratories may be of the "specially bred" variety, or they may have arisen from a "random source," that is, from dealers or pounds. In the case of dogs and cats, the great majority come from random sources and are typically stray, unwanted, or stolen pets. For example, of the over 300,000 dogs used in biomedical research in the US in 1969, only about 40,000 were bred specially for the laboratory. By the time a random-source animal reaches the laboratory, an average of one month has been spent either in pounds, with dealers, in transport, or in the wild. Within this period the animal is subjected to poor diet and shelter, and a variety of stressful, unhygienic conditions. In an article in *Life* (February 4, 1966), the outrageous conditions maintained by dog dealers were exposed. The article reports:

> Unscrupulous dog 'dealers,' taking advantage of the growing demand for dogs for vital medical research, are running a lucrative and unsavory business. Laboratories now need almost two million dogs a year. . . . Some dealers keep big inventories of dogs in unspeakably filthy compounds that seem scarcely less appalling than the concentration camps of World War II. Many do not sell directly to labs but simply dispose of their packs at auction where the going rate is 30¢ a pound. Puppies, often drenched in their own vomit, sell for 10¢ apiece.

Unsurprisingly, as one researcher testified, 40% of the dogs obtained this way die before they can even be used for research.

Hence, when such animals get sick or die upon receiving some treatment there is little reason to suppose the treatment was to blame. One researcher, testifying at a congressional hearing following the exposition in *Life,* told of the following case. A drug had been condemned on the basis of an experiment in which all the dogs receiving it died, but it turned out to be distemper and not the drug that was responsible for the deaths (U.S. House, 1966). Nor have conditions improved much since the passage of the Laboratory Animals Welfare Act of 1966, which requires licensing and inspections. The rate of animal deaths prior to experimentation is still high. A more detailed discussion of the humane animal laws may be found in Pratt, 1976 and Ruesch, 1978.

Animals that arrive at the laboratory with infections often render results equivocal. Experiments that are particularly vulnerable to the existence of intercurrent infections are immunology, radiation, and carcinogenesis studies. Carcinogenesis, for example, is affected by the efficiency of immunological responses and rate of cell turnover, and both of these are affected by indigenous pathogens.

To avoid the problem that the variability of the health of research animals presents, means by which animals are bred uniformly free of disease have been developed. Strains of bacteriologically sterile animals, most often rodents, are bred especially for use in the laboratory. These germ-free animals are born by cesarean section, raised in sterile surroundings, and fed sterile foods. The germ-free condition of specially bred animals is advertised as a major selling point in advertisements by suppliers of research animals. One recent ad (*Laboratory Animal Science,* February, 1977) boasts "*New* Cesarean derived, Barrier reared CAMM RATS- Certified Pathogen Free."

However, the resulting "uniform biological material" as these animals are often called, differs radically from normal animals. Animals raised in totally sterile conditions fail to develop immunologies that provide a natural defense mechanism against disease. Hence, they are likely to be far more susceptible to disease than their naturally bred counterparts.

By using specially bred animals that are biologically uniform there is less variability in response, and as such, the results are claimed to be more reliable. It is true that there will be more reliability in the sense of more agreement among successive experiments with uniform as opposed to non-uniform animals. However, there will not be more reliability in the sense of accurately representing natural, random-bred, heterogeneous animal and human populations. For this, uniformly bred animals provide extremely unreliable models. It is true that uniform research animals yield less variability in response, and hence an effect may be detected with fewer animals. However, the effect detected is not a reliable indicator of the effect that would arise in animals found in nature, much less in humans.

The demand for uniformity in research animals, is, according to a prominent researcher, M. W. Fox, often a "pseudo-sophistication." According to Fox, "Few investigators inquire or are aware of how the prior life history and environmental experiences of the animal may influence his experiment" (Fox, 1974b, pp. 96–97). However, a knowledge of how the background of the animal influences the experimental results is necessary to ensure the validity of the experiment. Even when not bred to be germ-free, the laboratory animal may fail to develop the resistance to various stresses that it would in its natural habitat where it would normally be faced with a number of stresses. An example of such a natural stress is brought about when baby animals are left alone while their mothers go in search of food. Hence, the laboratory animal is apt to be less hearty than its natural counterpart.

For the most part, research animals are chosen not on the basis of how appropriate they are for a given experiment—even when such information is available. Rather, they are selected for possessing such characteristics as taking up little space, being inexpensive, being nocturnal, being docile and adaptable to laboratory environments. An advertisement for Marshall beagles (*Laboratory Animal Science,* February, 1977) boasts of having designed their research animal as one might design a car model. The ad reads:

> TRY OUR 1977 MODELS. Large selection of COMPACTS, MID SIZE & FULL SIZE MODELS. The Marshall Beagle is built with you in mind. All models feature sturdy unitized body construction and Easy Handling.

As a result, inappropriate animals are often used for detecting the effect of an experimental treatment. One way in which an animal may be inappropriate is if it tends to spontaneously generate the effect in question even when the experimental treatment is absent (i.e., in control animals).

Rats and mice, for example, tend spontaneously to develop a high incidence of tumors. This renders them unsuitable for detecting tumors. Still, no animals are used as often as rats and mice for assessing the effect of numerous experimental treatments upon the production of tumors. The reason is that they are inexpensive and take up little room. Examples of recent experimental treatments tested this way are saccharin and oral contraceptives. The *British Medical Journal* (October 28, 1972, p. 190) made the following remark concerning the report of the Council on Safety of Medicines on tests of oral contraceptives:

> The tables in the report show incidences of 25% of lung tumours and 17% of liver tumours in *control mice* and 26% adrenal tumours, 30% pituitary tumours, and 99% mammary tumours in *control rats*. It is difficult to see how experiments on strains of animals so exceedingly liable to develop tumours of these various kinds can throw useful light on the carcinogenicity of any compound for man.

In this experiment, rats exposed to high dosages (up to 400 times the human use) of contraceptives were found to have more tumors than those not exposed. But this does not mean that the additional tumors were caused by the contraceptives, particularly given the high rate of tumors in untreated controls. The question of how many more tumors must be observed to conclude that treated rats differ significantly from untreated rats poses an important statistical problem. Given the high rate of tumors in controls, a rather high rate of tumors in a large sample of treated rats should be required. The report, however, does not indicate how many more tumors were observed. The Committee's report itself (Committee on Safety of Medicines, 1972) concludes that:

> . . . although a carcinogenic effect can be produced when some of the preparations are used in high doses throughout the life-span in certain strains of rat and mouse, this evidence cannot be interpreted as constituting a carcinogenic hazard to women when these preparations are used as oral contraceptives.

The report also notes that many animals died from the high dosage of the compound given.

The conclusion of this committee contrasts with the conclusions of the FDA with respect to the banning of DDT and the recently proposed banning of saccharin. In both cases the evidence consists of an increase of tumors observed in rats given extremely high doses of the substance in question. Without additional evidence, conclusions about the danger of these substances are ill-grounded, both because of the high doses and the high spontaneous rate of tumors in rats.

Even if the experimental effect observed does not result from a condition already present in the animal before entering the lab, it may arise from nontreatment factors introduced after entering the lab. Claude Bernard, the founder of modern animal experimentation, himself admitted that "The experimental animal is never in a normal state. The normal state is merely a supposition, an assumption." The animal is placed in an abnormal state because of the host of stressors it is confronted with.

Animals in the lab are deprived of their natural habitat, and often are terrorized by what they see even before they themselves are experimented upon. It is not uncommon for animals, particularly dogs, to be subject to devocalization prior to experimentation (sometimes referred to as "the anesthesia of the public"), which is itself a trauma. Also common is for an animal to be used for additional research after it has already undergone one or more experiments. (This practice, illegal in Britain, is legal in the US.)

When an animal is in fear or pain, all its organs and biochemical systems are affected. Measurement parameters such as blood pressure, temperature, metabolic rate, and enzyme reactions have been found to vary up to three times their normal value as a result of pain and stress

(Hillman, 1970). Hence the value of any of these parameters measured following an experimental treatment may result not from the treatment, but from unrelated pain and stress to which the experimental animal has been subjected. As such, any experiment attempting to detect variations smaller than those already known to be attributable to experimental stress is invalidated. As one physiologist studying the effects of fear and pain notes, "It [experimental stress] is almost certainly the main reason for the wide variation reported among animals upon whom painful experiments have been done" (Hillman, 1970).

Animals may be stressed simply by being handled by the experimenter, or by the order in which they are taken to the experimental room from the cage. When animals are taken one by one from a cage or pen, the last animals taken differ markedly from the first few taken because of the stress brought about from the changing composition of the cage or pen. As one researcher (Magalhaes, 1974, p. 103) states:

> Even apparently minor procedures such as successively removing rats from their colony cage for decapitation and subseqeunt biochemical analysis can have a marked influence on the results. One experimenter found that the corticosteroid measures obtained from the first few rats taken out of their cage were very much lower compared to those from the last few to be taken.

To avoid this problem, animals are sometimes experimented upon in their usual animal facility instead of removing them to the laboratory. But this gives rise to a number of other variables that tend to bias the experimental result, as the following notes (Magalhaes, 1974, p. 104):

> . . . the distress vocalizations, struggling and release of alarm odors during restraint (while taking blood samples, for example) may affect other animals that are to be sampled later, and from such animals 'later down the line' the samples might be qualitatively very different.

Experimental effects are also influenced by the manner in which the experimental treatment is administered. For example, different results are likely to be obtained if a substance is force-fed to an animal by a stomach tube as opposed to having it be eaten naturally. As we have already noted, the response may arise not from the substance administered, but from the large quantity the animal is forced to consume. As the toxicologist Dr. Leo Friedman points out: "We know that administration of *enough* of any substance in high enough dosage will produce some adverse effect" (Friedman, 1970). For example, massive amounts of common table salt cause birth defects in pregnant rats. However, there is no evidence that low doses of salt cause birth defects in humans.

There are a number of less obvious background variables that can arise to confuse the effect of the experimental treatment, and hence lead to invalid inferences. Such variables include the time of day, the temperature of the room, and even the color of the clothing worn by the experimenter. Drug toxicity, for example, has been found to vary with the time of day at which animals are given the drug. Other time-dependent reactions are susceptibility to seizures and irradiation. Animals apparently have certain 24-hour rhythms, including periods of maximum activity and periods of minimum activity (Ader, 1974). Animals react differently to test treatments according to whether they occur during their period of maximum or minimum activity.

For example, it is known that restraining rats causes them to suffer gastric lesions. However, when restraint occurs during the rats' periods of maximum activity, the rats are significantly more susceptible to the development of gastric lesions than those restrained during periods of minimum activity (Ader, 1974, p. 111). Presumably, being restrained during a period of maximum activity is perceived by the rat as being more stressful than being restrained during a period of minimum activity. It would seem likely, then, that rats restrained (as they are) for the introduction of some experimental treatment may show disturbances that are largely the result not of the treatment, but of the time of day. The influence of time also serves to explain some of the variation found between labs. After observing the significant effects of time, one researcher (Ader, 1974, p. 120, emphasis added) was led to conclude:

> . . . it does not seem facetious to ask how many discrepancies in the literature are attributable not to who is right and who is wrong, but to *when* the behavior was sampled.

CONCLUSION

The considerations I have discussed provide a strong argument against a scientific justification of a great deal of animal experimentation; for they have identified numerous sources that often prevent such experiments from being scientifically relevant or valid. Often, an animal experiment is rendered irrelevant because the question it seeks to answer is trivial or obvious, or has already been answered countless times. Invalid experiments frequently arise from the disparity between artificially induced conditions in the laboratory and natural ones, from differences within and among species, and from a number of background variables whose effects are confused with experimental variables.

A few of these problems can be avoided to some extent, for example, by being careful to control a large number of background variables. For the most part, however, these problems exist as limitations in principle to the beneficial use of animals in research. For instance, animals will never be able to produce tumors, ulcers, or a number of other pathological conditions in a manner that is similar to humans. The complex psychological problems humans face are not the sorts of things that can be induced in laboratory animals. The differences in drug toxicity in animals cannot be overcome, and accounting for the vast number of background variables is practically impossible.

Admittedly, it is possible to claim that animal experimentation is justifiable despite the fact that it is of no practical value. Ruesch (1978, p. 144) cites one such attempt made by Professor Leon Asher. According to Asher:

> . . . one might ask oneself whether it isn't a sacred case of conscience to follow the call toward the solution of the mysteries of life, and whether man shouldn't consider it a *religious duty* to satisfy the desire for exploration that Providence has placed in our hearts, without asking whether our research on life has any value for medical science or any other practical value.

However, not all actions that satisfy a "desire for exploration" may be considered permissible (much less a religious duty). Experimenting on humans may also satisfy such a desire, and one is likely to learn much more from humans than from animals. Surely, pointless inquisitiveness does not justify the infliction of pain and suffering, and both humans and animals have the capacity to suffer.

Increasing emphasis on research has also done harm to patient care. As Ryder notes (1975, p. 75):

> A doctor's merit is no longer rated on how many patients he cures but on how many papers he publishes in learned journals. This has encouraged a great deal of trivial research and often there has been a decline in the standards of clinical care.

The employment of human data (which is available in great abundance) is one way to replace a good deal of medical, psychological, and toxicological research on animals with a more scientifically sound alternative.

Other promising alternatives to animal experiments are the use of tissue cultures and mathematical models. Tissue culture involves cultivating living cells outside the organism, and has the advantage of permitting the growth of human as opposed to animal cells. Mathematical models, derived from theory or from past experiments, may be used to predict responses or effects from a variety of experimental conditions that are formulated mathematically. The strongest argument in favor of developing these alternatives is that much of animal experimentation is scientifically, and hence morally, indefensible.

"Animal Liberation" as Crime
The Hawaii Dolphin Case

Gavan Daws

This article discusses a court case in which some ideas about animal liberation and animal rights were subjected to the test of existing criminal law.

Between midnight on 29 May and dawn on 30 May 1977, two captive female Atlantic bottlenose dolphins held under the control of the University of Hawaii's Institute of Marine Biology at the Marine Vertebrate Laboratory of Comparative Psychology, Kewalo Basin, Honolulu, were lifted from their circular concrete isolation tanks, carried on stretchers to a panel van fitted with foam padding, driven to a fishing and surfing beach an hour from the city, and turned loose before sunrise in the Pacific.

The two men who organized the release of the dolphins, Kenneth W. Le Vasseur and Steven C. Sipman, lived on the laboratory premises at Kewalo. They had been students of the laboratory's director, Dr. Louis M. Herman, in his animal psychology classes at the University of Hawaii. Herman, whose research at Kewalo was funded by the National Science Foundation, originally made use of Le Vasseur and Sipman, among many other students, to assist in carrying out behavioral experiments on the two dolphins. One of the dolphins, called Puka, had been in her tank at Kewalo since 1969; before that, she had been for many years an experimental subject in Navy and Navy-sponsored research and operations-related programs. The other dolphin, Kea, also came to Kewalo by way of the Navy, in 1972. (She replaced a dolphin named Nana who had been found dead in her tank one morning, of causes not altogether satisfactorily explained by the autopsy results.) In 1975 Le Vasseur moved into a room on the laboratory premises. Sipman became his immediate neighbor there in 1976. By the start of 1977 both Le Vasseur and Sipman, on the basis of what they saw happening at the laboratory, had become disenchanted with the idea of using dolphins in close

361

captivity for rigorous scientific experimentation, and at the time they released the dolphins it had been many months since they had taken any part in the experimental program; they were doing maintenance work, principally cleaning the dolphins' tanks, in return for free rent.

Once the dolphins were in the ocean, Le Vasseur and Sipman went back to Honolulu and called a press conference to announce what they had done. They realized that the release of the dolphins might appear on the face of things to be criminal. But they did not consider themselves to be criminals. In fact, they took the view that if there was a crime, it was the crime of keeping dolphins—intelligent, highly aware creatures, with no criminal record of their own—in solitary confinement, in small concrete tanks, made to do repetitious experiments, for life.

Le Vasseur and Sipman were not systematic animal rights philosophers. They were students of dolphins whose personal experience at Kewalo led them to become animal rights activists. Neither were Le Vasseur and Sipman systematic thinkers about the law. Indeed they had not so much as consulted an attorney before they released the dolphins. Their view was that the release was morally right, and they believed the force of their act would compel the law to recognize as much. What they had done was intended to be exemplary. They wanted to convince other people—the world—that dolphins should be free. In their view, no matter what rights humans might claim over animals, no matter what assertions scientists might make about the importance of extending human knowledge, nothing was of more importance than the right of the dolphins Puka and Kea—and, by extension, all dolphins—to be free.

Le Vasseur and Sipman, in short, saw themselves as liberators. When the dolphins were taken from the tanks, a message was left behind identifying the releasers as the 'Undersea Railroad,' a reference to the Underground Railroad, the abolitionists' slave-freeing network of pre-Civil War days. Along the Underground Railroad in the 1850s, it sometimes happened that juries refused to convict people charged with smuggling slaves to freedom. That was the kind of vindication Le Vasseur and Sipman were looking for. If they were to be charged with a crime, then—so they believed—the evidence that could be brought forward about the nature of the dolphin would be persuasive that there was no crime, could be no crime, in freeing a captive dolphin.

What happened was the Le Vasseur and Sipman were charged with first-degree theft, a felony. In substance the indictment ran that on or about 29 May 1977 they obtained or exerted unauthorized control over property valued at more than $200 belonging to the University of Hawaii Institute of Marine Biology, to wit, two dolphins, with the intent of depriving the owner of the property.

As a result of legal maneuvering that is not of interest for present purposes, their trials were severed, Le Vasseur's being brought on first. It is with Criminal 50322, State v Le Vasseur, heard in First Circuit Court at

Honolulu before Judge Masato Doi, that this paper is principally concerned.

Prosecuting attorney Stephen B. Tom did not have to address the question of dolphin rights, much less argue a negative case on the issue; he had merely to prove beyond a reasonable doubt that the elements of theft set out in the indictment were satisfied. Le Vasseur had never denied taking the dolphins from the laboratory and putting them in the ocean. He had even agreed to be photographed for a magazine article, between the date of the release and the date of his trial, standing in the water where the release took place, with Sipman and the attorney who took his case. So there was the most public of admissions from the defendant. No one could argue that the dolphins were worth less than $200. There were perhaps some minor uncertainties about where ownership of the dolphins finally rested; with the Institute of Marine Biology, the National Science Foundation, or the Navy. No matter—if Le Vasseur had taken the dolphins with intent to deprive, this was theft. With the possible exception of intent to deprive, the major elements seemed to be incontrovertibly present; Tom could approach the case in a straightforward way.

The defense, perforce, would have to be innovative. This was a test case, but it had not been legally researched and prepared as a test case. Getting ready for trial, Le Vasseur's attorney, John F. Schweigert, canvassed several lines of argument (of varying potential usefulness) in an attempt to develop a workable strategy:

For example, could Le Vasseur's release of the captive dolphins, the message left by the 'Undersea Railroad,' and the press conferences be sustainably presented as actual and symbolic *free speech,* constitutionally guaranteed and protected? The analogy here would be with acts of protest during the Vietnam war, such as the burning of draft cards or unorthodox display of the American flag. There was case law to indicate the possible availability of such a defense, but of course no certainty that it would prevail.

Another approach would be to argue that Le Vasseur was *authorized* to act as he did, at least implicitly. If, for example, the rules and regulations under which work proceeded at the laboratory specified that at all times the well-being of the dolphins was paramount, and their well-being came to depend upon their being released, then in a sense this would constitute authorization.[1]

Again, might it be argued that Le Vasseur acted out of *necessity?* Le Vasseur would testify to his belief that the continued incarceration of the dolphins at Kewalo in bad physical conditions, and their subjection to a demanding experimental regimen, was damaging to their health and well-being and might even result in their death. Though taking them out of this

---

[1]Here Schweigert was looking toward the Federal Laboratory Animal Welfare Act, the Marine Mammal Protection Act, and related regulations.

situation was in a strict sense a criminal act, leaving them would have been the greater evil. This 'choice of evils' defense is not widely reported in case law, but it is contained in the Model Penal Code that has been enacted in a number of states, and was made part of the Hawaii penal code when it was revised during the early 1970s (Hawaii Revised Statutes § 703–302).

Before the choice of evils defense could stand, a number of tests would have to be met. It would have to be shown that the threatened harm to the dolphins was imminent, and that no lawful alternative was open to Le Vasseur. Then it would be for the court to agree that the evil Le Vasseur committed in breaking the law to release the dolphins was less than the evil that would have been perpetuated had he *not* released them.

The defense of necessity or 'choice of evils' emerges from conduct that the actor believes to be necessary to avoid an imminent harm or evil to himself or to 'another.' Who, in this case, might be 'another'? The definition in the Hawaii Penal Code includes under the category of 'another,' where relevant, the United States, the State of Hawaii, other states and their political subdivisions [Hawaii Penal Code § 701–119(8)].[2] So if it could be shown that what was happening to the dolphins in captivity at Kewalo constituted a violation of federal or state statutes or regulations, then that was a harm or evil to 'another,' so defined, and Le Vasseur's release of the dolphins might be seen as a legitimate choice of evils.

The more central definition of 'another' is 'any other person'; and this, of course, was the consideration that bore most closely on questions of animal rights. Could it be sustainably argued that a dolphin by its nature and characteristics deserved to be treated under the law as a *person?* Le Vasseur clearly thought so. Here he was on ground mapped out by those who hold that any useful definition of 'person' framed so as to include *every* human being without exception must at the same time include *some* animals. One way commonly used to illustrate this proposition is to compare—in point of intelligence, communicative ability, and so on—a massively retarded and disabled human infant, incapable of speech, social activity, or even self-preservation, with a physically self-reliant, intelligent and sociable chimpanzee able to use one of the languages developed for two-way human–primate communication. Clearly in this instance, the chimpanzee would qualify as more 'human' than the human infant in every important respect but the purely biological. Le Vasseur would be able, through expert witnesses, to bring evidence of the exceptional size and structure of the dolphin brain; the remarkable communicative abilities of the dolphin; and the sophisticated, apparently mindful individual and social behavior of the dolphin, especially in its relation with humans. On the basis of this sort of evidence, the dolphin would seem to rank with the primates as possessing characteristics that might qualify it as a person for purposes of a choice of evils defense.

[2] See also *State v Marley,* 54 Haw.

If a dolphin could be defined as a person, it could not be defined as property. Le Vasseur could not be convicted of theft. And the legal situation with respect to animal rights, at least the rights of dolphins, would be radically altered.

What, then, happened to these lines of argument in court? Before trial proper began, Judge Doi made rulings that determined the course of everything that followed. He defined dolphins as property, not as persons—as a matter of law. A dolphin could not be 'another person' under the penal code. Accordingly, Doi refused to allow the choice of evils defense: 'This is going to be tried as a theft case, pure and simple,' not 'this choice of evil . . . is a dolphin going to be considered a person? This is not going to be the area of trial at all as far as the court is concerned, and I will rule right now that it is irrelevant . . .'[3]

Schweigert moved to have Doi disqualified for prejudice because he would not even allow an offer of proof on the choice of evils defense. The move failed. Schweigert then asked leave to go to federal court for injunctive relief on the basis that Thirteenth Amendment rights in respect of involuntary servitude might be extended to dolphins. This failed. Judge Doi said: 'We get to dolphins, we get to orangutans, chimpanzees, dogs, cats. I don't know at what level you say intelligence is insufficient to have that animal or thing, or whatever you want to call it, a human being under the penal code. I'm saying that they're not under the penal code, and that's my answer. I may be in error, but that's my legal proposition here at this level.'[4]

During Schweigert's conduct of voir dire of the jury panel, Doi ruled that questions on potential jurors' attitudes to dolphin rights were irrelevant. This provoked Schweigert to move—unsuccessfully—for a mistrial because Doi had foreclosed questioning on the moral issue.

Throughout the subsequent days of testimony, Schweigert kept trying to open the trial up, to get to the issue of rights of dolphins, looking for a way around Doi's rulings on dolphins as property and on the choice of evils defense. This resulted in innumerable bench conferences. (The court reporter, a man of many years' experience, could not recall a trial in which the attorneys and the judge spent so much time in colloquy out of hearing of jury, spectators, and the press.)

Given the unsatisfactory way the courtroom trial took shape from the defense point of view, Schweigert and Le Vasseur had to make a decision about 'trying the case in the media' in order to make their point about rights for dolphins. In a way, this had been the intention of Le Vasseur and Sipman from the beginning: on the very day of the release they had contacted TV, radio, and newspaper reporters in Honolulu. But they had

---

[3]Trial transcript, Criminal 50322, *State v Le Vasseur,* 28 November 1977, p. 14.

[4]Trial transcript, 29 November, 1977, p. 32. Schweigert also moved at this stage to quash the indictment on the ground that the University of Hawaii could not have property rights in dolphins that originated in the wild. This failed.

no well-thought-out strategy for handling the media. Their belief was that their act would speak for itself.

This turned out not to be so. The release of the dolphins attracted a great deal of attention locally, nationally, even internationally. A number of the expert witnesses called by each side to testify on the characteristics of the dolphin got into print on the subject before, during, and after the trial (though the judicial ruling that defined dolphins as property meant that none of the experts took the stand).[5] Assertions and counter-assertions about the actual condition of the dolphins at Kewalo were freely published. And a verbal war of personal reputations was fought in the media between Louis Herman and Le Vasseur and Sipman. On balance, it would be fair to say that opinion in the media ran quite strongly against Le Vasseur and Sipman.

Even this unfavorable turn of events for the defense might, under certain circumstances, have been useful in the courtroom as a means of forestalling a conviction for theft. Again the analogy was with something that happened during the Vietnam war, when defendants in a protest case deliberately hurried to trial while controversy over their act was at its height, then claimed that publicity made a fair trial impossible, and were successful in having their case dismissed.[6] In Le Vasseur's case, though, six months elapsed (to the day) between the release of the dolphins and the opening of the trial, so that such a tactic could scarcely have been used. Then, too, Le Vasseur and Schweigert continued to believe up to the last moment that they would be able to use the courtroom as a forum for issues. When Doi defined dolphins as property, refused to allow the choice of evils defense, and announced that he would be hearing a theft case pure and simple, the defense once more considered attempting to take the issues direct to the media. But Doi gagged Schweigert—if he spoke to reporters, he would be held in contempt of court.

Trial, from the empaneling of the jury to the handing down of the verdict, extended over eight days of court time. Prosecutor Tom's presentation of the state's case was clear, economical, and precise, hewing closely to establishing the elements of theft. Schweigert, in his opening statement for the defense,[7] spoke about the exceptional nature of dolphins as animals; bad and rapidly deteriorating physical conditions at the laboratory; a punishing regimen for the dolphins, involving overwork, reductions in their food rations, the total isolation they endured, deprived of the company of other dolphins, even of contact with humans in the tank, deprived of all toys that they had formerly enjoyed playing with—to the point where Puka, having refused to take part consistently in experimental sessions, developed self-destructive behaviors symptomatic of deep

[5]With one exception, to be discussed later.
[6]Personal communication from attorney E. Cooper Brown.
[7]Trial transcript, 6 December 1977, pp. 2 ff.

disturbance and unhealth, and finally became extremely lethargic, 'coma-tose.' Le Vasseur, seeing this, fearing that death would be the outcome, and knowing there was no law he could turn to, believed himself author-ized, in the interest of the dolphins' well-being, to release them. The re-lease was not a theft in that Le Vasseur did not intend to gain anything for himself. It was intended to highlight conditions at the laboratory. It would not permanently deprive Louis Herman and the Institute of Marine Biol-ogy of the use of the dolphins—Schweigert proposed to show that Puka and Kea could be recovered from the ocean, and that they were in fact alive to be recovered.[8]

Once again Schweigert was talking in terms of choice of evils. Doi called him to the bench to present his offer of proof on this and other pro-posed defenses, asking him to specify what his witnesses would speak to. Schweigert listed the following topics: conditions at the laboratory; evi-dence from experts as to self-damaging, self-destructive behavior in cap-tive dolphins; questions of the true ownership of the dolphins; Le Vasseur's status as a state employee, which conferred authority on him to act as he did in coming to the defense of 'another,' in this case the United States, whose social values were injured by what was being done to the dolphins; the humane way in which the release was effected; the release as symbolic speech, a way to speak and be heard; the lack of lawful alter-natives; the necessity to act, from which emerged the choice of evils de-fense.

Doi rejected in general testimony about conditions at the laboratory as irrelevant to the matter of theft, and cautioned Schweigert in passing that if he pressed the point about bad conditions he might in fact be reinforcing the idea that Le Vasseur's intent was to deprive. Symbolic speech was ruled irrelevant. Doi also ruled that choice of evils was not established in Schweigert's offer of proof. As for the notion of harm to 'another,' meaning the United States, Le Vasseur should have gone to a United States district attorney and said: Herman is violating the law (by harming dolphins)—prosecute him. Doi ruled in sum that he was re-jecting all evidence except what related to the taking, transportation, and release of the dolphins; the question of the defendant's intent to deprive; and the question of implied authorization.

On the point of intent to deprive, Schweigert was able to bring an expert witness to the stand—the only one to testify on either side. This was Michael Greenwood, an ex-CIA and Navy scientist. Greenwood had worked on developing military systems involving dolphins at the Naval Undersea Warfare Center at Kaneohe, Hawaii (where Louis Herman had

---

[8]Louis Herman made effective use of the media to counter every one of those asser-tions, as they bore upon his conduct of the laboratory. The issue of the survival or death of the dolphins after their release was, and continues to be, one that arouses the strongest of emotions. It is a complex issue, deserving of extended treatment—which it did not get in the courtroom: the life or death of 'property' was not relevant to a theft charge.

done research work on Navy contract, where Puka had undergone training for a long period, and where Kea had been briefly before being
transferred, like Puka, as inventory from the Navy to the laboratory at
Kewalo). Schweigert was permitted to question Greenwood only on
points relevant to the intent of Le Vasseur to deprive the dolphins' owners
of their property. Greenwood's testimony[9] had to do with the demonstrated ability of trainers at the Naval Undersea Warfare Center to condition captive dolphins released in the open ocean to return to a given location on command, the command being the sound of an underwater pinger.
The point the defense was attempting to make was that Navy-trained dolphins such as Puka and Kea were not unrecoverable from the ocean; Le
Vasseur knew this; thus he did not have the intent to deprive the owners
of the dolphins of their property. (On many counts this was a strained
argument. As Tom pointed out at one stage, if someone steals a car it is
no less a theft if the car is later left where the owner can recover it.)

Le Vasseur took the stand in his own defense.[10] Doi had ruled that
he could testify to 'all the things he observed and made him believe that
he had to do what he did.'[11] Within the constraints of a theft trial, this
gave Le Vasseur and Schweigert considerable latitude, and Le Vasseur
stayed on the stand most of one day, going into minute detail about his
observations of deteriorating conditions at the laboratory; increased stress
on the dolphins, particularly Puka; Puka's increasingly worrying
behavior—erratic participation in experiments, interspersed with periods
of total noncooperation, neurotic jaw-snapping, tail-slapping, coughing
for hours at a time at night, beating her head against experimental appara-
tus and drawing blood. Schweigert took Le Vasseur at length through the
decision to release the dolphins, the planning for the release (which took
weeks to organize), and the way the release was carried out.

Testimony completed, the attorneys for state and defense submitted
their proposed instructions to the jury. None of the substantive instructions proposed by Schweigert were used. In other words, he had made no
impression on Judge Doi with his arguments about the theory of the case.
The instructions to the jury were basically those of the prosecution, of a
sort that might have been given in any theft case where the elements of
theft had been well established.

After eight days of hearing testimony and argument, the jury took
less than an hour to convict Le Vasseur of the felony of first-degree theft.

Schweigert moved for a new trial. He alleged that Doi, by curtailing
the defense's voir dire of the jury, had violated due process and violated
Le Vasseur's constitutional rights; by failing to allow testimony regarding
conditions at Kewalo, he had deprived Le Vasseur of the ability to dem-

[9]Trial transcript, 2 December 1977, pp. 163 ff.
[10]Trial transcript, 7 December 1977, pp. 3–195.
[11]Trial transcript, 6 December 1977, p. 42.

onstrate to the jury his state of mind; and he had deprived Le Vasseur of the choice of evils defense, in violation of his Fifth Amendment rights. Schweigert placed most importance on the argument that Doi had committed reversible error by instructing the jury as a matter of law that dolphins were property. Schweigert read the relevant statute [Hawaii Penal Code § 708–800(15)] as laying down factual criteria for property; it followed that the prosecution was required to adduce evidence to substantiate that a dolphin was property. 'This being *a question of fact* for the jury to resolve, it constituted an essential element of the crime for which the Defendant was tried. Accordingly it was error for the Court to instruct the jury that a dolphin was "property" '.[12]

Tom, opposing the motion for a new trial, [13] argued that on the procedural points raised by Schweigert, Doi had acted reasonably. On the question of marine animals as property, courts across the country had made this ruling as a matter of law. On choice of evils, Tom argued that there was no harm imminent to the dolphins at the time Le Vasseur took them, and that in any case, dolphins as property were properly excluded from the definition of 'another.' Finally, if Le Vasseur truly believed that the dolphins were being abused and mistreated, and that releasing them was the only remedy, then he would be acting to relieve cruelty to animals, and under the penal code cruelty to animals was a misdemeanor. But by releasing the dolphins he committed a felony. So he was reversing the logic of the choice of evils defense—relieving a lesser evil by committing a greater.

The motion for a new trial was denied.

Le Vasseur could be sentenced to anything between a period of probation and five years' jail. In arriving at sentence, Doi spoke of the necessity of deterrence, the defendant as an individual being submerged in the 'overall picture' because of 'larger considerations that came into play: the seriousness of the offence involved,' and the 'deterrent or nondeterrent effect of any sentence that may be imposed on him.'[14] Le Vasseur had no criminal record of any kind. On the other hand, 'release of the dolphins was an act which caused irreparable injury to many who were directly involved in dolphin-research and learning which caused great monetary loss and which may have . . . sentenced the dolphins, themselves, to death in . . . unfamiliar waters.' In sum, said Doi, 'there is nothing in the Defendant's personal background which indicates that he should receive harsh treatment. On the other hand, even assuming that he had good motives in doing what he did, the seriousness of his conduct in taking vigilante action cannot be condoned nor encouraged by this Court.'[15]

[12]Criminal 50322, Motion For New Trial, 19 December, 1977, p. 1.
[13]Criminal 50322, Memorandum of Points and Authorities in Opposition to Motion for New Trial, 10 January 1978.
[14]Trial transcript, 2 February 1978, p. 21.
[15]Trial transcript, 7 February 1978, pp. 75–76.

Doi sentenced Le Vasseur to five years' probation, as a special condition of which he was to spend six months in jail.

Schweigert moved for a reconsideration of sentence. This was denied; Doi said that he had been inclined to give Le Vasseur five years' jail. Schweigert gave notice of appeal. Le Vasseur was freed on his own recognizance pending the outcome of the appeal.

Schweigert's opening appeal brief[16] referred to errors he alleged Judge Doi had committed in the conduct of the trial, among them: advancement of the trial date, thus depriving le Vasseur of his right to effective counsel; failure to allow adequate voir dire of all jurors; failure to allow introduction of exculpatory evidence as to authorization and as to the state of mind of the defendant; failure to present the defense case affirmatively to the jury in the judge's instructions. And once again Schweigert returned to the choice of evils defense, arguing that Doi erred in not allowing it. (It is a curious fact that Doi, as a circuit court judge, has been reversed by the Hawaii Supreme Court only twice, both times in cases where choice of evils was advanced as a defense and disallowed by Doi.)

The appeal process took more than two and a half years to conclude. Appeal in the first instance was to the Supreme Court of the State of Hawaii; but between filing and hearing a new Intermediate Court of Appeals was established, and it was there that the appeal was heard, in May, 1980.

The court found none of Schweigert's arguments persuasive. On the all-important question of whether a dolphin might qualify as 'another' for a choice of evils defense, the court ruled negatively. And as for the idea that Le Vasseur's action protected the United States as 'another' from the 'evil' of keeping dolphins in captivity at Kewalo, the appeal court agreed with the criminal court, holding that "appellant's action in removing the dolphins from their tanks . . . and . . . releasing them into the ocean, thereby committing the crime of theft, was at least as great an evil as a matter of law as that sought to be prevented."[17]

Schweigert moved for reconsideration; the motion was denied without comment. Subsequently, the Supreme Court of the State of Hawaii declined to hear the appeal and so, in turn, did the United States Supreme Court. Le Vasseur's conviction stood, and, with it, the ruling that as a matter of law, dolphins are property.

The question of sentencing remained. Le Vasseur had been sentenced to six months' jail as part of five years' probation. The Intermediate Court of Appeals considered this not excessive, but pointed out that

---

[16]Criminal 50322, Appellant's Opening Brief, 15 January 1979 (Hawaii Supreme Court 6930, October Term 1977).

[17]Intermediate Court of Appeals of the State of Hawaii, No. 6930, *State v. Le Vasseur*, Appeal from First Circuit Court, 27 June, 1980, Opinion of the Court by Padgett, J., p. 9.

the sentence could be reduced if the criminal court saw fit to do so. In August, 1980, Judge Harold Y. Shintaku, having heard from both prosecution and defense, reduced the sentence so that Le Vasseur, as part of his probation, would serve no time in jail, but would instead be required to do four hundred hours of community service.

This outcome could be seen as less than satisfactory for defense and prosecution alike. The defense failed to make its point—that, for legal purposes, dolphins should not be regarded as property but should be assigned 'person-like' rights. The prosecution got a conviction that was upheld on appeal, but failed, at final sentencing, to make its point—that 'vigilante' action to release captive dolphins was a serious matter, deserving of jail.

Both sides had a second chance to make their case. The trial of Le Vasseur's codefendant, Steven C. Sipman, was eventually brought on at the end of November, 1981, four years almost to the day after Le Vasseur's trial. The principal argument of the defense went to the question of *intent:* Sipman's state of mind was such that he did not consider dolphins to be property; thus in removing the dolphins from the tanks at Kewalo he could not have intended to commit the crime of theft, one element of which is depriving an owner of property. Sipman was able to testify at length, as Le Vasseur had been able to; but neither judge nor jury was persuaded by the defense argument, and Sipman was found guilty of first degree theft, as Le Vasseur had been. Interestingly, some members of the jury remarked afterwards that they sympathized with Sipman, and said that in his place they might very well have acted as he did; but the law was the law, and under the law they had to find him guilty. In their discussions, they even mentioned the analogy between dolphins in capitivity as property and slaves in Southern states as property, and concluded that however reprehensible slavery might have been, they would have had to convict a slave-liberator just as they had convicted Sipman—obeying the provisions of existing law.

In pre-Civil War days, juries sometimes refused to convict those who took it on themselves to liberate slaves by removing them from the control of their owners. This came to be known as "jury nullification" of what was seen to be bad law. This doctrine was discussed again at the time of trials of anti-Vietnam protestors in the 1960s, but has not been brought to bear in any significant way in the courtroom.

Since Le Vasseur and Sipman released Puka and Kea from Kewalo, there have been other instances of "vigilante" action to release captive marine mammals in the United States and in Japan. And the animal rights or animal liberation movement continues to search for the best legal forum for presenting its case.

Fighting for Animal Rights

Issues and Strategies

Henry Spira

Animal liberation is also human liberation. Animal liberationists care about the quality of life for all. We recognize our kinship with all feeling beings. We identify with the powerless and the vulnerable—the victims, all those dominated, oppressed, and exploited. And it is the nonhuman animals whose suffering is the most intense, widespread, expanding, systematic, and socially sanctioned of all. What can be done? What are the patterns underlying effective social struggles?

All experience indicates that struggles for justice are not won, nor are significant reforms achieved, by politely limiting oneself to rules laid down by the power structure. The reason is obvious. The powerful do not design rules that encourage outsiders to take away or share their power. In addition, those who make rules can unmake and circumvent their own rules to suit their convenience and interests. As long as one remains boxed into their game, those who run the power system know they need not take one seriously and therefore will pay little attention to pleas for change.

Power concedes nothing without effective struggle. It was only after a determined effort by a well organized group of the physically handicapped that bureaucratic inertia was broken. They occupied the Health, Education and Welfare (HEW) offices, conducted an all night vigil at the home of the HEW Secretary, and demonstrated at a presidential function. Within ten days, a do-nothing policy was forced to give way to new regulations forbidding discrimination against the physically handicapped. An organizer commented, "We won because we had strong political connections, and mostly because we took matters into our own hands."

And only after the women's movement became mass action oriented, fighting its way into the political arena, prepared to make it costly to those who would keep them out, did women win any political and economic concessions.

The meek don't make it. But audacity must be fused with meticulous attention to program, strategy, and detail.

TARGETING

Before taking an action, we must consider what our priorities are, what we are most concerned about. The priorities of animal liberationists should be based on the number of victims, the intensity of their suffering, and our ability to effect change. The main bases of animal slavery, as analyzed in Peter Singer's *Animal Liberation,* are exploitation for food and as "lab tools."

We are surrounded by systems of oppression, and they are all related. But in order to influence the course of events, we must sharply focus on a single significant injustice, on one clearly limited goal at a time.

One criterion for choosing a target is that the mere statement of the issue tends to place the adversary on the defensive: Is another shampoo worth blinding rabbits to you? Do you want your tax monies spent to mutilate cats deliberately in order to observe the sexual performance of crippled felines?

And the goal must be winnable, with the possibility of expanding ripple effects. Success is then used as a stepping stone toward still larger struggles and more significant victories.

The animal research industry cloaks itself in the noble cause of protecting health and saving lives. We wanted to spotlight the lack of real scientific payoff in make-work, tax-supported bizarre horrors. At the American Museum of Natural History in New York, there was a 20-year history of deliberately mutilating cats and kittens in order to observe their sexual performance that we believed needed to be stopped. The public came to perceive the Museum experiments as a gross, grotesque, crude, useless perversion of science. We created a broad base of support, and in the process, helped change the ideological climate so that animal experimenters can be held accountable to an increasingly critical public. The campaign began in June 1976. By December 1977, the American Museum of Natural History's cat-sex experiments had stopped and the labs were dismantled.

WE CAN HAVE IMPACT

Basically the strategy in all struggles for freedom is similar. Generally, the other side has all the power, and the oppressed have only justice and the capacity to mobilize people on their side. To succeed, we need much expertise and credibility, and must carefully work out partial, short-term goals to reach people and effect the changes we desire.

Through meticulous preparation, a small group can release an enormous amount of energy. After all, the power structure has problems and weaknesses that render it susceptible to successful attack, and these offer us the needed openings. The task is to find these openings and to move in boldly.

To create bridges connecting us with the public's current awareness, we must first check an issue out with a wide variety of people, and listen carefully to their responses. How do they react? Can they feel themselves in the place of the victim? Are they outraged? What in particular do they focus on? Will this action propel their consciousness forward?

THE CAMPAIGN

We need personally to research and analyze all available verifiable materials, to find patterns and connections, and to see the vulnerable points that offer us maximum leverage.

We provide a framework for people to organize themselves through articles and activity sheets that must be clear, logical, and defensible in public debate.

Our current campaign concerns the Draize Test: the routine use of the eyes of unanesthetized rabbits to test hazardous substances. This test is vividly painful. We all know what it feels like to get a little soap in our eye. In addition, it is also unreliable, with extreme variation in the results. It could be replaced. And finally, the goal is trivial and frivolous: blinding rabbits for the sake of yet another mascara, yet another oven cleaner.

We went as far as we could within the system. For 18 months prior to our public campaign, we sought a meaningful dialog leading to a collaborative approach. We presented Revlon with scientific suggestions on alternatives. It was only after Revlon refused to be responsive to our concerns that we launched demonstrations and a boycott of Revlon products.

COALITION POLITICS

We organized a single-issue coalition that now includes over 400 different organizations, with constituencies in the millions, banded together to abolish the Draize Test. When enough of us band together around a single issue, the tide can be turned.

A coalition can assemble different elements to maximize pressure and destabilize a target by approaching it from every direction on every front. It is an orchestrated, purposeful action.

Participation should be possible at whatever level allows organizations and individuals to feel comfortable. Some will boycott, demonstrate, and march, others will not. Organizations can get media publicity. The Millennium Guild placed full page ads in major papers and thereby created a turning point. Individuals can write letters to the editor where one can reach millions without spending a penny.

Within nine months of publicly launching our anti-Draize Test campaign, Revlon pioneered the funding for non-animal alternatives with a $750,000 Rockefeller University project, followed by Avon, Estee Lauder, and Bristol-Myers. Within one year, over $2,500,000 was committed by multibillion dollar corporations towards phasing animals out of the laboratories, and the Johns Hopkins Center for Alternatives to Animal Testing was established.

Bills were introduced in the Senate and House encouraging government agencies to promote alternatives to Draize testing, and the issue was spotlighted on network TV and featured in science-oriented publications. Ideas, combined with clout, sensitized much of the research community, industry, and government to our concern.

ACCENTUATING THE POSITIVE

We have to work with people's perceptions. Whether they are true or not, we cannot ignore them. They are part of the political reality. One perception is that those who challenge live animal research are anti-intellectual, anti-science sentimentalists. But, we accentuated the positive. We offered realistic options that, from the scientists' own perspective, are more productive.

We had suggested earlier that the American Museum of Natural History's death and pain-dealing methods of research were turning curious and sensitive youth away from biology; but that elegant and humane research would inspire and intrigue imaginative youth with its creative beauty and challenge.

And in our current anti-Draize Test campaign, we have urged industry to unleash its scientific creativity to develop effective and reliable non-animal tests.

EFFECTIVE TACTICS FOR CATALYSTS

Do not start off being personally hostile to your potential adversary. Suggest reasonable options, realistic collaborative approaches. Such an approach might work, and if it does not, you are still ahead. You have gained extra credibility for having tried, and have very likely blunted in advance much of the criticism sure to result when you then move from words to action.

Thus, we need always to emphasize that there is a better way. That non-intrusive science is more imaginative, more elegant, and likely to produce more relevant data. And that scientists need to initiate productive new departures in biology for ethical and scientific reasons.

By tackling one issue at a time, we can isolate our opponent rather than ourselves. We are not confronting a monolith. Some animal researchers have scruples against painful experiments that have no serious application, and may join their voices with ours.

And our adversaries can also be split. Thus, the chief Museum cat-sex experimenter publicly bemoaned being abandoned by his peers. They did not want to be identified with a grotesque, indefensible experiment while it was being spotlighted, though their own research efforts might be similar.

No congressional bill, no legal gimmickry, by itself, will save the animals. The courts can, at best, open up the possibility for us to intervene in defense of animals. But, the courts will not act until effective protests disrupt the system's orderly operation. Laws function to maintain and justify the status quo. In movement-related issues, laws are changed to keep disturbances at a minimum. And the legal process often deflects struggles into parliamentary gabbery and inaction. The legal front is no substitute for action. It must be organically connected with mass struggles to produce social change.

THE ROAR OF MANY WATERS

Society programs us into inconsistency, into being kind to household pets while other animals suffer from birth to death. And society also programs us into not focusing on the true levers of power. We therefore need to study the realities consciously, in a detached way, as a guide to action. Who is profiting and who is calling the tune? And how does the rest of the world perceive our concerns?

The majority of people would certainly prefer that animals not suffer, were the matter brought to their attention. Yet the intense pain of billions of animal victims continues unabated.

To fight successfully for the rights of animals, we need priorities, programs, effective organization, imagination, tenacity, expertise, and a good sense of strategy and tactics as we create bridges with the public's current awareness and move forward.

And, we need to remember the words of the Abolitionist leader Frederick Douglas—''If there is no struggle, there is no progress. Those who profess to favour freedom, and yet deprecate agitation, are people who want rain without thunder and lightning. They want the ocean without the roar of its many waters. Power concedes nothing without a demand. It never did and it never will.''

Epilog

The papers in this volume connect with one another in many different ways, and therefore many alternate groupings are possible. One might group together the contributions of Fox, Scanlon, Wenz, and Tideman on the importance of the ecosystem-wide perspective. Or one might put into a single section those papers dealing with the variability of moral and legal standing (including, at least, those of Baier, Rachels, and Rollin). A section on rights might include the papers of Fox, Regan, and Rollin, in addition to those of Frey and Rachels. A section on utilitarianism would contain the papers of Cargile, Gruzalski, and Narveson, and perhaps others. Yet another plausible grouping, on the subject of sympathy and its limits, would include Baier, Buchanan, Clark, Fox, Tideman, and perhaps Daws, Gross, and Spira.

There remain at least three other themes that might provide illuminating principles of organization for this collection. One is what Gruzalski calls the 'animal husbandry' argument and what Henry Salt (1914) called the 'logic of the larder.' The crux of these arguments is the claim that the present situation is better for animals than that which would otherwise exist. Both Salt and Narveson express great reluctance to compare any sort of existence with sheer nonbeing, but such a comparison seems required by the argument. Cargile compares the actual (short) life of the particular pigs he buys with the (no longer) life those animals would have had had he not bought them. Gross's paper, though clearly relevant to this issue, makes no such explicit comparison.

Another theme is the relation between the good of individual animals and the good of the larger wholes of which the individuals are parts, such as species or Nature. Several papers, among them those by Clark, Gross, VanDeVeer, and Wenz, address aspects of this relation.

One last possible organizing principle, perhaps the most inclusive of all, is the tension between the actual, the possible, and the natural. The importance of the distinction between what is natural and what is possible is made clear by Rumbaugh and Savage-Rumbaugh. Baier, Becker, Buchanan, Clark, and Scanlon all draw conclusions (different conclusions) from considerations of the nature of our species. Benson, Fox, Gruzalski, and Wenz argue that it is possible for us to change what some would claim is our nature; and Gross is engaged in examining the possible

379

limits of the nature of the chicken species. Clearly we cannot simply equate the good and the natural; for racism is probably natural and tuberculosis certainly is. But neglecting the natural is very dangerous prudentially, intellectually, and morally. If we humans, or some of us, hope to become (even approximately) excellent, it is not excellent books or trees or stones that we wish to be, but excellent humans. It is human excellence, the full development of the natural potentiality of humanity, that we need to understand. That some of the things we naturally do (such as overeating) are contrary to our natural excellence is only superficially paradoxical.

Knowing how we should treat other animals is part of knowing how we should live. We already know some relevant factual truths, some relevant moral truths and some relevant truths that cannot be forced into either of those old pigeonholes. But we do not know enough. The authors in this collection cannot (as a matter of logic) all be right. In fact, probably most of them are wrong. Yet both individually and collectively they advance our knowledge by sharpening questions and by propounding and criticizing arguments. It is progress to become clearer about the boundaries of our ignorance.

Works Cited

A NOTE ON BIBLIOGRAPHIES

A very full bibliography of the English-language literature on human treatment of nonhuman animals has been prepared by Charles Magel. The title is *A Bibliography of Animal Rights and Related Matters,* and it was published by University Press of America in 1981. A substantially shorter bibliography, also prepared by Magel, is appended to the 1980 edition of Henry Salt's *Animals' Rights,* published in the US by the Society for Animal Rights and distributed in the United Kingdom by Centaur Press. The special issue of the journal *Inquiry* (Oslo: Universitetsforlaget) for Summer, 1979, contained a select bibliography jointly compiled by Charles Magel and Tom Regan.

REFERENCES

Adams, R. 1972. *Watership down.* New York: Macmillan.

——. 1978. *Plague dogs.* New York: A. A. Knopf.

Ader, R. 1974. Environmental variables in animal experimentation: the relevance of 24-hour rhythms in the study of animal behavior. In Magalhaes, 1974.

Alexander, S. 1939. The mind of a dog. In *Philosophical and literary pieces.* London: Macmillan. pp. 97–115.

Allport, G. W. 1958. *The nature of prejudice.* Garden City, NY: Doubleday Anchor.

Altman, N. 1973. *Eating for life: a book about vegetarianism.* London: Theosophical Publishing House.

Amory, C. 1974. *Man kind? our incredible war on wildlife.* New York: Harper & Row.

Anderson, K. V., Pearl, G. S., and Honeycutt, C. 1976. Behavioral evidence showing the predominance of diffuse pain stimuli over discrete stimuli in influencing perception. *Journal of Neuroscience Research* 2: 283–289.

Aries, P. 1962. *Centuries of childhood.* New York: Vintage Books.

Aristotle: references to Aristotle are to the standard Bekker pagination.

Arrow, K. J. 1965. *Social choice and individual values.* 2nd Edn. New Haven and London: Yale University Press.

Augustine, works cited as: CD: *Civitas Dei* (City of God) and DMM: *De Moribus Manichaeorum*

Austin, P. 1885. *Our duty to animals.* London: Kegan Paul & Co.

Baier, A. 1979a. Master passions. In *Explaining emotions,* ed. A. Rorty. Berkeley: University of California Press.

381

——. 1979b. Hume on heaps and bundles. *American Philosophical Quarterly* 16: 285–295.

——. 1981. The rights of past and future persons. In *Responsibilities to future generations: environmental ethics,* ed. E. Partridge. Buffalo: Prometheus Books.

Barrett, J. E. and Witkin, J. M. 1976. Interaction of D-Amphetamine with Pentobarbital and Chlordiazepoxide: effects on punished and unpunished behavior of pigeons. *Pharmacology, Biochemistry, and Behavior* 5: 285–292.

Barrett, W. F. 1926. *Death-bed visions.* London: Methuen.

Becker, L. C. 1975. The neglect of virtue. *Ethics* 85: 110–125.

Bennett, J. 1978. On maximizing happiness. In *Obligations to future generations,* ed. R. Sikora and B. Barry. Philadelphia: Temple University Press.

Bentham, J. 1789. *Introduction to the principles of morals and legislation.*

Berdyaev, N. 1934. Man and machine. In *The bourgeois mind and other essays.* New York: Books for Libraries Press.

Berger, J. 1979. *About looking.* New York: Pantheon Books.

Bergstrom, L. 1966. *The alternatives and consequences of actions.* Stockholm: Almqvist and Wiksell.

Berkson, G. and Becker, J. D. 1975. Facial expressions and social responsiveness of blind monkeys. *Journal of Abnormal Psychology* 84: 519–523.

Berry, W. 1978. *The unsettling of America: culture and agriculture.* San Francisco: Sierra Club Books.

Bider, J. R. 1979. Implications of blackbird and starling control in North America. Working document prepared for presentation at the 44th North American Wildlife and Natural Resources Conference, Toronto.

Bishop, R. C. 1978. Buck economics. *Wisconsin Natural Resources* 2(6): 10–11.

Blake, W. 1966. The marriage of heaven and hell. In *Blake: complete writings,* ed. G. Keynes. Oxford: Oxford University Press.

Bohaman, L. and Bohaman, P. 1953. *The Tiv of central Nigeria.* London: International African Institute.

Bradley, E. W., Zook, B. C., Casarett, G. W., Bondelid, R. O., Maier, J. G., and Rogers, C. C. 1977. Effects of fast neutrons on rabbits. *International Journal of Radiation, Oncology, Biology and Physics* 2: 1133–1139.

Bradley, F. H. 1897. *Appearance and reality: a metaphysical essay.* Second Edition. Oxford: Oxford University Press.

Brandt, R. B. 1979. *A theory of the good and the right.* Oxford: Clarendon Press.

Broadie, A. and Pybus, E. 1974. Kant's treatment of animals. *Philosophy* 49: 375–383.

——. 1978. Kant and the maltreatment of animals. *Philosophy* 53: 560–561.

Broome, J. 1978. Trying to value a life. *Journal of Public Economics* 9: 91–100.

Bruce, A. B. 1899. *The moral order of the world in ancient and modern thought.* London: Hodder & Stoughton.

Buchanan, J. M. 1978. Markets, states, and the extent of morals. *American Economic Review* 68: 364–368.

Capra, F. 1977. *The tao of physics.* New York: Bantam Books.

Carrier, L. S. 1975. Abortion and the right to life. *Social Theory and Practice* 3: 381–401.

Carroll, L. 1872. *Through the looking-glass.*

Clark, K. 1977. *Animals and men.* New York: William Morrow and Co.

Clark, S. R. L. 1975. *Aristotle's man.* Oxford: Oxford University Press.

——. 1976. God, good and evil. *Proceedings of the Aristotelian Society* 77: 247–264.

——. 1977. *The moral status of animals.* New York: Oxford University Press.

——. 1979. The rights of wild things. *Inquiry* 22: 171–188.

(U.K.) Committee on Safety of Medicines. 1972. *Carcinogenicity tests of oral contraceptives.* London: H.M.S.O.

Dallery, C. 1978. Thinking and being with beasts. In Morris and Fox, 1978. pp. 70–92.

Darling, F. F. 1970. *Wilderness and plenty*. Boston: Houghton Mifflin.

Darwin, C. R. 1859. *The origin of species*. Everyman's Library edition 1928.

———. 1871. *The descent of man, and selection in relation to sex.*

———. 1872. *The expression of the emotions in man and animals*. University of Chicago Press edition, 1965.

Dawkins, R. 1976. *The selfish gene*. Oxford: Clarendon Press.

Descartes, R. 1970. *Philosophical letters*. Oxford: Oxford University Press.

Devine, P. E. 1978. *The ethics of homicide*. Ithaca, N.Y.: Cornell University Press.

Diamond, C. 1978. Eating meat and eating people. *Philosophy* 53: 465–479.

Diner, J. 1979. *Physical and mental suffering of experimental animals: a review of scientific literature 1975–1978*. Washington: Animal Welfare Institute.

Domalain, J-Y. 1977. *The animal connection: the confessions of an ex-wild animal trafficker*. New York: William Morrow.

Donne, J. 1610. To Sir Edward Herbert at Julyers. (poem)

Dover, K. J. 1975. *Greek popular morality in the time of Plato and Aristotle*. Oxford: Basil Blackwell and Berkeley: University of California Press.

Dowding, Lady M. 1971. Furs and cosmetics: too high a price? In Godlovitch and Harris, 1971.

Dubos, R. J. 1972. *A god within*. New York: Scribner.

Dworkin, R. 1977. *Taking rights seriously*. Cambridge, Mass.: Harvard University Press.

Ehrlich, P. R., Ehrlich, A. H. and Holdren, J. P. 1973. *Human ecology: problems and solutions*. San Francisco: W. H. Freeman and Co.

Eliade, M. 1961. *The sacred and the profane*. Trans. by W. R. Trask. New York: Harper & Row.

Epictetus, works cited as: D: *Discourses,* Loeb edition trans. W. A. Oldfather and E: Encheiridion.

Fox, M. W. 1974a. *Concepts in ethology: animal and human behavior*. Minneapolis: University of Minnesota Press.

———. 1974b. Space and social distance in the ecology of laboratory animals. In Magalhaes, 1974, pp. 96–105.

———. 1976. *Between animal and man*. New York: Coward, McCann and Geoghegan.

———. 1978. Man and nature: biological perspectives. In Morris and Fox, 1978. pp. 111–127.

———. 1980a. *Toward the new Eden: animal rights and human liberation*. New York: Viking.

———. 1980b. *Farm animal science, ethology and human ethics*. m/s in press.

———. 1980c. *One earth, one mind*. New York: Coward, McCann & Geoghegan.

Frankfurt, H. G. 1971. Freedom of the will and the concept of a person. *Journal of Philosophy* 68: 5–20.

Freud, S. 1963. Reflections upon war and death. In *Character and culture*. New York: Collier Books.

Frey, R. G. 1980. *Interests and rights, the case against animals*. Oxford: Clarendon Press.

Friedman, L. 1970. Symposium on the evaluation of the safety of food additives and chemical residues: II the role of the laboratory animal study of intermediate duration for evaluation of safety. *Toxicology and Applied Pharmacology* 16: 498–506.

Gandhi, M. K. 1972. *Gandhi's selected writings*. Ed. by R. Duncan. New York: Harper and Row.

Giese, R. L., Peart, R. M., and Huber, R. T. 1975. Pest management. *Science* 187: 1045–1052.

Godlovitch, S. and R., and Harris, J., eds. 1971. *Animals, men and morals*. London: Gollancz and New York: Taplinger.

Goodrich, T. 1969. The morality of killing. *Philosophy* 44: 127–139.

Gouldner, A. 1960. The norm of reciprocity: a preliminary statement. *American Sociological Review* 25: 161–178.

Gray, J. G. 1967. *The warriors*. New York: Harper and Row.

Griffin, D. R. 1976. *The question of animal awareness*. New York: Rockefeller University Press.

Hacker, H. M. 1951. Women as a minority group. *Social Forces* 30: 60–69.

Hamermash, D. S. and Soss, N. M. 1974. An economic theory of suicide. *Journal of Political Economy* 82: 83–98.

Hare, R. M. 1973. Rawls' *A theory of justice*. *Philosophical Quarterly* 23: 144–155 and 241–252.

——. 1975. Abortion and the Golden Rule. *Philosophy and Public Affairs* 4: 201–222.

Harman, G. 1976. Practical reasoning. *Review of Metaphysics* 29: 431–463.

Harris, J. 1971. Killing for food. In Godlovitch and Harris, 1971.

Harrison, R. 1964. *Animal machines*. London: Stuart.

Harsanyi, J. C. 1976. *Essays on ethics, social behavior and scientific explanation*. Dordrecht: D. Reidel.

Hegel, G. W. F. 1952. *Philosophy of right*. Oxford: Oxford University Press.

Hicks, J. R. 1969. *A revision of demand theory*. Oxford: Clarendon Press.

Hillman, H. 1970. *Scientific undesirability of painful experiments*. Zurich: WFPA.

Holt, J. 1975. *Escape from childhood*. New York: Ballantine Books.

Hume, D. References to Hume are to the Selby-Bigge editions. 'T' refers to the *Treatise of Human Nature*, 'E' to the *Enquiries*.

Jeffrey, R. C. 1974. Preference among preferences. *Journal of Philosophy* 71: 377–391.

Jennings, H. S. 1906. *Behavior of the lower organisms*. New York: Columbia University Press.

Johnstone, H. W., Jr. 1970. On being a person. In *Essays in metaphysics*, ed. C. G. Vaught. University Park, Pennsylvania: Pennsylvania State University Press, pp. 127–138.

Kant, I. 1785. *Grundlegung zur metaphysik der sitten*.

——. 1963. Duties to animals and spirits. In *Lectures on ethics*. Trans. by L. Infield. New York: Harper and Row. Originally Methuen, 1930. Reprinted by Hackett, 1980.

Krieg, M. B. 1964. *Green medicine*. Chicago: Rand McNally.

Krishnamurti, J. 1970. *Talks and dialogues*. New York: Avon.

LaChapelle, D. 1978. *Earth wisdom*. Los Angeles: Guild of Tutors Press.

LaMettrie, J. O. de. 1748. *Man a machine*. La Salle, Illinois: Open Court Publishing Company. 1912 translation of 1748 edition of *L'homme machine*.

Lane-Petter, E. 1963. Humane vivisection. *Laboratory Animal Care* 13: 469–473.

Lappé, F. M. 1975. *Diet for a small planet*. New York: Ballentine Books.

Leiss, W. 1972. *The domination of nature*. New York: George Braziller.

Leopold, A. 1970. *A Sand County almanac, with essays on conservation from Round River*. New York: Ballantine Books.

Levin, M. E. 1977. Animal rights evaluated. *Humanist* 37: 12–15.

Lewinsohn, R. 1954. *Animals, men and myths*. New York: Harper, Inc.

Lewis, C. S. 1943. *The abolition of man*. London: Oxford University Press.

——. 1953. *The silver chair*. New York: Macmillan.

Linzey, A. 1976. *Animal rights: a Christian assessment of man's treatment of animals*. London: SCM Press.

Lippman, W. 1922. *Public opinion*. New York: The Free Press.

Locke, J. 1690. *An essay concerning human understanding*.

——. 1905. *Some thoughts concerning education*. 5th edition. London. Section quoted as reprinted in *The educational writings of John Locke*. J. Axtell, ed. Cambridge: Cambridge University Press, 1968.

Lockwood, M. 1979. Singer on killing and the preference for life. *Inquiry* 22: 157–170.

Loosli, R. 1967. Duplicate testing and reproducibility. In Regamey et al. 1967, pp. 117–123.

Lowrence, W. W. 1976. *Of acceptable risk*. Los Altos, California: William Kaufmann.

Lyons, D. 1965. *Forms and limits of utilitarianism*. Oxford: Clarendon Press.

——. 1977. Human rights and the general welfare. *Philosophy & Public Affairs* 6: 113–129. Reprinted in *Rights,* ed. D. Lyons. Belmont, California: Wadsworth, 1979.

MacCormick, N. 1978. Dworkin as pre-benthamite. *Philosophical Review* 87: 585–607.

MacIver, A. M. 1948. Ethics and the beetle. *Analysis* 8: 65–70.

Mackie, J. L. 1978. Can there be a rights-based moral theory? *Midwest Studies in Philosophy* 3: 350–359.

Magalhaes, H. ed. 1974. *Environmental variables in animal experimentation*. Lewisburg, Pennsylvania: Bucknell University Press.

Magel, C. and Regan, T. 1979. Animal rights and human obligations: a select bibliography. *Inquiry* 22: 243–247.

Majumdar, T. 1958. *The measurement of utility*. London: Macmillan. Reprinted 1975, Westport, Conn.: Greenwood Press.

Markowitz, J., Archibald, J. and Downie, H. G. 1959. *Experimental surgery*. Baltimore: The Williams and Wilkins Co. 4th Edition.

Maslow, A. H. 1968. *Toward a psychology of being*. Princeton, N.J.: Van Nostrand. 2nd edition.

Mason, J. and Singer, P. 1980. *Animal factories*. New York: Crown Publishers.

Matson, A. 1977. *Afterlife: reports from the threshold of death*. New York: Harper and Row.

Mayo, D. 1981. Testing statistical testing. In *Philosophy in economics*. ed. J. C. Pitt, pp. 175–203. Dordrecht: D. Reidel.

McCloskey, H. J. 1979. Moral rights and animals. *Inquiry* 22: 23–54.

McGilvary, E. B. 1956. *Toward a perspective realism*. LaSalle, Illinois: Open Court.

Melden, A. I. ed. 1950. *Ethical theories*. Englewood Cliffs, N.J.: Prentice-Hall.

Middleton, J. 1965. *The Lugbara of Uganda*. New York: Holt, Rinehart and Winston.

Midgley, M. 1976. The concept of beastliness. In Regan and Singer, 1976.

——. 1978. *Beast and man*. Ithaca, N.Y.: Cornell University Press.

Mill, J. S. 1852. Dr. Whewell on moral philosophy. *Westminister Review*. Reprinted in *Dissertations and discussions* (1859), Vol. II, pp. 450ff.

——. 1863. *Utilitarianism*.

Miller, H. B. and Williams, W. H. 1982. *The limits of utilitarianism*. Minneapolis: University of Minnesota Press.

Monod, J. 1972. *Chance and necessity*. London: Collins.

Moody, R. A. 1976. *Life after life*. New York: Bantam Books.

Moore, G. E. 1903. *Principia ethica*. Cambridge: Cambridge University Press.

Morris, R. K., and Fox, M. W. 1978. *On the fifth day: animal rights and human ethics*. Washington: Acropolis Books, Ltd.

Mufson, E. J. Balagura, S., and Riss, W. 1976. Tail pinch-induced arousal and stimulus-bound behavior in rats with lateral hypothalamic lesions. *Brain, Behavior and Evolution* 13: 154–164.

Nagel, T. 1978. Ruthlessness in public life. In *Public and private morality,* ed. S. Hampshire. Cambridge: Cambridge University Press, pp. 75–92.
———. 1979. *Mortal questions.* Cambridge: Cambridge University Press.
Narveson, J. F. 1977. Animal rights. *Canadian Journal of Philosophy* 7: 161–178.
Nozick, R. 1974. *Anarchy, state and utopia.* New York: Basic Books.

Ogilvie, W. H. 1935. Letter in *Lancet* (February 23), p. 419.
Orlans, F. B. 1980. Humaneness supersedes curiosity. Presented at a conference on The Use of Animals in High School Biology Classes and Science Fairs, 27–28 September 1979, at the Institute for the Study of Animal Problems, Washington, DC.
Ortega y Gasset, J. 1972. *Meditations on hunting.* New York: Charles Scribner's Sons.
Osis, K. and Haraldsson, E. 1977. *At the hour of death.* New York: Avon Books.

Passmore, J. 1974. *Man's responsibility for nature.* New York: Charles Scribner's Sons and London: Duckworth.
Paterson, D., and Ryder, R. D. 1979. *Animals' rights: a symposium.* London: Centaur Press.
Perry, D. 1967. *The concept of pleasure.* The Hague: Mouton and Co.
Plato: references to Plato are to Stephanus page numbers.
Pluhar, W. 1977. Abortion and simple consciousness. *Journal of Philosophy* 74: 159–172.
Portola Institute. 1971. *The last whole earth catalog.* New York: Random House.
Pratt, D. 1976. *Painful experiments on animals.* New York: Argus Archives.
Prawitz, D. 1968. A discussion note on utilitarianism. *Theoria* 34: 76–84.
Prosser, W. L. 1971. *Law of torts.* 4th ed. St. Paul: West Publishing Company.

Rachels, J. 1976. Do animals have a right to liberty? In Regan and Singer, 1976.
———. 1979. *Moral problems.* Third edition. New York: Harper and Row.
———. 1980. Euthanasia. In *Matters of life and death,* T. Regan, ed. New York: Random House.
Rawls, J. 1951. Outline of a decision procedure for ethics. *Philosophical Review* 66: 177–197.
———. 1971. *A theory of justice.* Cambridge, Mass.: Belknap Press.
Redding, W. T., and Stewart, J. 1977. *Traps and trapping, furs and fashion.* New York: Argus Archives.
Reeve, E. G. Speciesism and equality. *Philosophy* 53: 562–563.
Regamey, R. H., Hennessen, W., Ikic, D., and Ungar, J. eds. 1967. *International symposium on laboratory animals.* Basel: S. Karger.
Regan, T. 1975. The moral basis of vegetarianism. *Canadian Journal of Philosophy* 5: 181–214.
———. 1976a. Broadie and Pybus on Kant. *Philosophy* 51: 471–472.
———. 1976b. McCloskey on why animals cannot have rights. *Philosophical Quarterly* 26: 251–257.
———. 1979a. Exploring the idea of animal rights. In *Animal rights: a symposium* ed. D. Paterson and R. Ryder. London: Centaur.
———. 1979b. An examination and defense of one argument concerning animal rights. *Inquiry* 22: 189–219.
———. 1980. Utilitarianism, vegetarianism, and animal rights. *Philosophy & Public Affairs* 9: 305–324.

——. 1981. The nature and possibility of an environmental ethic. *Environmental Ethics* 3: 19–34.

Regan, T. and Singer, P., eds. 1976. *Animal rights and human obligations*. Englewood Cliffs, N.J.: Prentice-Hall.

Regenstein, L. 1975. *The politics of extinction*. New York: Macmillan.

Richards, D. A. J. 1971. *A theory of reasons for action*. Oxford: Clarendon Press.

Richardson, W. J. 1963. *Heidegger: through phenomenology to thought*. The Hague: Martinus Nijhoff.

Ritchie, A. M. 1964. Can animals see? a cartesian query. *Proceedings of the Aristotelian Society* 64: 221–242.

Roberts, C. 1980. *Science, animals, and evolution*. Westport, Conn.: Greenwood Press.

Rollin, B. E. 1978. Updating veterinary medical ethics. *Journal of the American Veterinary Medical Association* 173: 1015–1018.

Rollin, B. E. 1981. *Animal rights and human morality*. Buffalo, NY: Prometheus.

Rosenfield, L. 1968. *From beast-machine to man-machine; the theme of animal soul in French letters from Descartes to La Mettrie*. New York: Octagon Books.

Roszak, T. 1975. *Unfinished animal: the aquarian frontier and the evolution of consciousness*. New York: Harper and Row.

Rosseau, J-J. 1762. *The social contract*.

Ruesch, H. 1978. *Slaughter of the innocent*. New York: Bantam Books.

Rumbaugh, D. M. 1965. Maternal care in relation to infant behavior in the squirrel monkey. *Psychological Reports* 16: 171–176.

Rumbaugh, D. M., ed. 1977. *Language learning by a chimpanzee: the LANA project*. New York: Academic Press.

Russell, B. 1956. *Portraits from memory and other essays*. New York: Simon and Schuster.

——. 1975. *Autobiography*. London: Unwin Books.

Ryder, R. 1975. *Victims of science*. London: Davis-Poynter.

Sagan, C. 1977. *The dragons of Eden*. New York: Ballantine Books.

Santayana, G. 1913. *Winds of doctrine*. New York: Charles Scribner's Sons.

Savage-Rumbaugh, E. S., Rumbaugh, D. M., and Boysen, S. 1978. Linguistically mediated tool use and exchange by chimpanzees (*Pan troglodytes*). *Behavioral and Brain Sciences* 1 (4): 539–554.

Savage-Rumbaugh, E. S., Rumbaugh, D. M., Smith, S. T., and Lawson, J. Reference: the linguistic essential. *Science*, 1980, 210, 922–925.

Sbordone, R. J., and Garcia, J. 1977. Untreated rats develop 'pathological' aggression when paired with a mescaline-treated rat in a shock-elicited aggression situation. *Behavioral Biology* 21: 451–461.

Scanlon, P. 1973. *The tender carnivore and the sacred game*. New York: Scribners.

Scheffer, V. B. 1974. *A voice for wildlife: a new ethic in conservation*. New York: Scribner.

Schweitzer, A. 1965. *The teaching of reverence for life*. New York: Holt, Rinehart & Winston.

Selye, H. 1956. *The stress of life*. New York: McGraw-Hill.

Seton, E. T. 1901. *Lives of the hunted*. New York: New American Library.

Shaver, K. G. 1977. *Principles of social psychology*. Cambridge, Mass.: Winthrop.

Shepard, P. 1973. *The tender carnivore and the sacred game*. New York: Charles Scribner's Sons.

Sherif, C. W. 1976. *Orientation in social psychology*. New York: Harper and Row.

Sidgwick, H. 1907. *Methods of ethics*. London: Macmillan and Co. 7th edition.

Singer, P. 1975. *Animal liberation*. New York: A New York Review Book distributed by Random House (Also New York: Avon Books, 1977).

——. 1978. The parable of the fox and the unliberated animal. *Ethics* 88: 119–125.

——. 1979. Killing humans and killing animals. *Inquiry* 22: 145–156.

——. 1980. Animals and the value of life. In *Matters of life and death*, ed. T. Regan. New York: Random House.

Skinner, J. D., Von La Chevallerie, M. and Van Zyl, J. H. M. 1971. An appraisal of the Springbok for diversifying animal production in Africa. *Animal Breeding Abstracts* 39(2): 214–215.

Smith, J. M. 1978. The evolution of behavior. *Scientific American* 239(3): 176–192.

Sobel, J. H. 1970. Utilitarianisms: simple and general. *Inquiry* 13: 394–449.

Spinoza, B. 1677. *Ethics proved in geometrical order*.

Spotte, S. 1976. The "rights" of animals. *Zoo Review*, No. 7, p. 5.

Stiller, H., and Stiller, M. 1976. *Tierversuch und Tierexperimentator*. Munich: Hirthammer.

Stone, C. D. 1974. *Should trees have standing? toward legal rights for natural objects*. Los Altos, California: Kaufmann.

Stoner, H. B. 1961. Critical analysis of traumatic shock models. *Federation Proceedings* 20, Supplement 9, pp. 38–48.

Sumner, L. W. 1976. A matter of life and death. *Nous* 10: 145–171.

Swenson, R. M., and Randall, W. 1977. Grooming behavior in cats with pontile lesions and cats with tectal lesions. *Journal of Comparative and Physiological Psychology* 91: 313–326.

Talbot, L. M. 1963. The biological productivity of the tropical savanna ecosystem. *Proceedings of the IUCN 9th Technical Meeting* (Nairobi, September 1963) IUCN Publication N.S. No. 4, Part II "Ecosystems and biological productivity" pp. 88–97.

——. 1966. *Wild animals as a source of food*. U.S. Department of the Interior Bureau of Sport Fisheries and Wildlife Special Scientific Report Wildlife No. 98. (16pp).

Talbot, L. M., and Talbot, M. H. 1963. The high biomass of wild ungulates on East African savanna. *Transactions of the North American Wildlife and Natural Resources Conference* 28: 465–476.

Taylor, C. 1976. Responsibility for self. In *The identities of persons*, ed. A. O. Rorty. Berkeley: University of California Press. pp. 281–299.

Taylor, R. 1970. *Good and evil*. New York: Macmillan.

Taylor, T. 1792. *A vindication of the rights of brutes*. Reprinted 1966, Gainesville, Florida: Scholar's Facsimiles and Reprints.

Thomas, L. 1980. Ethical egoism and psychological dispositions. *American Philosophical Quarterly* 17: 73–78.

Tompkins, P., and Bird, C. 1974. *The secret life of plants*. New York: Avon Books.

Tooley, M. 1972. Abortion and infanticide. *Philosophy and Public Affairs* 2: 37–65.

Turner, E. S. 1964. *All heaven in a rage*. London: Michael Joseph, Ltd.

Ulrich, R. E. 1978. Letter in American Psychological Association *Monitor*. March, p. 16.

VanDeVeer, D. 1979a. Of beasts, moral persons, and the original position. *The Monist* 62: 368–377.

——. 1979b. Interspecific justice. *Inquiry* 22: 55–79.

van Lawick-Goodall, J. 1971. *In the shadow of man*. London: Collins.

von Neumann, J. and Morgenstern, O. 1946. *Theory of games and economic behavior*. Princeton, N.J.: Princeton University Press, 2nd edition 1953.

Vyvyan, J. 1971. *The dark face of science*. London: Michael Joseph.

Wallace, J. D. 1974. Excellences and merit. *Philosophical Review* 83: 182–199.

Wallach, M. B., Dawber, M., McMahon, M., and Rogers, C. 1977. A new anorexigen assay: stress-induced hyperphagia in rats. *Pharmacology, Biochemistry and Behavior* 6: 529–531.

Wambach, H. 1979. *Life before life*. New York: Bantam Books.

Watson, G. 1975. Free agency. *Journal of Philosophy* 72: 205–220.

Watts, A. 1961. *Psychotherapy east and west*. New York: Pantheon Books.

——. 1973. *This is it and other essays on Zen and spiritual experience*. New York: Vintage Books (reprint of 1960 edition).

Weeden, R. 1977. Nonconsumptive users: a myth. *Alaska Fish and Game* 10(4): 9–10.

Weil, M., and Scala, R. 1971. Study of intra- and inter-laboratory variability in the results of rabbit eye and skin irritation tests. *Toxicology and Applied Pharmacology* 19: 276–360.

Wenz, P. S. 1979. Act utilitarianism and animal liberation. Paper read at the Eastern Division meetings, American Philosophical Association, December 1979.

Westacott, E. A. 1949. *A century of vivisection and anti-vivisection*. London: Daniel.

Whewell, W. 1852. *Moral philosophy*.

Wilkes, B. 1977. The myth of the non-consumptive user. *Canadian Field Naturalist* 91: 343–349.

Wilson, E. O. 1975. *Sociobiology*. Cambridge, Mass.: Harvard University Press.

Wittgenstein, L. 1958. *Philosophical investigations*. 2nd edition, translation G. E. M. Anscombe. New York: The Macmillan Company.

Wollstonecroft, M. 1792. *A vindication of the rights of women*.

Subject Index

Abortion, 6, 136–137, 148, 156, 288
ACTH (adrenalcorticotropic hormone), 330
Activists, animal, 327
Act-utilitarianism and rule-utilitarianism, 184, *see also* Utilitarianism
Agreements ('conventions') with animals, 72
Alexander, Samuel, 131
Alienation, human from nature, 309
Aliens, animals as, 80–81
Alport, Gordon, 79
Alternatives to animal experiments, 352, 359
Altruism and evolution, 181
American Museum of Natural History, 374, 376–377
Anarchy, 101
Animal,
    as child, 81–83
    children and animals, 174–175
    as demon, 85–87
    as hero, 83–85
    and human pleasures, relative values, 256–257
    husbandry argument, 253–255, 379
    husbandry, humane and nonhumane, 259n9
    industry, consequences of collapse of, 32
    liberation, 361–377
        strategies and tactics, 373–377
    -loving and human-hating, 76
    mistreatment and bad character, Kant and Hume on, 72
    as paragon, 83–85
    random source, 353–354
    reverence for, 84
    rights, *see* Rights
    sentience, denial of, 110–111
    specially bred for research, 353–355
    state of when experimented on, 356–357
    as toys, 175
    welfare statutes, limits of, 114
Anthropocentrism, 307–310, *see also* Homocentrism
    in legislation, 112–114
Anthropomorphism, ineradicable, 89–90
Anticipation, 145
    moral importance of ability, 256–257
Antivivisectionists, 340
Ape language, 207–215, 217
Aquinas, St. Thomas, 25, 108
Argument, place of in moral disputes, 243
Aristophanes, 179
Aristotle, 1–2, 177, 181–182
Asceticism, 171, 239–240
    and attitudes to animals, 171
Attitudes, human, and animal welfare, 337

Background variables, confusion of, 353–358
Beastliness, 86–87
Bellarmine, Cardinal, 107
Benefits to animals of raising them for food, 222–223, 249, 253–255
    see also Replaceability argument
Bennett, Jonathan, 145
Bentham, Jeremy, 11, 25, 37, 104
Berdyaev, Nicholas, 88
Berger, John, 80, 81
Bernard, Claude, 343, 356
Bibliographies, 381
Biography, *see* Life, a
Biotic pyramid, 185
Blake, William, 124
Blood clots, 348
Bores, obligations to, 247
Bradley, F. H., 175
Breeding humans for use, 157, 161

391

Bringing creatures into existence, 49, 50
Brutalization, 113
    effect of mistreatment of animals, 108,
        see also Aquinas; Kant
Buddhism, see Eastern philosophies
Burdens, any moral system brings some,
    247

Cancer cells, analogy of humans to, 192
Cancer research, see Experimentation,
    animal
Capability,
    of distraction, 257
    of doing damage, 57
    for entering into agreements, 57
    of utility, 51–54
Cargile's argument, see Animal
    husbandry argument
Causation, contributory, 264–265
Cetaceans, see Whales and Porpoises;
    Dolphins
Character, 28
    mistreatment of animals shows
        defective, 3
Charity and rights, 279
Chickens, 329–337
    chicken industry, 334–335
    effects of human handlers, 333,
        336–337
    light level and stress, 335–336
    raising practices,
        humanity of, 325
        stressfulness of, 334–336
Chimpanzees,
    behavior of, 177
    temperament, 213
Christianity, attitude to animals, 3
Civil order, 101
Comas, persons in, 282
Common conceptions, importance of, 138
Communication, nonlinguistic, 111
Compensation, and value of life, 150
Competition, human with other species,
    200, 203–205
    and vegetarianism, 166
Complexity, mental, 121, 130–133
Conception, avoiding, 156
Concern for animals, causes of recent,
    4–7
Conditions for being party to morality, 56
Confinement, 159
    frustrations of, 258
Consciousness,
    and death, 318–321

degrees of, 146
extent of, 318
good in itself, 145
and object of consciousness, 144
reflexive, 121, 146
and self-consciousness, 131
simple, 135, 139
    grounds of obligation not to kill, 135
transformation in, 309–310
value of, 121, 136
Consent, 49–50
    hypothetical, 50
    implicit or tacit, 50
Consequences, 252
Conservation ethic, 202
Contraceptives, oral, testing of, 355–356
Contract view of morality, see
    Contractarianism
Contractarianism, 56–59
    and animals, 58–59, 67–68
Contractualism, see Contractarianism
Corticosterone, high plasma (HPC) and
    low plasma (LPC), 330
    effects of, 331–332
Cosmogonies, animals in, 86
Creation Requirement, 156
    and breeding animals, 158–160
Creatureliness, 87
Cruelty, 27–28
    to animals and to humans, 313–314
    psychological, 113
Cruelty account, 26–29
Cuteness and subjugation, 83

Darling, Fraser, 181
Darwin, Charles, on man as animal, 3
Dawkins, Richard, 180–181
DDT, banning of, 356
Death,
    concept of, 128
    evil of, 281–282
    irrelevance of in evaluation of animal
        husbandry, 257–258
    nature of, 318–321
    significance of, 305–306
    as transition, 321
Death rate, in wild and domestic animals,
    259
Defenses, legal, 363–364
Deontological theories and social
    distance, 230–231
Dependency and social distance, 234
Descartes, René, 3, 5–6, 110, 132, 165,
    176, 346

Desirability, expected, 251–252
Desires,
first-order and second-order, 127
satisfaction and elimination, 129–130
Desire for exploration, as justification for research, 359
Devine, Phillip, 125–126, 133
Discrimination,
and differences in standing, 75–76
objectionable and non-objectionable, 240–241
Diseases, artificial and natural, 343–349
Distraction, moral relevance of capacity for, 257
Diversity, 186, 190
Dog dealers, 353–354
Dogmatic beliefs, 66–67
Dolphins, 361–362, 364–365
survival of released, 367, 367n8, 369
Domestication, 200
advantages of, 159
domestic animals set loose, 50
Domination, 202
Draize tests, see Rabbit eye tests
Dualities, inapplicable to reality, 308
Duty,
to act, 43
and interests, 276
positive and negative, 187
Dworkin, Ronald, 34–35, 104–105, 297–299

Eastern philosophies, 309–310
Eco-ethics, see Ecological ethics
Ecological awareness and hunting, 193–194
Ecological ethics, 308–310
Ecology, see Ecosystems
Economic effects of stopping hunting, 196–197
Ecosystems, 166, 183–194
duties to, 308–309
good, 190
healthy, 186
humans in, 199–201, 203–205
human destruction of, 191–192
Empathetic identification, see Empathy
Empathy, 137, 221, 231–233, 317
failure of, culpable and nonculpable, 240–241
limits of, 317–318
and moral status, 244–245
and utility calculus, 232
Environment, 183–185

history of damage of, 191–192
and stress in chickens, 332–336
Epictetus, 172, 173
Epicurus, 144, 281–282
Epilepsy, 348
Equality principle, 160–161
Erect posture and infant dependency, 208–209
Euthanasia of homeless animals, 116
Evolution and status of animals, 176
Evolutionary scale, relevance of place in, 108
Experimentation, animal, see also Research; Alternatives
aggression, 347–348
cancer research, 344–345
cat sex, 374
dolphins, 361–362, 366–368
induced stress, 346–347
invalid, 343–358
irrelevant, 340–343
justification of, 339–340
shock, electric, 347–348
trivial, 341–343
value of, 339–359
Extrapolation, 326, 350–352

Factory farming, see Intensive farming
Factual knowledge and moral attitudes, 283
Fate and interest of species, 77n8
Federal Animal Welfare Acts, 114
Food animals, numbers, 23
Frankfurt, Henry, 127
Friendship,
a part of virtue, 228
and reciprocity, 232
Future generations, 73, 148

Gandhi, Mahatma, 312
Generosity, and rights, 279
Genetic package, of species, 329
Genetic plan, 329
Goodrich, T., 123
Greenwood, Michael, 367–368
Growth rates, as measure of welfare, 330

Haraldsson, Erlendur, 320
Hare, R. M., 285–286, 298
Harman, Gilbert, 128
Having a life, see Life, a
Hedonism, radical, 132
Hegel, G. W. F., 127, 138
Heidegger, Martin, 310

Holism, ecological, 305, 308–310
Holt, John, 82–83
Homeostasis, of ecosystems, 190
Homocentrism, 183–184, *see also*
  Anthropocentrism
HTS (Hegel/Tooley/Singer) view,
  139–140
Humanism, *see* Homocentrism
Humans, *see also* Priority, of human
  interests
  eating, on libertarian grounds, 49
  as hunters, 180
  as natural predators, 195–196
Human contact with nature, 203–204
Human handlers, effect on chickens, 333,
  336–337
Human life, mechanization of, 89
Human nature and animal nature,
  170–173, 178
Hume, David, 64–65
  on animal and human nature, 68–71
  right to life of, 277
Hume's theory, 68–77
  attractions of, 75–76
Hunting, 9–10, 166, 191–197, 201–205
  and balance of nature, 193–194
  and ecological damage, 192–194
  as harvest, 202
  permissible cases, 197
Husbandry, 336–337
Hypothesis, time of formulation, 340

Imagination, 137
  and morality of killing, 137
Impartiality, 152
  and perfect virtue, 239–240
  and private virtue, 239
  and 'saintliness', 239–240
  and social distance, 239
Impersonalism, *see* Objectivity, scientific
Imported animals, numbers, 23
Inconsistency in attitudes to animals, 7,
  377
Insurance, and value of life, 149–150
Intelligence, as morally relevant, 53
Intensive farming, 88, 158
Interanimal communication, in trained
  chimpanzees, 214–215
Interests, 128n1, 140–141
  basic and peripheral, 161
  common of humans and animals, 226
  criteria for like, 160–161
  equal consideration of, 161
  equal moral weight of, 161

equality of, *see* Equality principle
equivalence of, 237
negative and positive, 129
of plants, 153
priority of human, 225–249
significant, 236n11
International Whaling Commission, 19
Interspecific conflicts, 147–148
Intuitionism, 285, 288–289, 290–292,
  299
Intuitions,
  appeal to, 187–191
  concerning animals, suspiciousness of
  opposed, 67
  moral, 295–296

Jennings, H. S., 131
Johns Hopkins Center for Alternatives to
  Animal Testing, 376
Johnstone, Henry, 127
Judeo-Christian ethic, 202
Judging from one's own point of view,
  125–126
Justification,
  conceivable and actual, 42–43
  onus of, 42

Kant, Immanuel, 25, 89, 108, 127, 170,
  276–277
Kantian account, 25–26
Kantian theory, 77
Killing,
  animals and humans, 269–270
  morality of, 319–322
  negative utility of, 142–143
Kubler-Ross, Elizabeth, 319

La Mettrie, Julien Offray de, 111
Land, the, *see* Leopold, Aldo
Land ethic, 183
Language, 207, 209–217
  in higher primates, 116, 167
  and moral standing, 5–6, 167
  and other forms of communication,
  109
  relevance of, 108–110
Language acquisition and social behavior,
  in apes, 215
  in severely handicapped children, 216
Language learning,
  in apes,
    accomplishments and limits, 217
    difficulty of, 213–214
    keyboard description, 211–213

keyboard contrasted with manual signing, 210
social factors, 209–210
in severely disadvantaged children, 215–216
Lappe, Frances Moore, 254n5
Law,
and moral principles, 104–105
and morality, 103, 118, 371
natural, 103, 177, 178–180
of harmonized restraint, 313
LD50, 349–350
Legal positivism, 104
Legal protection of animals, 311
Legislation, inadequacy of, 115
Leopold, Aldo, 183, 185–186, 193–194, 202
Le Vasseur, Kenneth W., 361–371
Levin, Michael E., 5
Liberation, animal, 327
and human liberation, 5, 362, 371, 373
Libertarianism, 46–51
Life, a, 38–39, 273, 280–282
biological determinants of, 280–281
and memory, 281
and personhood, 281
Life,
after death, 318–321
before birth, 320–321
contrast between human and animal, 129
interest in, 128–133
meaningfulness of, 47–48
purposes of, 320–321
relative importance of human and animal, 123–133
value of, 121, 161
worth living, 155–156
Life Preferability Requirement, 155
and breeding animals, 158–159
Lippmann, Walter, 79
Locke, John, 25, 139
Lyons, David, 264

Machines, animals as, 3, 18, 87–89
MacIver, A. M., 123
Mackie, J. L., 294–299
Mammals, lives of, 283
Management, of wildlife, 202
Marginal humans,
argument from, 226
justification for different treatment from animals, 58
and rights, 311

Mason, Jim, 88
Maximax and maximin, 150–151
McCloskey, H. J., 280, 292–294, 298, 300n9
McGilvary, E. B., 125, 133
Meat eating,
engrained habit, 260–261
inefficiency of, 55
pleasures of, 259–260
Mental complexity, 145
significance of, 284
Mental States, see also Cruelty
as directly recognized, 176–177
inference to, 176–177
Mice, see Rodents
Microcephalic animals, 117
Midgley, Mary, 87
Mill, John Stuart, 11, 51–52, 123–125, 251
Millenium Guild, 376
Monkeys, lives of, 282–283
Monod, J., 178
Monotonous diets, frustrations of, 260
Moody, Raymond, 319
Moore, G. E., 144
Moral agents, 12–14, 109
animals as, 84
Moral argument and conflicting intuitions, 296–297, 299
Moral capacity, human, 95
Moral community, 13, see also Social distance
bounds of, 96
determined by evolved characteristics, 96
distinct roles in, 76
and ecological community, 310
extension of, 312
and genetic lottery, 167
in Hume, 71
inherent limits of, 98–100
and moral order, 93, 95–102
tribal basis, 96–97
Moral demands as agreement-based, 73–74
Moral excellence, see Virtue
Moral order,
and animals, 101–102
as escaping limits of moral community, 100–102
Moral patient, 12–14, 109
Moral philosophers, task of, 308–310
Moral philosophy, political importance of, 14

Moral principles,
  criteria for acceptability, 288–289
  and feelings, 241–242
  and virtue, 241–242
Moral priorities and special
  commitments, 268–269
Moral priority of humans, 245
Moral problem of world hunger, 247, see
  also Moral community
Moral reasoning and self-interested
  reasoning, 74
Moral relevance, 41
Moral standing, 2
  moral agents and moral patients, 12–14
  and phylogenetic scale, 283–284
  of the unlovable, 244
Moral status of animals, options, 45
Moral theory,
  and convictions, 61–63
  point of, 62
Moral virtue, see Virtue
Morality,
  and amorality, 100–101
  and animal nature, 178
  in animals, 177
  and character, 269
  and human self-interest, 65
  and law, 103, 118, 371
Morally relevant characteristics, 226,
  226n3
Morally relevant differences, 108
  absence of between humans and
  animals, 111–112, 245
Muscular dystrophy, 348–349
Mythologies, 97–98

Nationalism, 97–98
Natural and possible, 379–380
Neighbor, see Moral Community
Noble-Collip Drum, 346–347
Non-priorism, 246–248
  compatible with empathy and
  reciprocity, 247
Nonsentient creation, status of, 314
Nozick, Robert, 46, 47–48, 50, 280
  on animals, 48

Object of moral concern, see Moral
  patient
Objectivity, scientific, 175, 176
Obligation,
  not to kill conscious beings, 145
  prima facie, 187
Octopus, 283

Opportunity costs of being killed,
  156–157
Organizations concerned with animals,
  'activist' and
  'establishment', 4
Origin of a tendency and justification of
  tendency, 245–246
Original contractors, 153, see also Veil
  of ignorance
Original position, see Veil of ignorance
Ortega y Gassett, José, 195
Osis, Karlis, 320
Oysters, mental life of, 132

Pain, freedom from, 143–144
Panmictic ethic, 202
Panpsychism, 318
Parfit, Derek, 55
Peritonitis, simulation of, 345
Personhood, 127
Persons, 145, 294
  as beings with reflexive consciousness,
    139–140
  distinction between 'person' and
    'human', 13–14, 139
  dolphins as, 364–365, 370–371
  legal definition, 364–365
Pets, 10
  vegetarian feeding of, 263n13
Philosophy, limits of Western, 307–310
Pigs, Cargile's described, 249
Plants, status of, 140, 283–284, 305, 318
Plasticity of behavior,
  human and animal, 207
  of squirrel monkeys, 208
Plato, 1–2, 132, 179
Pleasure, 144, 144n15
  capacities for, different, 222, 256–257
  human and animal, 124–125
  of meat eating, 259–260
  quantities of, 222–223
Pluhar, Werner, 136–138
Plutarch, 177
Population control, 256n6
Positivism, legal, 297
Poultry, see Chickens
Precision gained by calculus, value of, 74
Predation, 179, 270
Predators, humans as, 166, 200–203
Preferences, 229–231
  and actions, 238–239
  human and animal, 269
  preferring someone's company and
    giving moral preference, 244

frustration of in humans and animals,
    257–258
and moral priority, 267–269
Pre-original position, 153–156
Primates, wild, 20–21
Principle of equality, 29, 31
Principle of utility, 29, 32
Priority,
    giving and claiming, 244–245
    of human interests, compatible with
        virtue, 249
Product testing, 21
Projection, 174
    of human dark side, 87
Property, animals as, 112, 369, 370
Public morality, 239

Quality of life, 159

Rabbit eye tests, 21, 350–351, 375–376
Racism, 33, 90, 133
    and social distance, 240–241
    and speciesism, 126
Radical zoophily and radical feminism,
    180
Rats, see Rodents
Rawls, John, 56, 61–62, 121, 147,
    152–154, 298
    difference principle, 154
Rawls's contractors, heads of families, 73
Reason,
    relevance of ability to, 108–110
    and sentiment, 76, 179
Reciprocity, 231
    in family, 232–233
    and friendship, 232
    and intimacy, 232
    and social distance, 232–233
    and virtue, 231
Reflective equilibrium, 288–289, 298
Reflexiveness, 127–130, 135–136, 139,
    140
Religions, 97–98
Replaceability argument, 138, 142–143,
    254n4, see also Utility to
    animals of eating them
Research, see also Experimentation
    animal, 326
    inference from and status of animals,
        176
    justification of, 326
    numbers of animals, 23
    value of, 326
    worthlessness of much, 114

Restrictions of animal husbandry,
    frustrations of, 258
Revlon, 375–376
Rhythms, effect on experimentation, 358
Rights, 11–12, 141, 184, 273–301
    accorded and intrinsic, 292
    acquired and unacquired, 274, 299–301
    analysis of concept, 34–36, 273,
        278–279
    animal, 34, 241, 285, 292, 294,
        300–301, 305, 310–314
        benefits for humans, 312
        dolphins, to be free, 362
        on libertarian view, 50–51
        not to be tortured, 279
        ridicule of, 117
    attribution to animals, 47
    Bentham on, 37, 104
    of breeders over animals bred,
        157–158, see also Replaceability
        argument
    and characteristics, see Rights, grounds
        of
    and charity, 279
    conflicts between, 289–294, 298
        compromise as resolving, 294–295
        importance, 241–242
    to develop natural potential, 313
    as distracting from central issues, 299
    and duties, 278–279, 291
    ecological argument for, 310–314
    to equal concern and respect, 297
    existence of and treatment of animals,
        287–288
    forfeiting of, 300–301
    fundamental, 294, 297–298
    and general welfare, 107
    and generosity, 279
    grounds of, 276, 278, 280, 295–296
        inherent value, 38
        self-direction, 292
    institutional, 287, 287n3
    and intuitions, 296, 298
    legal,
        conferral of, 116–118
        extension of, 115–116
    to life, 166, 273, 280, 282–284
        practical importance of, 275
    moral,
        functions of, 105–106
        as principles in law, 105
    and moral principles, 274, 288–289
    natural, 103
    negative and positive, 46–47

not to be harmed, 36–39
not to be tortured, 278
and obligations, 190
and other moral concepts, 286–288
overriding, 35–36
and permissibility, 279n4
postulation of, 295–298
prima facie, 290–292
to property, 299–300
and propriety of protest, 279
and self-consciousness, 276–277
speculative nature of, 286
stringencies of, 292–294
superfluity of, 289
and third parties, 279
usefulness of talk of, 273–274,
    285–301
and utilitarianism, 40, 297
welfare-rights and action-rights, 47
to worship, 278
Rights account, 41
Roberts, Catherine, 84
Rodents, tumors in, 355–356
Rodeos, 10
Ross, W. D. (Sir David), 293
Rousseau, Jean-Jacques, 125
Ruddick, William, 280n5
Ruesch, Hans, 82
Russell, Bertrand, 124
Ruthlessness, 239
Ryder, Richard, 25, 139

Saccharin, banning of, 356
Safari parks, see Zoos
Saints, and sexuality and vegetarianism,
    249
Saintliness, 246–248
  consistency with virtue, 248
  contrasted with genuine virtue, 240
  in good times, 268
Salt, Henry, 379
Santayana, George, 132–133
Satisfied pig and dissatisfied human, 154
Scientific education, need for changes,
    115
Scientific language, 276, see also
    Objectivity, scientific
Search for truth, as justification for
    experiments, 359
Self-assertion, as basis for ethics, 133
Self-consciousness,
  Hume on, 70
  and rights, 276–278
Self-direction, as ground of rights, 292
Self-evaluation, 127–128

Self-evidence, 293–294, 298–299
Selye, Hans, 346
Sense of the future, 53
Sentience, relevance of, 314–315
Sentiment, 174–175
  as influence on conscience, 76
  and rise of humane movement, 81–82
Sexism, 33, 90, 180
  and social distance, 240–241
Sexuality, animal and human, 170
Shepard, Paul, 191, 195–196
Sidgwick, Henry, 144n16
Simple consciousness, see Consciousness
Singer, Peter, 25, 88, 130–132, 138–139,
    141–142, 147–148, 158, 160–161,
    253, 254n4, 287–288
  argument criticized, 30–34
  a utilitarian, 11
Sipman, Steven C., 361–362, 371
Slaughter, terror of, 258–259
Social distance, 221, 229–241, see also
    Moral community
  and animals, 234–235
  contrast between actual and necessary,
    240–241
  and dependency, 234
  and distribution, 232
  and empathy, 233
  and interaction, 234
  and preference, 232
  and reciprocity, 232–233
  and species differences, 233–235
Socrates and the fool, 124–125
Soul, immortal, 132
  alleged lack of by animals, relevance
    of, 107
Sources of error in moral conviction,
    63–67
Special interests, 63–65
Species,
  differences, 349–352
  point of view, 50
  preferences in law, 114
  protection of, 311
Speciesism, 25, 33–34, 139, 148,
    160n12, 178, 180–182, 221,
    310–311, 314
  absolute, 235–236
  moderate to strong, 237–238
  and racism and sexism, 126, 240–241
  resolute, 236–237
  varieties of, 235–238
  and virtue, 238
  weak, 237
Spinoza, Baruch, 170

Standing, of natural objects, 183–184
Stereotypes, 79–80
   of animals, 174
   and moral standing, 79, 89
   selective application, 82, 85
   unreality of, 83, 85
Stewardship, 181, 313–315, 336–337
Stoics, 173–174, 181
Stone, Christopher, 183–184
Stress,
   experimental, 356–357
   factors in chickens, 330
   results in chickens, 331–333
   results in other animals, 331
   social, in chickens, 332–333
   and welfare, 325
Suffering, 143
   and right not to be tortured, 278
Suicide, rationality, and value of life,
      150–151
Sumner, L. W., 143–144
Symbolic action, 172
Sympathy, 71, see also Empathy

Tastes, 66
Taoism, see Eastern philosophies
Taylor, Charles, 127–128
Taylor, Richard, 127, 131–132
Telos (animal nature), 116–118, 179
Teratogenic substances, 352
Terror, death and, 259
Thalidomide, 352
Theism, 179, 184, 202
Tooley, Michael, 138, 277–278
Toxicology, 349–353
Traditional farming, 158
Transportation of animals, 259n9
Tumors, 344
   in rodents, 355–356
Typical behavior and possible behavior,
      208

Ulcers, 346
Ulrich, Robert, 347–348
Uses of animals,
   in education, 9
   inventory of, 8–10
Utilitarianism, 11, 51–55, 105–106, 112,
      141–144, 165, 169–170, 174,
      183–184, 197, 222, 228, 251–265,
      285, 288, 291, 295–297
   on animals, 67
   classic, 146, 251–252
   hedonistic vs. preference, 141

   and objections to preference by social
      distance, 230
   preference, 145
   and rights, 40
   rule-utilitarianism, 245
   and social progress, 170
   total and average, 55
   utilitarian account, 29–34
Utility,
   of animals, 51
   cardinal, 52
   downrating animal, 52
   as pleasure or preference, 52, 54
   to animals of eating them, 54–55

Value,
   inherent, 37–39
      as ground of rights, 38
   of life, 149
      not infinite, 151
Variability, of individuals, 329
Veal calves, 22, 158
Vegetarianism, 30, 50, 52, 54, 222–223,
      317
   challenge to vegetarians, 269–270
   and competition with other species,
      166, 204
   conflicts with natural impulses, 249
   defended against animal husbandry
      argument, 254–255
   delights of, 261
   health implications, 261
   individual and meat industry, 262–265
   and land use, 254
   less helpful to animals than some meat-
      eating, 249
   limited diet, 260
   market effects of, 262–264
   pleasures of, 260–261
   practical effects of individual, 222–223
   and purity, 171
   requires changing habits, 260–261
   and starving humans, 255, 256n6
   utilitarian grounds for, 252
Veil of ignorance, 56, 121–122, 147, 152
Veterinary medicine, ethics in, 115–116
Virtue, 174, 221–222, 225
   in animals, 69, 73
   as basis for treatment of animals,
      71–73
   a matter of character, 227
   coherence of traits constituting, 228
   empathy and social distance, 243–245
   friendship a part of, 228
   and moral principles, 241–242

and moral priority, 268–269
and priority of human interests,
    246–247
produces spontaneous conduct, 228
and speciesism, 238
stability of traits, 228
and vice, theory of, 74–75
Vocalization, indicator of condition in
    chickens, 332, 334, 336–337

Watson, Gary, 129
Welfare,
    of animals, measure of, 325
    evaluation of, 329

Welfare economics, 141, 141n9
Whales and porpoises, 6, *see also*
    Dolphins
Whaling, 19–29
Whewell, William, 169
Wild animals,
    and domestic animals, 270
    as food supply, 201
    pleasure of, 254–255
Wilderness, value of, 194
Wittgenstein, Ludwig, 80
Wolf, Susan, 129n2
Writings on animals, 5

Zoos, 10, 81